The Best-Dressed Miners

LIFE AND LABOR IN THE MARYLAND

COAL REGION, 1835–1910

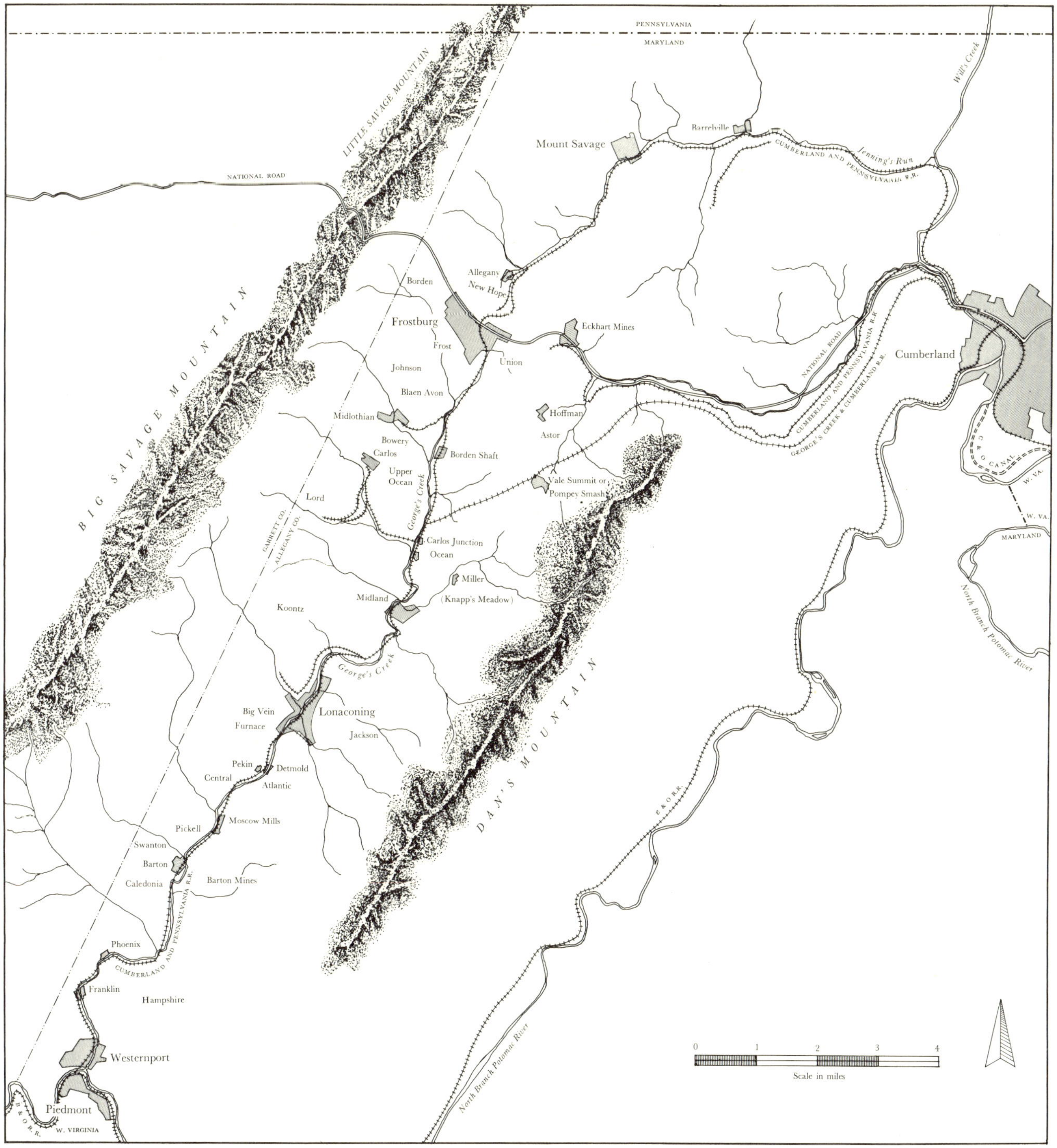

Map of George's Creek Valley, Allegany County, Maryland, showing the approximate locations of coal mines mentioned in the text

The Best-Dressed Miners

LIFE AND LABOR IN THE MARYLAND COAL REGION, 1835–1910

By Katherine A. Harvey

Cornell University Press

ITHACA AND LONDON

First published 1969

Standard Book Number 8014-0494-0

Library of Congress Catalog Card Number 71-78924

PRINTED IN THE UNITED STATES OF AMERICA
BY THE VAIL-BALLOU PRESS, INC.

To O. L. H.

"Taken as a whole, this is a community of the best-dressed miners I have ever seen."

W. M. WARDJON, Organizer
United Mine Workers' Journal
May 15, 1905

Preface

This is an unorthodox picture of a coal-mining region. There are no dreary rows of tumble-down houses set in a gray landscape. There are no hard-core unemployed, no hordes of ragged, hungry children facing a life without a future.

Covering the last three quarters of the nineteenth century and running into the first decade of the twentieth, this is the story of a small coal-mining valley in western Maryland. Some, not aware of Maryland's past importance as a coal producer, may find the setting obscure. Yet the western counties of Maryland for decades successfully competed for markets with the great Pennsylvania anthracite and bituminous fields, supplying fuel for the change-over from water power to steam power in the New England mills. Maryland coal was bunkered in steamships of the early commercial fleets; the preferred fuel of the United States Navy, it was tested in the 1870's and 1880's and proved against the Spanish fleets at Santiago and Manila Bay.

A wealth of evidence suggests that in the period of which I write, tumble-down houses, hungry children, and grimy landscapes were common to many coal-mining towns in the eastern seaboard competitive fields. Despite the absence of detailed studies to provide a valid comparison, I nevertheless believe that the Maryland region was in many respects atypical. Its towns do not fit the description of other mining towns of the times; its miners seem to have been economically better off than those in the directly competitive areas; its population up to 1910 certainly contained only a trace of the "new immigration" from eastern and southern Europe; and many sources comment on the intelligence and stability of the mining population.

If the region had been in all ways atypical, its story would add little to the annals of nineteenth-century labor history. The significance of this regional study is that it provides on a small scale a typical example of the vicissitudes of coal miners' unions throughout the country as they struggled for improved working conditions, for recognition, and for status. Without the drama and violence of Molly Maguireism, Maryland miners had their own union in the early 1870's, following the peaceable teachings of John Siney. In the 1880's under the Knights of Labor the Maryland miners associated themselves with a national (but not trades union) labor movement and for the first time met massive resistance from the operating companies. In the 1890's and the early 1900's, Maryland miners were briefly and in varying degrees allied with the United Mine Workers of America. In 1909 the national officers of the UMW acknowledged their failure to organize Maryland, as well as Virginia and West Virginia.

Perhaps somewhat atypical is the attitude of the employers. Until as late as 1905 Maryland operators tended suavely to override rather than aggressively to suppress labor organization. But as the operating companies increased in size and capitalization and combined in ever larger corporations, their resistance to labor progressively increased. Thus the labor policy of the Consolidation Coal Company changed radically in 1905, when its West Virginia officers, installed after a merger of Maryland, West Virginia, and Pennsylvania interests, began to impose methods of espionage and intimidation. For the most part, between the anti-labor activities of Maryland mine owners and those of coal mine owners elsewhere the difference was only one of degree.

This is a chronicle of people, not of corporations. The industrial story has been told in part in a few published "histories" of various coal companies and of the Maryland coal region. These accounts in general tend to neglect the interests and contributions of labor. I have tried to correct this imbalance by concentrating on the social and economic problems of the miners. Thus the coal companies are considered primarily as employers of labor, and their financial structure, interrelationships, and political influence are merely touched upon. The initial chapter, dealing with the development of the mining industry in the region, is therefore frankly summary in nature.

In order to provide the fullest possible account of the Maryland miners I have presented their story in two parts. The first describes the men and their environment: their national backgrounds, their personal characteristics, their homes and manner of living, their working conditions, their means of cultural expression and recreation. The second consists of a chronological account of the miners' relationships with their employers, with labor unions, and with the state. The two sections of the narrative should not be viewed as separate entities, for each is necessary to the other. For example, up to 1905 the absence of feudalistic company control and the relatively high standard of living probably were in large measure responsible for the continuing failure of union organization. Similarly the presence of an industrious, thrifty, and stable labor supply undoubtedly inhibited (at least until 1905) the development of harsh and oppressive practices on the part of the employers. The settled condition of the miners and their acquisition of real estate made them reluctant to risk strikes which might result in the loss of all they owned. On the other hand, the fact that the miners owned their homes made it difficult for the operators to dislodge any considerable segment of the working population.

Historians who have dealt with Maryland have barely touched upon economic and social aspects in the western part of the state during the nineteenth century, and primary source materials are scattered and fragmentary. I have tried to reconstruct the pattern of life of the George's Creek miner before memories of his trade have vanished. In pursuit of the many threads woven into the narrative I have been greatly aided by the staffs of the Social and Economic Division of the National

Archives, the Library of Congress, the Archives of The Catholic University of America, and the libraries of the United States Departments of Commerce, the Interior, and Labor. I am particularly indebted to Miss Margaret Brickett, librarian of the Department of Labor, for locating and making available the microfilms of early labor newspapers. Miss Hester Rich of the Maryland Historical Society library at Baltimore made a painstaking search of uncatalogued materials and supplied me with many nineteenth-century coal company reports. Other annual reports were secured from the Maryland Department of the Enoch Pratt Library at Baltimore, which also possesses the folios of the *Frostburg Mining Journal.* Archivists at the Hall of Records at Annapolis were helpful in locating the files of Executive Papers, and the University of Maryland's library at College Park (particularly the Maryland Room) furnished Maryland House and Senate journals and documents and the volumes of laws passed at each session of the legislature. The University also possesses an extensive collection of microfilmed newspapers. I am indebted also to the office of the *United Mine Workers' Journal* for the use of its microfilms; to the office of the Superintendent of Schools of Allegany County; to The Times and Alleganian Co., Cumberland, for the use of its exceptional collection of microfilms of newspapers dating back to 1845; and to the office of the Maryland Bureau of Mines at Westernport. I should like to thank individually Glenn F. Massay, of Fairmont State College, Fairmont, West Virginia, for his generous contribution of material from his dissertation in progress, and Mrs. R. L. Stallings, Hugh Fazenbaker, A. F. Greene, Thomas W. Richards, and John P. Stevenson, all of whom contributed information leading to the finding of the George's Creek Coal and Iron Company journals. Lastly, I express my gratitude to Lloyd R. Stallings, owner of the journals, for permission to make use of them.

K. A. H.

Silver Spring, Maryland
May 1968

Contents

Illustrations

Tables

PART I

Portrait of a Coal Region

Chapter 1

The Stable Bulwark

Today, when even a second-growth forest seems a formidable barrier, we marvel at the speed with which our forebears moved back from the Atlantic seaboard in pursuit of Lebensraum and riches. Before the French and Indian War the Maryland frontier had pushed from Frederick to Hagerstown and on to Cumberland, where a tributary of the Potomac cuts through the hills and offers a way west. After 1750 the Cumberland area was traversed by trappers, traders, land buyers, road, railroad, and canal surveyors and engineers, and a host of western travelers both foreign and domestic. Gist's Trace passed through western Maryland, as did Braddock's Road into western Pennsylvania, and in the infancy of the United States this was the passage chosen for the National Road. Here, too, lay the logical route for the railroad and the canal which were planned to bring Baltimore businessmen a share of the western trade.

Carrying out their various commissions or seeking their individual fortunes, a diversity of shrewd and observant adventurers swarmed over mountainsides and along stream beds, and at least some among them must have speculated as to the existence of exploitable minerals. George Washington, who was no stranger to the territory, was one of the first to comment on its natural resources, writing to the Colonial Governor of Maryland in 1755:

"This wonderful country impresses me more and more each time I go through it. I am sure, Hon'd Sir, that it is destined to out-rival some day and outgrow His Most Christ'n Majesty's home country—England. These mighty forests of soft and hard woods will furnish the ships of the world, and with the native stones build the residences of future generations.

"Again, what is beneath the soil? There may be stored mineral wealth which will astonish the countries of the Old World, while the fuel of the future may also be found therein." [1]

At least three eighteenth-century maps indicated the presence of coal in Maryland. Two Frye and Jefferson maps of Virginia, dated 1751 and 1755, show the words "Coal Mine" near George's Creek,[2] and a French map of 1782 entitled "Carte générale des Treize Etats Unis et Indépendants de l'Amérique Septentrionale d'apres M. Bonne" shows a "Mine de Charbon" at the mouth of George's Creek.[3]

The first discoveries of coal and iron may have been accidental, made easy by the nature of the terrain. Deposits of both minerals lay close to the surface and were exposed by spring floods on the swift-moving streams which sliced through the mountains to join the Potomac. A typical discovery of this sort was described on September 2, 1789, by an anonymous Virginia gentleman who had been investigating the navigability of the wild upper reaches of the Potomac. His letter, published in the *Alexandria Daily Advertiser* (reprinted in the *Virginia Gazette* of October 1, 1789), said in part:

"[We] went up the river in a boat from the Great Falls to twelve miles above Fort Cumberland, and continued on foot to the mouth of Savage river. . . . About the mouth of Savage river and George's Creek are inexhaustible beds of coal; some of which the river has laid bare—we found them easy to dig, and the coal of good quality. This range or bed of coal extends along the first ridge of the Allegheny mountain to Will's Creek, above Cumberland; from which place we saw them carrying coal for the nail manufactory, at Hager's town." [4]

This letter is probably the first written report that coal from the so-called Cumberland region was being used commercially. It is also a sort of pre-prospectus for investors, for it carries the magical words which were to appear and reappear in praise of Maryland's coal resources: "inexhaustible beds," "easy to dig," "coal of good quality." Only one catch phrase remained to be added with the advent of the canal and the railroad: "unusual transportation facilities . . . as compared with other coal regions." [5]

The mines opened in the first quarter of the nineteenth century were small, individual operations, and the difficulties of marketing were very great. The sparse local population could absorb only a limited amount of the output as domestic fuel. Manufactories of any sort were also small and few in number in these early days. A glass works established in Cumberland did try coal from Frostburg but found that it gave the glass a greenish tinge not popular with customers; the makers therefore abandoned coal and began to use wood.[6] However, this George's Creek coal proved to be excellently suited for blacksmithing, and smiths were a necessity in frontier settlements. Forges in Winchester and Romney, Virginia, purchased coal laid bare by a flood near Lonaconing in 1810. Some of this same coal, "mined" with mattocks, was taken to Westernport and loaded on flatboats and rafts to be sent down the Potomac to Washington.[7] To dispose of coal discovered in Eckhart in 1814 or 1815 during construction of the National Road, the entrepreneurs had to haul it in wagons to Cumberland and Baltimore.[8]

In spite of its seasonal nature and in spite of the transportation difficulties, by 1820 the Maryland coal trade had become commercially important.[9] The total output came from a few mines at Eckhart, Pompey Smash and Frostburg, and large-scale transportation was limited to periods of high water, when boats could be taken down the Potomac.[10]

During the winters the coal trade became a part-time occupation for those who had other means of livelihood in more clement seasons. Once the coal was dug, it was loaded into wagons and pulled by teams of two, three or four horses to Cumberland, where it was piled up beside the river.[11] In the meantime other men were busy building boats, using the abundant white pine from the forests west of Cumberland.[12] The coal transports were usually flatboats eighty feet long, thirteen feet wide, and three feet deep. Each could carry 1,500 to 1,800 bushels of coal (fifty to sixty tons).[13]

When the river had reached the "boating stage," hundreds of men and boys with wheelbarrows loaded the boats with the coal piled on the river bank.[14] The departure of the fleet—sometimes as many as twenty boats, each manned by a crew of four—was an annual diversion for the townspeople. In addition to the coal, the boatmen carried provisions for a rather limited diet of bread, bacon, coffee, and whisky. If they arrived safely, as many did not, they sold boat and cargo at their destination somewhere between Cumberland and Georgetown and returned home on foot. No one ever brought a flatboat back up-river, but occasionally a crew poled a keel boat back with a load of salt, fish, and other goods. For their dangerous journey the boatmen received wages of ten to twelve dollars per trip. The coal they carried, worth six to eight cents a bushel at Cumberland, brought thirty-five to forty-five cents at Harper's Ferry and fifty to sixty cents at Georgetown. From the first three or four boats on the Cumberland-Georgetown run, the trade grew to fifty or sixty boats a year before the completion of the Baltimore and Ohio Railroad to Cumberland in 1842 all but put an end to the river trade.[15]

These were the haphazard beginnings of the Maryland coal region,[16] an area roughly five miles broad from east to west and twenty-five miles long from the north branch of the Potomac at Piedmont, West Virginia, to the Pennsylvania state line,[17] in a great basin between Dan and Savage Mountains, drained by Jenning's Run and George's Creek,[18] and their tributaries. The deposits in Maryland were confined to Allegeny and Garrett counties. As the area eventually developed, it included the villages of Mount Savage and Barrelville northeast of Frostburg (which lay on the old National Road), Eckhart Mines, a little east of Frostburg, and down the length of the George's Creek valley to the southwest of Frostburg, Borden Shaft, Midlothian, Carlos, Klondike, Lord, Vale Summit (Pompey Smash), Ocean, Midland, Lonaconing, Pekin, Moscow, Barton, and Westernport.

In 1828, stimulated possibly by the successful carriage of coal down the Potomac and almost certainly by the beginning of work on both the Chesapeake and Ohio Canal and the Baltimore and Ohio Railroad,[19] the first of the coal-mining companies sought incorporation under the laws of Maryland. From this time on, unincorporated companies engaged in mining only on a very small scale.[20] In terms of legislation, the Maryland Mining Company was the prototype coal corporation in the George's Creek region.[21] The company was authorized to hold up to 5,000 acres of land and to build a railroad from its mines to Cumberland or to some point on the Chesapeake and Ohio Canal. Originally the Maryland General Assembly approved a stock issue of $200,000, but in 1835 authorized the increase of stock to $1,000,000.

Most mining companies incorporated in Maryland before the Civil War were given essentially the same privileges as the pioneer company. Each was allowed to hold a liberal amount of land, to issue an equally liberal amount of stock, and to build a railroad. However, one opportunity extended to the president and directors of the Maryland Mining Company was unique. These gentlemen were permitted to open a banking business in Cumberland, the Mineral Bank, which ostensibly would have no connection with the mining company. Nevertheless, the boards of directors of the mining company and of the bank were identical.[22] In the charters of all subsequently-incorporated mining companies, banking powers were specifically denied.

In all cases, the Maryland Mining Company included, the acts of incorporation carefully protected the interests of the Chesapeake and Ohio Canal and the Baltimore and Ohio Railroad in the sole, and extremely narrow, passage through the mountains near Cumberland. The supplementary act for the Maryland Mining Company in 1835 contained the following provision:

> It shall not be lawful for the said Maryland Mining Company to occupy or use any portion of the lands in the gorge of the mountains, from the mouth of Braddock's Run towards Cumberland, that may be necessary for the accomodation of the canal, and works of

the Chesapeake and Ohio canal company, or for the main route of the Baltimore and Ohio Rail Road.[23]

A similar restriction, worded in a slightly different fashion, was contained in the act incorporating the Allegany Mining Company.[24] Occasionally the limitations were not spelled out, and certain mine railroads were merely "vested with the same rights and privileges as have heretofore been granted to the Maryland Mining Company by the legislature of this State." [25]

Between 1828 and 1850 the Maryland legislature passed acts incorporating thirty coal or coal and iron companies.[26] Although political quarrels and financial difficulties delayed progress on the transportation projects begun in 1828, no one doubted that both the canal and the railroad would in time serve the coal fields. Indeed, the defenders of the waterway used the potential coal trade as an argument for their cause, shrugging off the railroad as a mere carrier of passengers. One of the chief engineers of the canal stated that it "would pass through coal banks where coal could be thrown from the mines into the boats with a shovel"; that at the Savage mines "the vein was from ten to twenty feet in thickness, over one-fifth of the whole field of two hundred square miles"; and that "each square mile of the principal vein would supply five hundred tons of coal per day for four hundred years." [27]

Prospective investors did not have to rely entirely on enthusiastic hearsay. Beginning with the 1830's a number of scientific writers dealt with the Maryland coal resources.[28] Among them were Philip T. Tyson, later the state's agricultural chemist, and J. T. Ducatel, the first state geologist. Undoubtedly a great impetus to English investment came with the publication of Sir Charles Lyell's *Travels in North America*, which included that eminent geologist's observations on the Maryland coal fields. Of great importance was the establishment of the Maryland Geological Survey in 1833, with J. H. Alexander as engineer and J. T. Ducatel as geologist. Ducatel's first reports on the coal region were published in 1836 and 1840.[29]

The Ducatel survey and others which followed pointed out that the coal resources were not only extensive but also unusually accessible. One writer remarked, "The great bed in which mining is done in the Cumberland coal region is of the extraordinary thickness of 14 feet. Scarcely anywhere in America is there a seam of bituminous coal of this size, extending, as this does, throughout the region, and maintaining its size wherever it occurs." [30]

This was the famous "Big Vein." The Cumberland field also included the "Six-foot Vein" and the "Four-foot Vein," besides a number of smaller veins not thought worth naming.[31] Estimates of 1854 placed the content of the three principal veins alone at more than 1.2 billion tons of coal and the total coal resources in all veins at 4 billion tons. Between 1854 and 1891 about 64 million tons were mined, mostly from the "Big Vein." [32]

By no means all of the companies incorporated went into business. The Baltimore *Sun* of August 7, 1841, reported that twelve mining companies with a chartered capital of $6,700,000 were waiting to begin operations upon the arrival of the canal or the railroad.[33] Nevertheless, in 1850, when both the canal and the railroad had reached Cumberland, only four companies were engaged in the regional coal trade.[34]

In spite of the grandiose statements of their charters, most companies before 1870 held only a few hundred acres each.[35] Many of the early companies failed; many were absorbed by the Consolidation Coal Company, established in 1864, which eventually became the largest bituminous coal company in the United States.[36]

English investors were among the earliest to support Maryland coal companies; the George's Creek Coal and Iron Company was formed by a London and Baltimore syndicate in 1835,[37] and even as late as 1882 much of its stock was held in England.[38] The Maryland and New York Iron and Coal Company, incorporated in 1837, secured subscriptions in England to the amount of £120,000.[39]

Local men also had faith in the region. The original incorporators of the George's Creek Mining Company (later renamed the George's Creek Coal and Iron Company) were John Alexander and Philip Tyson, who at the time of incorporation owned 11,000 acres in Allegany County, valued for capitalization purposes at $800,000 by J. T. Ducatel.[40] By 1848 northern and eastern capitalists were turning their attention to the Maryland coal fields,[41] gradually absorbing the holdings of the pioneer coal companies.[42]

New coal corporations were being chartered even in the twentieth century. Between 1900 and 1910 the Maryland Bureau of Industrial Statistics and Information reported twenty-six new coal corporations with authorized capital stock of $1,861,250.[43] Its annual report for 1908 listed forty coal companies operating seventy-one mines, not including purely local operations, or so-called farmers' diggings.[44]

As later chapters will show in more detail, the absentee ownership predominant before the Civil War resulted in a pattern of close company control in the early mining villages and a lack of communication between employers and employees. When the region entered its major period of development after the war, freely expressed community hostility toward "New York capitalists" prompted Consolidation, which had already become the leading producer, to move its offices to Baltimore and to elect a respected Baltimore citizen as its president. By 1880 only two of the companies maintained New York offices; all the others conducted their administrative affairs in Baltimore or in Cumberland. Moreover, the company presidents not only kept in close touch with their superintendents, but also made it a point to visit the mines in person whenever labor trouble threatened.

The potential uses of bituminous coal in manufacturing and transportation drew the attention of some of the nation's shrewdest businessmen to the western Maryland region. Erastus Corning, the New York railroad magnate, was one of the owners of the Mount Savage Iron Company.[45] This company, which had the distinction of rolling the first heavy iron rails in the United States,[46] produced its own coal in what has been called "an early captive mine." [47] Corning participated in the iron venture with John Murray Forbes and John F. Winslow of Boston.[48] Forbes and Corning were already associated in railroad enterprises at the time of the purchase of the iron company in 1847.

William H. Aspinwall, one of the founders of the Pacific Mail Steamship Company and of the railroad across the isthmus of Panama, urged the combinations which resulted in the organization of the Consolidation Coal Company.[49] Howland and Aspinwall's Pacific Steamers used Maryland coal exclusively, even carrying it around Cape Horn to supply their fueling stations.[50] It was Aspinwall and Forbes, incidentally, who were sent to England during the Civil War to see if they could buy for the Union the iron-clad rams being built there for the Confederacy.[51] Corning, Forbes and Aspinwall were among the first directors of the Consolidation Coal Company.[52]

Warren Delano, maternal grandfather of Franklin Delano Roosevelt, was trustee for two pieces of coal property transferred to the Consolidation Coal Company in 1871.[53] He was also a director of the Consolidation Coal Company from 1864 to 1875,[54] and one of the first directors of the Cumberland and Pennsylvania Railroad.[55] Franklin Roosevelt's father, James, was a director of the Consolidation Coal Company from 1868 to 1875.[56] *

The Borden Mining Company, incorporated in 1846, was linked with New England. The Borden family had earlier founded the town of Fall River, Massachusetts, where it con-

* It is perhaps of passing interest that F.D.R.'s fateful voyage to Campobello in 1921 was made aboard the yacht of Van Lear Black, member of a Maryland family which operated five companies in the George's Creek region and was one of the largest stockholders of the Consolidation Coal Company. See Arthur M. Schlesinger, Jr., *The Crisis of the Old Order* (Cambridge, 1957), p. 366; B. F. Johnson, *Men of Mark in Maryland* (Baltimore, 1910), II, 331; and *Coal Age* XVII: 232–233.

trolled an iron works, a steamboat line, a calico printing works, a gas company, and a railroad line to Boston.[57] All of these enterprises used coal.

Among the incorporators of the Lonaconing Ocean Coal Mining and Transportation Company in 1853 were Edward Cunard and August Belmont, the New York banker.[58] Since 1849, at least, Cunard's British West India and New York to Liverpool steamships had consumed only Cumberland coal on their eastbound trips. This endorsement of the Maryland product was significant because Cunard could have obtained a drawback of duty on English coal at New York and used it exclusively.[59]

The rapid growth of the Maryland coal industry was due in large part to the early development of a system of rail and water transportation. In November and December of 1842 the region shipped via the newly arrived Baltimore and Ohio Railroad more than 1,700 tons of coal. The following year over 10,000 tons went out by this route. The Chesapeake and Ohio Canal carried during its first full operating year (1851) almost 83,000 tons; in the same year the Baltimore and Ohio Railroad hauled approximately 175,000 tons. The Pennsylvania Railroad began to serve the region in 1872, and in that year these three carriers transported close to two and one-third million tons.[60] Several of the mining companies had exercised their privilege of building railroads from their mines to some convenient point on through transportation. Many of these small lines were absorbed by the Cumberland and Pennsylvania Railroad in the 1850's and 1860's.[61] Another important regional railroad which began to carry coal in the 1880's was the George's Creek and Cumberland.[62]

The intercorporate holdings of coal companies and railroads were too intricate to summarize here, but it should be pointed out that from 1876 to 1906 the Baltimore and Ohio, through its stock ownership and its interlocking directorates in the Consolidation Coal Company, had a practical monopoly of the Maryland coal trade.[63] In 1906 the railroad sold all of its Consolidation stock, which then amounted to more than 52 percent of the shares. This sale was made during the course of an Interstate Commerce Commission investigation of the relations between the coal companies and the coal-carrying railroads.[64]

With capital and transportation available, the mining enterprise was guaranteed almost certain success because of the superior quality of the coal from the Cumberland field. Technically semibituminous, it is one of the "supercoals" found in a few isolated sections of the East.[65] Tests by the United States Navy Department in 1844 proved that "a pound or a bushel of coal from this region will generate more steam than the same amount of coal from any other mines in the country." [66] A committee of the Maryland General Assembly reported in 1865 that the coal was particularly suited for blacksmiths and was the best fuel in use for producing steam for ocean steamers, for locomotives, and for steam machine shops.[67] Scharf, commenting in 1882 on the changeover from sail to steam by the world's navies, predicted an increasing demand for Allegany County coal.[68]

Maryland coal maintained its reputation as a quality product. The *Coal Trade Journal* said in its issue of June 15, 1904:

> In all discussions of the Bituminous situation [at that time depressed] the fact must not be overlooked that genuine Georges Creek coal occupies a class by itself and is in a much firmer condition than any of the other Bituminous products. Like Anthracite, Georges Creek coal is now becoming a specialty. The fields are clearly defined and have been exploited to such an extent that no large increase in annual output is probable, and at the same time, like Anthracite, Georges Creek coal has its own particular market, a certain class of trade that will take nothing else, with a good competitive chance, on a large volume of business, so that this coal forms an exception to the general course of the Bituminous market.

Chronologically, Maryland was the third state to produce coal,[69] being preceded only by Virginia and Pennsylvania. The expansion of the coal industry in all three of these states must not, of course, be viewed as an isolated phenomenon, but as part of the national response to the arrival of the age of steam. The availability of water power made possible to a large extent the initial growth of manufacturing in the northeastern United States, and up to the 1840's factories did not commonly use steam power.[70] In the 1840's the proponents of steam were successfully demonstrating its flexibility and energy potential and were establishing steam-driven textile mills in New England and the Middle Atlantic States.[71] A dependable supply of coal for the new mills was vital.

Iron furnaces had already found their charcoal sources dwindling before 1840 and were experimenting with coal, mainly anthracite. Railroads, too, realized that the supply of wood could not last forever, and ocean-going steamships needed a compact fuel to carry them over the seas. In terms of coal resources and transportation facilities the bituminous mines of Virginia, Pennsylvania and Maryland were admirably equipped to suit all these needs and were developed to meet them. (However, before the Civil War about half the coal mined in the United States was Pennsylvania anthracite,[72] which remained the chief smelting fuel until 1875.)[73]

Bituminous coal from all three of these states entered the "tidewater trade" for steamship bunkering, export, and coastal distribution.[74] Cumberland coal went to Cuba and to the French government in 1859;[75] in 1872 to the East Indies, Egypt, Rio de Janeiro, St. Thomas, Valparaiso, "and other distant points";[76] and in 1873 to the British government for supplying its naval stations in the West Indies.[77] In 1898 it was carried to the Klondike gold fields for blacksmithing, and in the same year, "It was our Maryland coal that Sampson and Schley had in the bunkers of their battleships when the whole of Cervera's fleet was destroyed and sunk out of the harbor of Santiago."[78] The principal consumers of Maryland coal, however, were the factories of New England and the steamships which took on fuel at eastern seaports.[79]

To a much smaller degree and for a relatively short period Maryland coal met the need for a suitable fuel for smelting iron. "The first notable success in the use of bituminous coal in the blast furnace in this country was achieved at three furnaces in Western Maryland." [80] These furnaces—one at Lonaconing and two at Mount Savage—were designed to use the coal in the form of coke, and were the first of their kind in the United States.[81]

The Baltimore and Ohio Railroad was using Maryland coal when most other railroads were still burning wood. After its chief engineer had carried out two years of experiments and had found that one ton of Cumberland coal would do the work of two and a half cords of wood in a locomotive engine, besides taking up less space in the tender, the Baltimore and Ohio progressively adapted almost all its locomotives to the new fuel.[82]

During the period 1850–1910 Maryland became one of the nation's principal coal producers. Capitalization of the industry rose from $605,000 in 1850[83] to $22,871,136 in 1909.[84] In terms of tonnage there was an increase from 196,848 tons (not specified whether short or long) in 1850,[85] to 4,039,386 long tons (4,524,112 short tons) in 1910.[86] As late as the 1880 Census, Maryland maintained its place in the top five bituminous coal producers of the United States, being surpassed by Pennsylvania, Illinois, and Ohio, with West Virginia a not very close fifth.[87] By 1889 Maryland had dropped to seventh place.[88] Maryland ranked eleventh among the coal producing states in 1900.[89] However, during the period 1816–1906, over 50 percent of the total coal produced in the United States came from six states, of which Maryland was one.[90] At the close of 1906 Maryland ranked thirteenth among the twenty-eight coal-producing states.[91]

It is understandable, therefore, that Scharf should have commented: "The stable bulwark of the county [Allegany] is its mineral wealth. The effort to develop this and to afford facilities for its introduction into market has absorbed much of the legislation of the state for years. . . . The extent of this mineral wealth, its value to the State, and its importance to the country at large in its commerce and manufactures, fully justify the consideration which has been given to it." [92]

Chapter 2

Intelligent and Fearless

What kind of men mined coal in Maryland? Regional histories, newspapers, and the reports of state officials and labor union organizers convey the impression that the Maryland miners formed an unusual laboring class; yet why this should be so is not easy to determine. In describing the Maryland miners no less an authority than the *United Mine Workers' Journal*, in an editorial dated November 9, 1899, recognized that "in some respects the Maryland mining field and the Maryland miner differ somewhat from those of other localities." The editor then expatiated on the areas of difference: habits, customs, long established local rules of government, reputed conservatism, intellectual abilities, and stable character. The editorial continued, "There is perhaps no mining field in the country which contains so small a per cent. of what is usually termed the 'floating element,' as in Maryland. If we are correctly informed, the greater majority of these miners are permanent fixtures, having resided in the one community and worked at the same mine for many years. They are noted for their morality and honor, and for their quiet, peaceful dispositions."

The only solid facts we have available are those derived from census data concerning place of birth, occupation, citizenship, literacy, and property ownership. Added to these facts is a mass of comment from persons who, for the most part, have no particular axe to grind. Recognizing that some of these comments are Victorian clichés, meaningless in themselves, we can reasonably assume that they were intended to express approbation, to sum up what was considered admirable in a man: honesty, loyalty, industry, sobriety, intelligence, manliness, and the like.

To ascertain why the Maryland miners may have been a unique group we must look beyond national origins, which were not initially different from those of Pennsylvania miners working in the same kind of coal. We must take into account the surprising ease with which the industrial workers merged with the rural community. And we must concede that the coal companies in Maryland were less repressive than those of Pennsylvania and West Virginia.

Allegany County, which until 1872 included the present Garrett County, was Maryland's frontier. Settlement around Fort Cumberland began to expand after 1750, once the French had been disposed of and the Indian danger had lessened. New settlement was limited during the Revolution, but after 1783 there was a renewed surge of population from other sections of Maryland, from Pennsylvania and Virginia, and from Europe.[1] Like much of the Appalachian region, Cumberland and its environs had a number of pre-Revolutionary German pioneers,[2] and some of the farmers who found coal on their lands were of German descent.

Except in Cumberland, where there were skilled craftsmen, merchants, and professional men, the early settlers were dependent to an enormous extent on natural resources, making their livings as farmers, hunters, woodsmen, cattle-grazers—in short, following the traditional pursuits of an agrarian frontier community. Had large-scale speculative development been possible as soon as coal and iron were discovered, this community would have been disrupted. Instead, the exploitation of mineral deposits came about so gradually that no indigestible industrial mass suddenly overwhelmed the rural population. It required seventy years for the number of miners to grow from 140 to roughly 5,800. Even in 1910 miners constituted only 7 percent of the population of the two western counties.

In 1840 Allegany County census takers listed somewhat less than 16,000 persons, including 800 slaves. They also provided a rough occupational breakdown: agriculture, 2,192; manufactures and trades, 648; mining, 140; commerce, 79; and learned professions and engineers, 66.[3]

The first official count of miners in what was to be known as the Cumberland coal region included some who were not coal miners, and some who had been on the scene during the preceding decade. We know that well before the census was taken, both the George's Creek Coal and Iron Company and the Mount Savage furnaces were in operation and employing ore diggers as well as colliers. These two companies probably established the pattern of the mining force in terms of nationalities.

The George's Creek Coal and Iron Company, which had an elaborate plant under construction in the late 1830's,[4] was among the first to import skilled hands for its operations. It brought in a superintendent and supervisory labor from Wales,[5] a founder from England,[6] at least four Scottish miners,[7] and from time to time hired workers from the Pottsville, Pennsylvania, coal and iron establishments.[8] In 1839 the directors of the company complained that miners were scarce and that the few men imported from England that year were not of the most desirable class of workmen.[9] The company also employed many Germans in various capacities during the construction of the works,[10] and found it necessary to have a storekeeper who could speak their language.[11]

The George's Creek Coal and Iron Company imported labor for an enterprise already under way, even though on a very small scale. At least one other company was optimistic enough to establish an advance cadre of workers. In 1843 a Scottish traveler in Maryland observed, "We visited the mines of the Boston and New York Company, and there we met an intelligent countryman of the name of Douglas, at the head of a small colony of Scottish miners, who have been for some time surveying, and are ready to commence operations as soon as the [Chesapeake and Ohio] canal is finished, of which there is still about ten miles to complete." [12]

The Frostburg region in general, according to Andrew Roy

in his *History of the Coal Miners of the United States*, was settled by Scottish miners and was still inhabited by their descendants in 1907.[13] A Scottish immigrant himself, Roy worked in the Frostburg mines for a number of years. The Cumberland *Civilian* for July 28, 1848, reported the recent arrival of coal miners from Cornwall at Eckhart Mines. Large numbers of unskilled Irish who had worked on the Baltimore and Ohio Railroad and the Chesapeake and Ohio Canal found it possible to become coal miners[14] because the early workings were so close to the surface that "a common laborer might here be immediately made competent to do the work of a miner without passing through the apprenticeship of a regular collier."[15] It should not be assumed, however, that all Irish were completely unskilled in mining, for many had migrated from Ireland to Scotland and Wales, where they in time became experienced colliers.[16]

By midcentury Scots, Welsh, English, Irish, Germans and Cornishmen formed the nucleus of a coal-mining class in western Maryland. In reply to the question, "Who are the miners?" a Frostburg correspondent of the Cumberland *Civilian* wrote, "Few if any Americans have given attention to this peculiar handicraft."[17]

Although the miners were not a conspicuous segment of the county population, by reason of either their numbers or their national origins, they were members of an elite class, as distinguished from other labor in and around the mines, and were often related by ties of blood. Pick mining was a handicraft brought to America from the coal fields of England, Scotland and South Wales.[18] As Maryland coal mining changed from a surface to an underground operation, the need for miners to have special skills went unquestioned until the great strike of 1882. In the antebellum period mine foremen were imported from the mining districts of Wales and Scotland, and these foremen were followed by relatives and friends from their old country neighborhoods.[19] The Lonaconing miners in particular were spoken of as being "all uncles and cousins."[20]

In the decade 1860–1870, the number of coal miners in

Table 1. National origins of coal miners of Allegany County, Maryland, at census years 1850–1870

Year	Tot.	Eng.	Welsh	Scot.	Irish	Ger.	Brit. Am.	Other for.	Am. (U.S.)
1850	370	42	36	143	65	46	2	6	30
1860	928	136	99	230	215	144	24	1	79
1870 *	2,463	469	282	654	369	137	56	3	493

* Taken from the three major mining districts only (Westernport, Frostburg, Lonaconing).

Sources: United States Census Bureau, Manuscript Population Census Returns, Allegany County, Maryland, 1850, 1860, and 1870, National Archives.

Maryland more than doubled and the numbers of English, Welsh and Scottish miners roughly tripled (Table 1). One of the reasons for the influx from Great Britain was that throughout this decade Alexander McDonald, the great leader of Scottish and English miners, urged his followers to go to America where they might eventually acquire land and become farmers.[21] Many Scottish miners who were acquaintances of his settled in Lonaconing, and he greeted them when he visited Maryland in 1867.[22] The Irish gained more slowly, and the Germans actually decreased. A substantial number of German-born lived in Allegany County in 1870, but few of these were miners (Table 2). Most German immigrants who entered the United States through the port of Baltimore after the Civil War were headed for the farm lands of the Middle West and stopped in Cumberland or the mining towns only because their original destination seemed too distant.[23]

It will be noticed that there is a discrepancy between total numbers of those who reported themselves "miners" in the population census (Table 1) and those listed as employed in the mines under the censuses of manufactures and mineral industries (Table 2). Employment figures include all mine labor, not merely miners. Furthermore, a "miner" in the population census could be a boy of nine to seventeen, irregularly employed, or an old fellow of eighty or more who no longer pursued his occupation.

Table 2. National origins of all inhabitants of coal mining areas of Allegany and Garrett counties in Western Maryland at successive census years

Year	Totals		Foreign born by national origin									Total mine workers
	All origins	Foreign born	English and Welsh	Irish	Scottish	German	Hungarian	Polish	Italian	Russian	Others	
1870	38,536	7,969	1,959	1,847	1,558	2,312	–	–	–	–	293	2,712
1880	50,178	7,781	1,845	1,834	1,643	2,107	–	1	–	–	351	3,677
1890	55,784	6,207	1,440	1,225	1,265	1,958	15	22	–	–	282	3,741
1900	71,395	5,176	1,254	871	1,121	1,490	21	35	–	–	384	4,936
1910	82,516	5,195	1,056	519	919	1,013	248	–	689	247	504	5,798

Sources: Ninth Census of the United States, I, 358; III, 762; *Tenth Census of the United States*, I, 64, 435, 513; XV, 681; *Eleventh Census of the United States*, XIV, 347; XV, 24, 632; *Twelfth Census of the United States, Population, Part I*, I, 23, 758–59; *Thirteenth Census of the United States*, II, 830–31, 842, 844; XI, 93; United States Bureau of the Census, *Special Reports, Mines and Quarries, 1902* (Washington, 1905), p. 384.

Taking into account mine workers' wives and children, almost the entire foreign-born population of the western counties must have had some connection with the coal mines (Table 2). Obviously, Allegany County miners as a group were not likely to arouse racial or nationalistic antagonisms.[24] Until 1910 the immigration to the Maryland coal region was of the "old" type. Although the strikebreakers imported in 1882 included Poles, Hungarians, and Swedes,[25] they did not remain long enough to change the basic character of the mining population.[26] By 1886 Welsh, Irish, Scots, English, Germans, and Americans again predominated in the mining force of some 3,500 men.[27] Ten years later a visiting newspaper editor remarked, "The mines in this region are free from the pestilential Hungarians, who afflict Pennsylvania, and as a result they are contented and happy." [28] Union organizers had no language problems here; a correspondent of the *United Mine Workers' Journal* commented in 1901, "Out of nearly 5,000 miners not more than 300 or 400 foreign speaking miners are among them." [29]

The Maryland coal region was scarcely affected by the massive exodus from Europe before the first World War. Instead, Allegany and Garrett counties found their foreign-born population steadily decreasing, both in numbers and in proportion to total population. In 1870 about 20 percent of the residents of Allegany County had been born outside the United States, in 1910 only 6 percent. This trend continued throughout the effective life of the coal field. In 1932 historians of the Maryland coal industry stated that nearly all mine employees were native-born white persons,* and that the Maryland field was perhaps the most completely Americanized coal-producing area in the United States.[30] Not until 1910 did immigrants from southern and eastern Europe make up a substantial proportion—22.8 percent—of the region's foreign born, and these 1,200 Italians,

* Few Negroes were employed; in 1892 Negroes composed less than 3 percent of the combined populations of Allegany and Garrett counties. Maryland Bureau of Industrial Statistics, *First Annual Report, 1892* (Baltimore, 1893), p. 212.

Hungarians, and Russians cannot definitely be linked to the coal mines. In any event, they must have caused only a weak dilution of the total population of 82.5 thousand.

In the meantime almost the opposite situation was taking shape in the West Virginia and Pennsylvania coal counties which were in the Maryland competitive orbit. In Clearfield County, Pennsylvania, the proportion of foreign born to the total population rose from a little over 8 percent in 1870 (when the mines opened) to nearly 20 percent in 1890, and dropped back to 17.6 percent in 1910. "New" immigration accounted for 9 percent of the county's foreign born in 1890, 20 percent in 1900, and 31 percent in 1910.

In Somerset County, Pennsylvania, which had less than 8 percent foreign born in each of the census years from 1870 to 1900 inclusive, the proportion of foreign born rose sharply to 20.6 percent in 1910. Of the approximately 11.4 thousand foreign born in 1910, more than half were immigrants of the "new" variety.

Until 1900 Marion County, West Virginia, was sparsely settled, and its population was more than 97 percent native born. The 1900 census showed a tremendous increase in the number of foreign born (from 168 in 1890 to 1,774 in 1900.) More than 60 percent of these new elements of the population came from Hungary, Poland, Italy, and Russia. Between 1900 and 1910 the number of Marion County's foreign born doubled, with the proportion of "new" immigration rising to 64.2 percent. Some of the increase may be attributable to the expansion of the Consolidation Coal Company into this field in 1902.

Contemporary newspaper accounts show that Pennsylvania operators sometimes took advantage of a strike to clear out their Anglo-Saxon and Celtic miners, whom they viewed as persistent trouble-makers, and substitute a new force which they hoped would from sheer poverty and ignorance prove more docile.[31] A complete labor turn-over was relatively easy to effect in areas where the company owned all the houses, ran all the stores, and paid in scrip or tokens, but it was next to impossible to accomplish in Maryland. As their properties ceased to be self-sustaining units remote from other settlement, Maryland companies gradually relinquished the complete feudal control of the environment [32] which Pennsylvania companies and others maintained and were perpetuating in the new coal regions in their states. Maryland miners could buy land and build their own houses. Company stores were outlawed in 1868, although they may have continued to operate in disguise, and many of the businesses were run by retired miners.[33] Cash payments of wages were mandatory by 1880.

In Maryland also an unbusinesslike touch of sentiment at times inhibited any radical change in the labor force. Consolidation, as it demonstrated in the 1882 strike, would probably have been willing to go a long way in importing a new type of labor. It represented for this area the big impersonal corporation as contrasted with smaller operations, not yet absorbed by it, whose Scottish managers or superintendents, and in at least one case whose president, had risen from the ranks. Some of these managerial figures had been officers of military units raised in the mining area at the outbreak of the Civil War and were sympathetic toward the men who had served under them. Although the companies stood together in the strike, the smaller operators had agreed to import English, Scots and Welsh, not Germans, Hungarians and Swedes, to break the strike and with it the Knights of Labor. At the end of the strike, when Consolidation prepared a roster of those whom it would not reemploy, Andrew Spier of Blaen Avon Mines took on as many of the blacklisted miners as he could.[34]

The Americanization of the Maryland coal area, mentioned above, began early. The diarist of the George's Creek Coal and Iron Company recorded on October 16, 1839, "Mr. Schmidt, Dr. Herman and other Germans go to Cumb[d] to declare their intention of becoming citizens." Although most persons looked upon naturalization as a normal and desirable process, the eagerness of the foreign-born to obtain American citizenship was a matter of grave concern to the anti-Catholic, anti-immigrant

American or Know-Nothing party, which became active after the 1852 presidential election. The Know-Nothing Cumberland *Telegraph* attributed the defeat of its party in the local elections in 1855, to the increase in the "anti-American" vote caused by the naturalization of some 180 foreigners during the October session of the Court.[35]

The Know-Nothing party, which was firmly entrenched in Maryland, in 1856 put before the state legislature a resolution to request of Congress a change in the naturalization laws. The proposed amendment would have made the residence requirement twenty-two years.[36] Possibly this agitation, combined with continuous editorial fulmination against foreigners, had a bearing on pre-election naturalization statistics of that year. During the autumn term of court, 140 persons (purportedly mainly Irish and German) were naturalized.[37]

A study of the manuscript population census schedules for 1870 shows that the desire of Maryland miners to become American citizens was a continuing one. Taking the census districts of Westernport, Frostburg, and Lonaconing, the principal coal-mining areas of Allegany County, we find that 1,970 reported themselves as miners and as of foreign birth. Forty percent of these had already become citizens, even though the first real explosion of the foreign mining population occurred in the decade 1860–1870, and probably in the second half of that decade.[38]

The older agricultural community in this area might have resented the intrusion of what was essentially a laboring class, and foreign at that; this does not seem to have been the case except during the brief ascendance of the Know-Nothing party in the 1850's. Like most Americans after the War of 1812, Allegany Countians dreamed of an expanding economy based on natural resources, transportation, and manufactures. They spoke of miners as "the valuable and industrious class . . . who devote their time to the development of the mineral wealth of our State," [39] or as "the stalwart operatives of the mines." [40]

Any "foreign problem" that existed could well be attributed to the Chesapeake and Ohio Canal Company and to the Baltimore and Ohio Railroad because they employed large numbers of Irish laborers, who were by no means as welcome in western Maryland as the transportation facilities they were constructing. The county press had great sympathy for starving and oppressed Irishmen in the abstract, but deplored the brawling, boisterous mob of humanity engaged in internecine warfare on the very outskirts of Cumberland. The "Irish problem" in Allegany County was entirely different from that in the anthracite region of Pennsylvania, where many Irish had migrated in order to work in the mines,[41] and where they carried on definitely nationalistic feuds against English, Welsh, and Germans.[42] There is no documentation of similar feuds among Maryland miners. On the contrary, "The various nationalities of Welsh, Scotch, Irish, English, German, Americans and Swedes fraternize with tolerable congeniality in their home life—the sense of common danger being a tie to good fellowship."[43] However, local newspapers reported numerous incidents of bloody dissension between factions of Irish on canal and railroad.

Most of the Irish who joined the western Maryland mining force between 1850 and 1860 had been canal or railroad laborers. They accepted the unskilled jobs around the mines, both above and below ground, and were not truly in competition with the trained craftsmen of England, Scotland, and Wales. In 1850 only 65 *miners* were of Irish birth; in 1860, only 215. It seems unlikely that these small numbers were entirely responsible for the anti-Irish sentiment expressed by Know-Nothing editors in the 1850's.

The Know-Nothing party in western Maryland was also anti-German. Germans had lived in the county since the very early days of settlement, but they had become thoroughly Americanized during the first half of the nineteenth century.[44] However, increasing numbers of Germans settled in Cumberland, Lonaconing, Frostburg, and Eckhart Mines around the middle of the century.[45] To the Know-Nothings, the new wave of German immigration seemed to stem directly from the uprisings of 1848, which were either ignorantly or deliberately misunderstood by the native descendants of other revolutionaries. The Cumberland *Telegraph* pursued an anti-foreign policy during the presidential year 1856, repeating with variations the theme that German adventurers were making "every possible endeavor to engraft upon our political system a spirit of radicalism and licentiousness." [46]

These criticisms were aimed at Irish and Germans in general, not at the mining population in particular, although, by putting two and two together, we can conclude that some of the *Telegraph's* post-election remarks in 1856 were intended to hit the mining districts. Fillmore, the Know-Nothing candidate, carried Maryland, but lost Allegany County to Buchanan by 310 votes. The mining districts, which cast a third of the total vote of the county, were responsible for 258 votes of Buchanan's 310–vote majority.[47] Later the *Telegraph* referred briefly to "The Irish Locofoco Demonstration [in celebration of Buchanan's victory] at Mount Savage," the "Limerick of Allegany." [48]

After a few feeble attempts at survival in the local elections of 1857, the Know-Nothing party quietly died, and violent attacks against the foreign population ceased. In 1860, when Germans were mentioned in the Cumberland press, it was as "our industrious and quiet Germans." [49] The last suggestion of serious hostility to foreigners occurred in the early part of the Civil War, when the region, strongly Unionist in sympathies, was tense and suspicious. One newspaper expressed a fear that many Allegany County residents of British birth would seek exemption from military service on the grounds that they were not American citizens.[50] There is no evidence that miners made such claims of exemption.

Almost from the beginning of corporate mining in Maryland, any number of persons felt impelled to comment on the characteristics of the miners. From this body of material it is possible to reconstruct a picture of how these workers appeared to contemporary observers.

The Maryland miner did not emerge as a stock character with generally agreed-upon traits until after the Civil War. Up to this time both mines and mining force had been going through a period of development in which employment conditions were anything but stable. Work was irregular, and miners of necessity moved from place to place looking for jobs.[51] In the 1840's and 1850's Maryland men tended to drift into western Kentucky for seasonal work in the mines there.[52] Some of the earliest Maryland miners engaged in their occupation only long enough to earn money for buying farms in the West.

Andrew Roy, who bought a farm in Arkansas just before the Civil War, noted in his *History of the Coal Miners* that many Maryland miners acquired land in Iowa or Wisconsin, where they farmed during the summers, returning to the mines after harvest in order to earn money for livestock and agricultural equipment.[53] Although not referring specifically to Maryland miners, Alexander McDonald told a Parliamentary committee that in the United States he had found thousands of young men who had left Scotland a few years earlier already possessing forty to a hundred acres. Many of them worked part time on the land and part time in the mines.[54] William Cullen Bryant, visiting the region in 1860, had remarked that the more provident miners were investing their savings in land, and predicted that they and their children would eventually move into the agricultural class.[55]

Not all were so prudent. In the Maryland coal field, "as is invariably the case in every mining region," there were "a number of thriftless men, who possessed a roving and adventurous disposition, and who spent their hard-earned money in soul-debasing pleasure." [56]

Bryant described the majority of the inhabitants of the mining villages as owing "little to the school-master" and as being not particularly interested in the education of their children; the farmers of the region also were "ill-schooled, if schooled at all." In his *History of Western Maryland*, Scharf published an expurgated version of Bryant's article, adding in a footnote, "Since then there has been a great change, and Mr. Bryant, if alive, would find the miners intelligent, surrounded with pleasant homes, and their children attendants at the free and Sunday-schools, than which none in Maryland are better than those of the mining regions of Allegany County." [57]

In any event, Bryant's sneer at the miners' indifference to education was not warranted. The manuscript population census schedules for the year of his visit show that of 928 miners, only 71 fell in the category of those "over 20 y'rs of age who cannot read & write." Furthermore, although schools were not free, the same schedules show that in the census districts containing the bulk of the mine population (Frostburg and the fourth district), more than half of the mine families with children of the appropriate age (six to sixteen) managed to scrape up enough money to send one or more of their children to school.

Miners who emigrated to Maryland brought their families with them.[58] As a result, the mining class which became established there was settled, not migratory as in some other districts. Furthermore, owning their homes, Maryland miners were reluctant to leave the region even in periods of economic crisis.[59] In contrast, the Pennsylvania Department of Internal Affairs reported that in the coal regions of that state thousands of Hungarians, principally men without families, had no intention of staying longer than was necessary to accumulate a few hundred dollars to take back to Hungary.[60]

Since the miner's way of living is discussed later, it will suffice at this point to establish him as typically a family man and worthy citizen: "The public and private morals of the mining community are excellent. Men, as a rule, marry young, and while there are no restrictions on the use of intoxicating drinks, intemperance is the exception. . . . Women and girls attend exclusively to household duties." [61] "Large families are the rule in the mining districts, six children to a household being a low average. The children are early initiated into the duties and responsibilities of life, both domestic and industrial." [62] Maryland miners "are kind to their wives and children, and provide them with plenty of wholesome food and comfortable clothing. Hospitality is another characteristic of the Maryland miner. He has always an extra plate for the friend or stranger that may drop in at meal-time, and to refuse to sit at his board would be a personal insult." [63]

The characteristics of this settled family class which their contemporaries stressed over and over in their writing were thrift, industry, and intelligence. Official note of the condition and character of the mining population began with the mine inspectors' reports in 1876 and was added to after 1884 by reports of the Maryland Bureau of Industrial Statistics and Information and its various successors.

The first mine inspector wrote in 1876, "In this teeming valley of Georges Creek and the adjacent mines, there is an air of thrift and industry. The population here partake of the hardy character of the mountains around them and are a healthy and industrious people. But few willing idlers are to be found and the hive is almost entirely free from the presence of drones." [64]

In 1885 the chief of the newly created Bureau of Industrial Statistics and Information, whose duty it was to describe the condition of all labor in the State, singled out the miners as "an intelligent and thrifty body of men," well informed and communicative.[65]

This type of description continued throughout the period under discussion, and is perhaps most adequately exemplified in the quotation from which the chapter heading is taken: "It is a mistaken idea that the coal miners are a rough, uncouth or lawless class. Their social condition compares favorably with that of any other part of the State. These men located in Maryland from twenty to forty years ago, and here they have made their home, steadily and industriously pursuing their chosen calling. Although composed of five different nationalities, under our remarkable and elastic system of government they have assimilated with our people in habit and custom, until today the race lines are almost obliterated in the common bond of American citizenship. My experience has been, that these men who, in the western mountains, have by their labor added so greatly to the wealth of this community, are an intelligent, sympathetic and fearless class." [66]

The contention that the miners were thrifty is supported by statistics on personal property holdings derived from the census manuscripts for 1860 and 1870. In the former year 30 percent of Maryland coal miners reported that they owned personal property in small amounts, usually between $50 and $150. In 1870 the proportion claiming personal property was about the same—roughly 34 percent—but the amounts held were larger, none less than $100, with a surprising number running to several hundred dollars and a few even into the thousands. As to thrift we learn further from a correspondent of the *Frostburg Mining Journal* on August 6, 1910, that at the close of business on June 30 the three Frostburg banks had on deposit $1,690,850, "the great bulk of this sum representing the earnings of miners."

It should not be assumed from the catalog of praise that the Maryland miners were a docile working class. Between 1847 and 1900 they engaged in twelve major strikes, to say nothing of a host of smaller ones; yet they did not incur public disapproval. The Baltimore *American*, which was outspokenly on the side of the mine owners during the bitter and hard-fought strike of 1882, remarked on March 15 of that year, "There is no doubt that the class of miners in this region is the best in any section of the country. The men are quiet, sober and hard-working, and what is even more unusual, and to their credit, are disposed to be saving and economical." Perhaps more significantly, the article went on to say, "A general agent of one of the largest companies when questioned on this point today, at once heartily admitted that the men were personally all that could be desired. He further said that he thought that the George's Creek region had the best men in the world."

Even a circuit court judge, in handing down a decision against a labor leader in an injunction case arising from a later strike, was moved to comment, "The miners of this region, as a class, have for years been noted for their intelligence as well as their orderly conduct. The Court has often noticed that many of its most intelligent and satisfactory jurors have been from those engaged in the mines." [67]

Many of the migrants from the British Isles came to America with ambition to rise "from the grade of a working miner." [68] In the Maryland coal fields the most industrious, thrifty, and intelligent found ample opportunity either to move upward themselves or to provide that their sons would not spend their lives in the pits. By the 1890's some mine boys had become "teachers, doctors, lawyers, preachers and priests." [69] An early twentieth-century history of the area, including biographical sketches, confirms the fact that mining families had produced lawyers, physicians, ministers, bankers, and merchants.[70] *

* In addition, it should be noted that three figures of national importance came from the Maryland mines: Andrew Roy, labor leader and historian of the coal miners; John Gardner Murray, Presiding Bishop of the Protestant Episcopal Church in the United States; and David J. Lewis, member of Congress and champion of social reform in the Woodrow Wilson and Franklin Roosevelt administrations.

Chapter 3

Sooty Demons of the Mine

Writing in 1860, William Cullen Bryant has provided us with a highly colorful account of his trip underground in a Maryland mine: "Our party made a visit to a coal-mine some three miles distant from Mount Savage. From one of the black entrances flowed a lively little stream with yellow waters, into which I dipped my finger to ascertain their flavor. It was acidulous and astringent, holding in solution both alum and copperas. Leaving the Stygian rivulet we came to another entrance, out of which a train of loaded trucks was passing, every one of which was attended by a miner blackened from head to foot with the dust of his task, and wearing in the front a small crooked lamp to light his way. As they emerged from the darkness they looked like sooty demons of the mine with flaming horns coming from the womb of the mountain. We now entered, each carrying a lantern, attended by a guide. The vein of coal is from eight to ten feet thick, and the passage is of that height, with a roof of glistening slate, propped in some places by wooden posts. Here and there, on each side of the passage, yawned chambers cut in the veins of coal, and extending beyond the reach of the eye in the faint light of our lanterns. At length we heard the sound of sledges, and proceeding for some distance farther came to the end of the passage, where the workmen, each with a lamp in his cap, were driving wedges into the cracks and fissures of the coal to separate it from the roof and walls. We saw several large blocks detached in this manner, the workmen jumping aside when they fell, and then we retraced our steps." [1]

From Bryant's description it is evident that Maryland coal-mining in 1860 had progressed far beyond its initial stage, whose primitive methods were comparable to those used in England up until the twelfth or thirteenth century.[2] The miners dug short tunnels, seldom more than a hundred feet long, into the side of a hill where there was an outcropping of coal. The timbering was so inadequate that the tunnels soon caved in. After 1840, tunnels were longer and better timbered, and might have small openings called rooms driven off them. Between the rooms enough coal was left standing in the form of pillars to support the roof. The length of the tunnel was limited by the problem of ventilation; a man could, after all, go only as far as he could breathe. The rooms were opened in a haphazard fashion and were only about fifty feet by twenty feet in size.

The location of the coal veins made mining in Maryland "extremely simple and inexpensive" and gave the region an economic advantage over competing coal fields. The veins lay in a nearly horizontal position high up on the hills. Thus they were easily accessible through adits, or tunnels, driven into the side of the mountain near the outcrop of the vein. The coal could be worked without any outlay for deep perpendicular shafts, elaborate drainage arrangements, or expensive hoisting and pumping machinery. By 1854 the mines were penetrating well into the hills, and were substantially timbered. The coal was brought out by horses or mules hauling small cars over railroad tracks laid into the mine.[3]

By the 1870's most Maryland mines had "two parallel headings, 9 or 10 feet wide, and from 30 to 100 feet apart, having a block of coal between them as a pillar. The headings may reach any distance up to a mile or more, according to the extent of the property on the line they are directed. The object of driving two together is that, by means of occasional cross-headings connecting them, a system of ventilation is obtained. This is often perfected by connecting one of the headings with an air-shaft, at the bottom of which a fire may be kept up to increase the upward current of air which is supplied from the heading, and its

View from the Eckhart Mines, 1854

Mount Savage Iron-Works, 1854

Mouth of Eckhart Mines, 1854

Borden Shaft, *ca.* 1870

"Strikers and Those Who Suffer: A Scene at Eckhart, Md.," 1882

Page from *Leslie's Illustrated Newspaper* showing four scenes from the 1882 strike

course regulated by tight partitions with doors when needed." [4] Another improvement was the opening of rooms in a systematic fashion to get out more of the coal.

When the mine inspectors refer to "furnaces" in their reports, they are speaking only of the large fires used for producing a current of air to aid in ventilation. The mine Bryant described had such a furnace: "Before returning to the entrance, however, our guides took us into a branch of the main passage, in which, after proceeding a little way, we heard a roar as of flames, and then saw a bright light before us. A furnace appeared, in which a fierce fire was blazing; the blackened workmen were stirring and feeding it, and a strong current of air rushing by us went with the flames up the shaft, which reached above to the surface of the ground."

Obviously this method of ventilating created certain hazards. Maryland was fortunate in having no deaths caused by fire in the mines. In Pennsylvania, on the other hand, a fire started by the furnacemen in Avondale Mine, Luzerne County, caused the death of 109 persons in 1869. Maryland miners on learning of the catastrophe contributed at least $450 to the victims' wives and children.[5]

Mining in Maryland until near 1900 was pick mining. "The roof in these mines is generally treacherous, and it is not allowable to use any powder for blasting." [6] Besides, blasting with powder shattered the coal and made it less marketable.[7] Even in 1902 the Consolidation Coal Company notified its miners to stop using powder because it crumbled the coal.[8]

James M'Killop, a Scottish observer of various American mining areas in 1869, reported, "The miner all over the state of Maryland is supplied with a set of tools (free) by the employer when he commences work; which set consists of two picks, two wedges, one shovel, and one sledge. This set is checked against him, so that he may return them to the owners when leaving his employment. If he fails to restore the same number, a deduction equivalent to their value is retained from his balance of wages." [9]

The method of mining from 1870 to 1910, with some modifications, was essentially that described by Peter Cain, the first Maryland mine inspector, in 1876.[10] In most of the mines the headings ran right through the mountain, "entering it on one side and coming out, so to speak, to daylight on the other side." [11] In fact, as a retired mine superintendent remarked to the present writer, the women at Hoffman Hollow used to walk through the mine to get to the store at Eckhart. At this period mines usually had three headings, the third being for the steel rail track which took the coal out. Miners opened rooms off the headings and then drove passages called galleries from one room to another to help in getting out the coal and also in improving ventilation. Temporary rails were carried right into the rooms and laid along one side between the wall and a row of props necessary to prevent the roof from falling on the men. These props were logs or pieces of round lumber five to seven inches in diameter, installed about three feet apart in a line through the center of the room. The other side of the room held the tracks and what the miners called "the gob," a deposit of waste matter taken out in working or hauled away to clear the working area after an accidental fall of coal and slate.

The men preferred to work two to a room. They had a working space six to seven feet high. Each miner would begin operations by digging with his pick a horizontal slot three to four feet into the solid coal from one side of the room to the other near the floor. This process was called undermining. Wedges put in at proper intervals prevented the coal from falling before the slot was finished. Undermining involved a considerable element of risk. When working in loose or yielding coal, the miner would cut in only about two feet and would remain in a standing position so that he would have a better chance to escape if the coal suddenly gave way. In solid coal he would work in a standing position until he could no longer reach under the coal. Then, relying on the "sound" of the coal as a measure of safety, he would lie on his side to complete the job, in a position which made him particularly vulnerable if the coal gave way sooner than he expected.

The next step was cutting or shearing, making vertical slots eight to ten inches wide and three to four feet deep on both sides of the room from the undermining to the roof. When the shearing was finished, the miner was ready to "take down" the coal, removing the wedges from the undermining and inserting wedges near the roof line. A skillful workman could bring the coal down in lumps weighing from one thousand to three thousand pounds, again a risky operation. The coal companies were particularly fond of exhibiting large blocks of coal at such public shows as the Philadelphia Centennial and the World's Columbian Exposition. It must have been deeply gratifying to six miners of the George's Creek Coal and Iron Company to receive "diplomas of honorable mention from the Board of Lady Managers of the World's Fair [Mrs. Potter Palmer *et al.*] for 'skill as expert artisans' in mining the big lump of coal exhibited . . . at the great Exposition." [12]

Next the coal was broken into lumps of a convenient size for loading into the small mine cars which carried about two tons apiece, and the cars were pushed by hand out of the rooms to the headings, where mules or horses (later locomotives) were attached to haul the cars to the mine entrance. From here the cars went down an inclined plane to the railroad which carried the coal to market. At the "dump house" on the railroad the mine cars were emptied into the railroad's coal hoppers. The cars on the plane were hitched to an endless wire rope, the loaded mine cars going down by their own weight and at the same time pulling the empty cars back up to the mine.[13]

Rumors spread in 1883 concerning the invention of a coal-cutting machine costing only $350 and able to do the work of thirty miners. The editor of a Frostburg newspaper thought it most unlikely that machinery would ever be used in the George's Creek region because of the nature of the coal.[14] However, in 1898 the Consolidation Coal Company put the first machine into its Ocean Mine.[15] The men did not like the machine because it left them all the hard work,[16] and by 1899 the mine inspector had come to the conclusion that the new method of mining would not be satisfactory in the area. The output per machine would not compensate for the cost of installation, and there would inevitably be wage disputes.[17] In 1910 the Maryland mining region had only forty-seven machines, all in Allegany County. Out of a total production of 4,716,382 tons in that year, only 124,399 were mined by machine.[18]

There were two reasons why no major mine disasters occurred in the Maryland coal region. First, there was, according to the report of the Bureau of Industrial Statistics and Information, "no inflammable gas or 'fire-damp' in the Maryland mines." [19] (A later writer said that it was not strictly true that the Big Vein contained no fire-damp, but that the amount given off was so small that only once in fourteen years had a detectable quantity accumulated.) [20] Second, Maryland mines had no deep shafts going great distances vertically into the earth. In the whole region there were only two shaft mines, the Borden Shaft and the New Shaft, and these were not continuously in operation during the period under discussion.

Maryland miners, therefore, were not trapped by the explosions, fires, and floods which took hundreds of lives at a time in mines elsewhere in the country. Only 70 miles from the George's Creek field, in one of the greatest disasters in American mining history, an explosion at the mines of the Fairmont Coal Company (an associate of the Consolidation Coal Company) at Monongah, West Virginia, killed 359 workers on December 6, 1907.[21] Roy's *History of the Coal Miners*, published shortly before this event, had commented on the West Virginia mines in general: "More miners are annually killed by explosions in West Virginia, man for man employed, or ton for ton mined, than in any coal producing State in the Union or any nation in the world." [22]

This is not to say that there were no accidents in the Maryland mines, but that the accidents which took place were not of a nature to create headlines. Both the owners and the mine inspectors were cognizant of potential danger in the Borden Shaft and were vigilant in seeing that conditions there were satisfactory. The inspector was able to report in 1881, for instance, that all due precautions were exercised in operating the cage which carried the men underground; that ventilation was sufficient, without a sign of black-damp; and, most important of all, "In case of fire or any mishap to machinery, there is a good horse road, by which the miners can leave the mine safely." [23] In the Avondale fire (in Pennsylvania), miners perished because they had no way of exit except the blazing shaft.[24]

The chief danger to Maryland miners lay in the characteristics of the coal itself. "Unlike anthracite coal, it does not bind in a solid mass, but is full of 'slips,' into which the miner may unwarily penetrate, and suddenly dislodge great masses of coal." [25] Many fatal accidents were caused by the "dull, heavy nature" of the coal, which gave no warning to the workman that a chunk of several tons was about to burst "out of the very solid." [26] A miner might be deceived by the sound of the coal and be crushed while lying down undermining, or the sound of the top coal might induce him to postpone setting a prop under a dangerous roof.[27]

There were human deficiencies, too. A legislative commission set up to investigate mining conditions in 1880 reported: "The accidents in these mines are in a great measure due to the carelessness of the men in moving about in the mines, and the neglect of the miner to set up his props at the proper time to secure safety, on account of his eagerness to send out all the coal possible upon the receipt of a barge [*sic*] order at the mine. Many more accidents would occur but for the constant vigilance of the master miner and his assistants in their daily rounds of the mines." [28] The Maryland Bureau of Industrial Statistics observed in its report for 1890–91, "The constant presence of danger makes these men indifferent to the risk they take in the prosecution of their industry." [29] Mine Inspector Thomas Brown, long a miner himself, said in his 1881 report that his experience of over forty years in the mines led him to conclude that in many cases men in the mines lost their lives because they were careless. Furthermore, overseers did not always insist upon the observance of elementary safety measures. "If overseers were held responsible for the lives of men under their charge, there would soon be a decrease in the number of lives lost. The present system requires every man to look out for himself. Some, through ignorance, and some, through neglect, do not use the precautions they ought to do." [30]

Apparently supervision and the habits of the miners did not improve much in the next two decades. Another mine inspector almost twenty-five years after Brown made essentially the same comments: the mine foreman or his assistants ought to visit every working place at least once every two days; he should report careless men to the mine inspector, who had the power to force them to adopt safety measures.[31]

Added risks came in the early twentieth century when miners using explosives, with seemingly no conception of the safe way to handle them, began to carry 25-pound cans of powder into the mines. "In making the different inspections I discovered the dangerous practice of making a cartridge from a large can of powder and very often I would find a small boy making a cartridge with a large [open] light on his head." [32] Safety lamps were not used in the Maryland mines. Each man worked by the unguarded light of an oil lamp carried in his cap,[33] a means of illumination not changed since the opening of these mines. "The miners in this region always use a common oil-lamp, which, being small, is fastened to the front of their caps. They can then see to work much better than if carried in any other way." [34]

Whether the cause might be assessed to their own carelessness or to physical forces beyond their control, the fact remains that men were constantly meeting with accidents "not only when actually at work, but while eating their meals in the mine, or when loading cars, or even in passing along the passages." In the 1880's it was difficult to find a miner thirty years of age who had not met with some "disaster." [35]

The statute which created the post of mine inspector required that the coroner of Allegany or Garrett County should notify the mine inspector when fatal accidents occurred in the mines, so that he might be present at the inquest.[36] Up to this time it had not been the practice to hold an inquest following the accidental death of a miner. The only notice taken of such cases was a brief statement in the papers: "Killed by a fall of coal." [37]

Except in 1894 and 1895, the number of fatal accidents per thousand miners in Maryland during the last decade of the nineteenth century was well below that of the states with which she competed for the eastern market (Table 3).[38] The remarkably good showing for 1898 represented a total of only four fatal accidents, one to each 1,143 men employed and to each 1,031,611 tons of coal mined.[39]

Table 3. Fatal accidents per 1,000 employees in bituminous coal mining, 1890–1899

State	1890	1891	1892	1893	1894	1895	1896	1897	1898	1899	Average
Md.	2.08	1.54	1.52	1.23	1.69	2.30	1.58	1.17	0.89	1.08	1.51
Penn.	2.18	3.21	1.69	1.60	1.44	1.83	2.14	1.72	2.38	2.82	2.10
W. Va.	—	3.16	2.76	4.20	2.98	3.97	2.68	2.89	3.86	3.55	3.34

By 1907 at least one company, the Consolidation Coal Company, had provided an ambulance at its largest mine to carry injured men to their homes. First-aid materials were also available, together with "such drugs as can be administered until the service of a physician can be secured." [40] The Miners' Hospital at Frostburg was not opened until October 4, 1913.[41]

In 1910 the area got its first visit from one of the mine rescue and training cars of the United States Bureau of Mines. Much of the equipment demonstrated was intended primarily for gas and smoke victims and was not applicable to George's Creek mining conditions. However, the demonstrations of first aid for injuries from a fall of coal were particularly valuable, and the manager of the Consolidation Coal Company planned to organize "a full-fledged corps of rescue men." [42]

Maryland had no semblance of a workmen's compensation law until 1902,[43] and that law was short-lived and ineffective. Other states also were remiss in this respect. West Virginia laws in 1907–08 were so worded as to give no recourse to survivors of the Monongah disaster victims. These widows and orphans received as compensation for their loss only what funds were voluntarily paid by the company or raised by public subscription. Maryland miners donated over $1,000.[44]

Although Maryland coal mines may have been considerably less dangerous than those in other regions which plunged deep below the surface of the earth or contained explosive gases, they were, as is any mining operation, far from ideal places in which to work. They were dark, illuminated as late as 1905 only by the naked flame of tin lamps carried in the miners' caps;[45] they were cool—about 60° F. summer and winter;[46] and some of them were wet. "In portions of some mines the men may be found working in 10 or 12 inches of water."[47] And although there was no fire-damp, "what is known as the 'black damp' is met with, and is very foul and injurious."[48] "It is not surprising, therefore," wrote Maryland's labor statistician in 1885, "that miners suffer both in health and body. Shut out from air and daylight, engaged in arduous toil, exposed to damp and danger, the miner who lives to 45 or 50 years of age is decrepit and broken down."[49]

Some of these conditions could be, and were, remedied. The *Coal Trade Journal* of March 6, 1895, reported that the George's Creek Coal and Iron Company was planning to use incandescent electric lights in the mines as far as the place where the horses were hitched to the mine cars, a distance of about a quarter of a mile. In remote parts of the mines oil lamps would still be used. In 1905 the Consolidation Coal Company was about to install near its Ocean No. 1 opening an electric plant to provide light and power for that mine as well as a lighting system for Midland and other towns.[50]

The mine inspectors' reports show continuous improvement in drainage. In 1903–04, wet conditions were reported in only two out of fifty-one locations visited.[51] In 1904–05, there were eight wet places out of fifty-four locations;[52] in 1905–06, two out of seventy;[53] in 1908–09, four out of sixty-seven;[54] in

1909–10, eight out of ninety-five;[55] and in 1910–11, seven out of one hundred five.[56]

By 1901 the Consolidation Coal Company had made some provision for the comfort of men coming out of the mine, installing three bathrooms at the New Shaft for the use of employees who had to work all day in wet places in the Hoffman mine. A miner could "go to the Shaft in his Sunday clothes, take them off, take a check for them, hand his valuables to the keeper, put on his work-a-day suit, go down into the shaft, spend the day, come out, take a bath, put on his Sunday clothes, collect his valuables and reach home looking like the presiding elder." Meanwhile, his wet suit was placed in a dryer until the next morning.[57] The Carlos Mine of the Barton and Georges Creek Valley Coal Company also provided a bathroom in its powerhouse so that the men could have a hot bath and change their clothing before going home.[58]

These facilities, however, were available only to the approximately five hundred men employed at Hoffman and the two hundred men employed at Carlos out of a total of nearly six thousand miners.[59]

So far as working conditions were concerned, the problem which most plagued operators and miners alike was that of ventilation. During the 1870's the principal spokesman for all American coal miners on this point was Andrew Roy, who had been a miner and mining geologist and was in 1874 mine inspector for the state of Ohio. Roy had strong views on ventilation. For example: "Constant labor in a badly-aired mine breaks down the constitution and clouds the intellect. The lungs become clogged up from inhaling coal dust, and from breathing noxious air; the body and limbs become stiff and sore; the mind loses the power of vigorous thought. After six years' labor in a badly ventilated mine—that is, a mine where a man with a good constitution may from habit be able to work every day for several years—the lungs begin to change to a bluish color. After twelve years they are black, and after twenty years they are densely black, not a vestige of natural color remaining, and are little better than carbon itself. The miner dies at thirty-five years, of coal-miners' consumption." [60]

Putting forward arguments of this sort, the miners of several states fought for laws requiring them to be supplied with the fresh air on which their health and lives depended. The operators uniformly resisted because of the expense involved. Maryland's first ventilation and inspection law for the coal mines was enacted in 1876,[61] during the same general period in which other states were securing similar legislation: Pennsylvania, 1869; Illinois, 1870; Ohio, 1874; Indiana, 1876; and West Virginia, 1879.[62] The Pennsylvania law applied only to one county in the anthracite region. Bituminous miners in Pennsylvania did not get the benefit of inspection laws until 1877.[63]

As one can see by reading the text of the Maryland law, it was not intended to force the coal companies into any radical change of ventilating methods. The mine inspector would have found it difficult to prove a violation under the statute, but he was perfectly free in his reports to the governor to describe conditions as he saw them and to make any recommendations he wished. The first inspector found that "the miners were somewhat timid and disposed to evade giving . . . information of irregularities or defects . . . through fear of being discharged from employment." [64] So far as his own observations were concerned, he noted in his second report that ventilation was good except in two mines where black damp was present in old workings, and that the companies had later remedied that situation.[65]

We find frequent references to black damp in reports from the mine inspectors and the Bureau of Industrial Statistics and in comments of the miners themselves. Black damp, as Roy describes it, "in its pure state . . . is a deadly lung poison, neither light nor life being capable of existing amongst it, and the miner's lamp, when placed in a solid stratum of it, becomes instantly extinguished as though it were plunged in water. When 10 per cent. of black-damp is diffused through the air of mines, a light cannot be maintained; and after a light ceases to burn it is never safe for a miner to trust himself for any length of time in such an atmosphere." [66]

In 1851, when the Maryland mines were in their early stage of development, a Cumberland newspaper published a straight-faced account of one solution to an accumulation of black damp which had temporarily closed one of the largest mines. "The foul air was let into the working chambers through holes in the solid coal, bored by those monster Norway rats that are so numerous in this region. It seems that these animals infest the mines in great numbers, seeking the fragments of food thrown away by the miners. In the case alluded to, they bored holes through the wall dividing the chambers in which the miners work from those that have been closed up. This foul air is carbonic gas and is so heavy that it settles on the ground like a weight. It was removed, we are told, by pushing a train of cars into the infected places, and then drawing it out loaded with the gas! Thus, an old miner told us, it was carted out." [67]

Unfortunately, not all problems of black damp could be handled so expeditiously. In the 1880's this gas was most troublesome in openings which were nearly worked out and which the owners refused to ventilate because of the cost.[68] Most of the mines at this time relied on natural ventilation, that is, air brought in by the headings, with circulation aided in a few cases by air shafts and furnaces.[69] The mine inspector in 1881 cited the Savage and Old Detmold mines as examples of the failure of natural ventilation, arguing that this method was ineffective when the air was the same temperature inside and outside the mines. In damp and rainy seasons air inside the mines became "stagnant, then impure and poisonous . . . , and therefore totally unfit for man or beast to breathe." [70] He urged that the mining law be amended to require some form of artificial ventilation, pointing out that the companies would profit and the stockholders benefit because good ventilation would enable the miners to take out more coal per acre.[71] The first ventilating fan in the George's Creek region was installed by the Consolidation Coal Company in its Hoffman Mine in 1885,[72] and by 1890–91 artificial ventilation was generally taking the place of natural ventilation.[73] Even so, black damp continued to be troublesome, particularly since many companies were reopening abandoned and supposedly worked-out mines to take out additional coal by new and improved methods.[74]

The mining laws were extensively revised in 1902.[75] By 1903–04 the mine inspector was able to report a marked improvement in ventilation. Although the volume of air delivered might not always be up to the requirements of the law, "at no time was any person in danger of being injured on this account." [76]

But progress in ventilation seems to have been offset by the use of inferior oils in the miners' lamps and by the increasing use of blasting powder.

In 1906 the mine inspector requested an amendment to that part of the mining law which prescribed the type of oil to be used for light in the mines; he felt that the standards were so high that they were generally disregarded. The matter was a serious one, involving "five hundred, six hundred, and in one case nearly one thousand lights burning in a mine, many consuming the cheapest oil to be bought, regardless of the menace of the poisonous smoke to the health, not only of the user, but of all other men employed where the smoke can befoul the atmosphere." [77] That the oil lamp continued to be an offender is evidenced by the complaint of the mine inspector in 1912 that for years the inspectors had been asking for a change in the law concerning oil. "With the inferior grade of oil and the amount of explosives used in our mines it has become a menace to the lives and health of those working underground." [78]

In addition to polluting the air, blasting brought other problems to the mines. The inspectors were concerned because the shooting was done carelessly and recklessly in an attempt to get out as much coal with as little work as possible. As one of them said in 1905, "Entering some mines for inspection, one will hear one blast after another, going off like an artillery duel." [79] The same inspector also feared that the practice of haphazard and unscientific blasting would eliminate the old type of miner who took pride in his work, and would bring an inferior class of labor to the Maryland coal fields.[80]

The mine inspectors' reports from 1905 on cite enough instances of indiscriminate blasting, lingering powder smoke, and dust in certain mines to make one feel that their concern was not without foundation. Nevertheless, the Bureau of Statistics and Information reported in 1908: "Careful laws have been enacted for the protection of miners, and the ventilation and safety of the mines. These laws are enforced, and in consequence thereby, and of the exceptional intelligence of the miners in the Georges [*sic*] Creek region, the Allegany coal fields are regarded as the most favored in the land." [81]

Chapter 4

Half Turns

Child labor in the Maryland coal mines was as undeniable an economic fact as child labor in most industries during the nineteenth century. Trapper boys of eight or nine were responsible for opening and closing the doors which helped to regulate ventilation in the mines. Other boys slightly older drove mine mules or worked as miners whenever they could secure a half turn. Because of the nature of the coal, there were no breaker boys in the state.[1]

Most boys inside the mines were employed under an arrangement known as the half turn. A turn entitled a miner to a car for removing the coal he had dug, and a boy's half turn gave his father or older brother the opportunity to dig and load extra coal. If the standard turn were two cars per day for each adult, the miner who had his son with him would receive three cars, and the family income would be considerably augmented in the course of a month.

The practice of giving boys a half turn was well established by the 1880's.[2] The Knights of Labor organization in the coal region formally recognized the right to a half turn in its published working regulations.[3] According to Rule No. 5, boys became entitled to a half turn at the age of twelve, and might go to the mine to claim it. Furthermore, at the discretion of the mine boss, the oldest son of a widow was to be allowed a full turn if his father had died while in the employ of the coal company. Apparently much depended on the individual company's willingness to hire workers below a certain age. The Consolidation

Coal Company, for example, announced in the spring of 1876 that it would not employ any one under twenty-one.[4] But this policy was only temporary, possibly designed to limit employment to family men during a period of serious depression. Children were employed in this company's mines in 1888 and 1889,[5] and in 1893 its superintendent fixed the minimum age at thirteen.[6]

Some adult miners expressed their disapproval of child labor. One miner replied to a questionnaire: "Boys employed around mines range from 10 years of age up. The majority of them work with fathers or brothers as miners claiming not less than half a turn, which is, I think, a detriment to all men who have a family to support and who have no boys. Thus, a boy who is able to do nothing, comparatively, will receive at pay-day half as much as an able-bodied man, which, I think, is not right. Not that I am opposed to boys working in the mines, provided that they are of suitable age and can read their own name when they see it written; and when I see men bringing boys to work regardless of the fact that they can neither read nor write, I think they should be compelled to send them to school until they are, say, 14 years old. . . . I am in favor of compulsory education. I know the need of education, as I did not go to school after I was 12 years of age. I never had the chance, being taken to work as lots are now, when they should be at school." [7]

By 1890 most operators would not permit boys to go into the mines as half turns until they were fourteen years old.[8] The minimum age of fourteen was not inflexible, as evidenced by Consolidation's establishment of thirteen in 1893. But at least the exclusion of boys of ten, eleven, and twelve was a step which the Bureau of Industrial Statistics viewed as beneficial, even though it might mean "a few more years of hard labor for many a father whose twelve-year-old son could materially assist him in providing food, clothing and all the necessities of life for a house full of younger children." [9]

An official of the same Bureau subsequently remarked on the large families of the miners—from four to twelve children—as a great drawback, and urged the establishment of factories to employ the girls, who "cannot work in the mines, and . . . must work at something when there are a dozen or more in the family." [10] He commented further, "In the mining regions the children between 12 and 16 years old whom we found working were bright, and nearly every one could read without difficulty. It was sad to see the little boys going down into the mine. The officers of the Consolidation Coal Company and several other mining companies said they preferred not to have boys, but that the miners took their sons with them. The work is done by the piece, and the father draws the weekly wage."

By 1898 half turns were not easy to secure, and many boys were near the age of manhood before they obtained regular employment. Again the Bureau of Industrial Statistics felt that a temporary hardship would in the long run be outweighed by the benefits. "The difficulty now experienced in getting employment for the boys enables them to devote a longer term at school, and this makes a greater likelihood of an increased percentage of those who incline toward the business or professional spheres of life." [11]

Circumstances in the individual mines made hiring children more expedient for some companies than for others. An investigator for the National Child Labor Committee wrote: "A coal mining community in Maryland presents a striking contrast between mines located on opposite sides of the same mountain. In one, an independent concern, it was estimated by several of the employees, and was the judgment of the investigator who saw all the people going into the mines, that not less than forty per cent of those employed were under sixteen years of age; while in the other, a large railroad mining company [in all probability the Consolidation Coal Company], six or seven of whose mines were visited, no boys under sixteen could be found. . . . In the first mine where the boys were employed the vein is from three and a half to four feet thick. Small mules are used for hauling in the mines, because the company does not care to excavate more earth than is necessary to get the coal out from the headings. For driving these mules boys are convenient, as also for tending doors and for loading cars. In the other mines the vein is a large one, running from eight and a half to nine feet in thickness. The work is very heavy. The cars are large, and the boys are not found profitable there. The differences were due entirely to natural conditions." [12]

At what age did boys in Maryland go to work in the mines? How many were included in the mine labor force? What were they paid? Were there statutory limits on their employment in this hazardous underground occupation? Information to answer these basic questions is scanty, especially before 1885, and is largely statistical. Any data for 1850 or 1860 must be gleaned from the manuscript population censuses for those years. The published census summaries for 1870 and 1880 provide employment and wage figures, and mine inspectors' reports give data on accidents. The Bureau of Industrial Statistics and Information began to publish its reports in 1886, giving the customary type of labor statistics and also furnishing a picture of social conditions which may have had bearing on the employment of children in the mines.

In 1850 the youngest persons enumerated in the census as "miners" were sixteen, and there were only two of them. In a mining population of only 350, most of them recent immigrants with presumably very small children, the absence of young workers is not surprising.

In 1860, of a mine labor force of roughly 900, only 34 were younger than sixteen: 2 were twelve; 3, thirteen; 6, fourteen; and 23, fifteen. (There were also 25 aged sixteen.)

The 1870 census was the first to attempt a breakdown of the labor force in terms of age. So far as mines were concerned, the classifications were "boys under 16 years" and "men 16 years and over." * The returns of "boys" in these statistics are not to be relied upon. The Bureau of the Census itself described its

* The term "boy" was not defined in the census volumes for 1870 and 1880, but the historical section of the 1902 *Special Report on Mines and Quarries*, Table 3, p. 6, indicates that these age classifications were used.

mining statistics as "distressingly inadequate," obtained "by correspondence with the proprietors of establishments so far as they were disposed to afford the information." [13] Since a boy's earnings as a half turn were credited to his father, it is likely that many company superintendents did not know how many boys were working underground. Under a system of payment by the ton of coal dug and loaded, a half turn was in effect employed by his father or by an older brother rather than by the company.

In any event, one must view with some skepticism the report that of the 2,712 workers in the Maryland coal mines in 1870, only 55 were boys younger than sixteen.[14] A much more probable figure is derived from the manuscript population census for this year, when the Census Bureau had been particularly anxious to secure a "true return of the occupations of the people." Enumerators were told that in most cases children should be entered as "at home" or "attending school"; but they were definitely instructed, "If a boy or girl, whatever the age, is earning money regularly by labor, contributing to the family support or appreciably assisting in mechanical or agricultural industry, the occupation should be stated." [15]

Selecting the three major mining districts of Allegany County —Westernport, Frostburg, and Lonaconing—the writer counted the number of children whose occupations were described in the 1870 manuscript census of population as "miner," "mine driver," or "work in mines." Since the count did not include other districts, it understates the situation. The exploration revealed a total of 221 boys aged nine through fifteen: 1 aged nine; 2 aged ten; 14 aged eleven; 30 aged twelve; 34 aged thirteen; 67 aged fourteen; 73 aged 15; and in addition 70 aged sixteen. This figure of 221 makes sense in terms of the proportion of child labor to total labor * and eliminates what would have seemed an inexplicable jump to the number of child workers reported in 1880 (Table 4).

* The figure for total labor in 1870 seems reasonable and is consistent with the number of miners turned up in the check of national origins prepared for Table 1, Chapter 2.

Table 4. Proportion of child labor (boys under 16) to total labor, Maryland coal mines, 1860–1910

Year	Total workers	Child workers	% child workers
1860	928	34	3.7
1870	2,712	221	8.1
1880	3,677	512	13.9
1890	3,702	300	8.1
1902	4,936	129	2.6
1910	5,798	164	2.8

Sources: For 1860, count from manuscript census of population, Allegany County, Md., 1860; 1870, count from manuscript census of population, Allegany County, Md., 1870, census districts of Westernport, Frostburg, and Lonaconing; 1880, *Tenth Census of the United States*, XV, 681; 1890, *Eleventh Census of the United States*, XIV, 349; 1902, United States Census Office, *Special Reports, Mines and Quarries, 1902*, p. 712; 1910, *Thirteenth Census of the United States*, XI, 93, 95.

Census techniques had improved by 1880; there is no reason to suspect the accuracy of the statistics in the *Report on the Mining Industries*, which showed that the Maryland coal mines in 1880 had 3,677 employees, including 475 boys working underground and 37 above ground.[16]

While the Bureau of the Census continued to use the term "boys under 16 years," other sources of employment figures in the 1880's had different definitions. Thus we find the *Coal Trade Journal* reporting that in 1884 the George's Creek region had 453 boys under seventeen mining coal, and 17 younger than seventeen were driving mules or horses in the mines.[17] The report of the Bureau of Industrial Statistics and Information for 1884–85 also drew the line between boyhood and manhood at the age of seventeen. The Bureau indicated that boys' wages were 75 cents a day. During this two-year period boys worked on the average 254 days a year and earned on the average just under $200 per year.[18] In 1888–89 the Bureau had still another definition of boy, reporting that during this period 11 coal mines in Maryland employed a total of 836 boys from the ages of twelve to twenty-one.[19] Their average wage was $6.75 a week, which the Bureau says appears high because "in this Trade boys over seventeen years receive men's wages." The average wage of those under seventeen was "about $5.25" a week. The working day in eight of the mines was 10 hours; in two, 9 hours; and in one was not specified. The Bureau listed the "nationalities" of these boys as: American, 502; Bohemian, 9; Colored, 4; English (including Scotch and "Welch"), 164; German, 59; Irish, 96; Poles, 2.

The 1890 census reported that on the average throughout 1889, 284 boys under sixteen worked underground in Maryland for a wage of $1.06 per day.[20] In Pennsylvania bituminous mines at the same period boys underground got 78 cents a day; in West Virginia, 66 cents.[21] The same source listed 16 boys under sixteen working above ground around the Maryland mines for 90 cents a day. Wages above ground for boys in Pennsylvania and West Virginia were 83 cents and 76 cents per day, respectively.[22]

In 1902 only 129 out of 4,936 coal mine workers in Maryland were boys under sixteen.[23]

In 1907 110 boys under sixteen were employed in the Maryland mines.[24] While conforming to the census definition in giving this employment figure, the Maryland Bureau of Statistics and Information in the same report stated that the average weekly wage of "children" sixteen years of age in coal mining was $4.46.[25]

In 1908 the mines employed 10 children "between" the ages of twelve and fourteen; 95 "between" fourteen and sixteen. Again this is a Bureau of Statistics report, and it does not indicate the precise limits of the age-intervals.[26] At this time the general minimum age for coal mining throughout the United States was fourteen, and John Mitchell, president of the United Mine Workers, would have liked it to be sixteen. Only Pennsylvania, Alabama and Maryland allowed the employment of boys of twelve in the mines.[27]

And finally, in 1910, the last year covered by this study, 164 boys under sixteen were employed in the Maryland mines.[28]

The highest incidence of child labor occurred during the late nineteenth century. The employment of 512 boys in 1880 may have reflected a period of recovery from the hard times of 1877–79; boys' wages were desperately needed for feeding and clothing large families and for replacing depleted savings accounts. Moreover, the Knights of Labor were beginning to insist upon the right of a miner to his son's half turn. Although Chapter 8 presents evidence that Maryland mine workers would have preferred to see their children in school, the financial advantage which the half turn gave the miner with sons was a decisive factor in the persistence of child labor in the mines. Mine workers desired not only to rise above the subsistence level but also to move into the ranks of small merchants and farmers, and considered it only natural that a boy should contribute his part to the family's social and economic move upward.

Table 5. Proportion of child labor to total labor, bituminous coal mines of Pennsylvania and West Virginia, 1880–1910

	Pennsylvania			West Virginia		
Year	Total workers	Child workers	% child workers	Total workers	Child workers	% child workers
1880	33,248	2,209	6.6	4,497	203	4.5
1890	53,132	2,351	4.4	9,778	293	2.9
1902	92,095	1,612	1.7	23,914	849	3.5
1910	184,408	1,138	0.6	69,666	837	1.2

Note: Based on same census sources as Table 4, but excluding 1870 because of the questionable accuracy of the statistics on child workers.

Although many more children were employed in the bituminous mines of Pennsylvania and West Virginia (Table 5), the proportion of children in the mining force was smaller than in Maryland, not exceeding 6.6 percent in Pennsylvania in 1880 and 4.5 percent in West Virginia in the same year. (Without being specific as to industry, the Pennsylvania Department of

Internal Affairs commented in its report for 1886 that the 1880 census figures for child labor in Pennsylvania were too low.)[29]

Two statistical factors should be noted in making comparisons between Maryland on the one hand and the West Virginia and Pennsylvania bituminous fields on the other. First, between 1880 and 1910 the Maryland work force did not even double, while in Pennsylvania it increased more than five-fold and in West Virginia more than fifteen-fold. Thus, in the latter states, even a large increase in child labor would still be reflected as a small proportion of the total. Second, while the Pennsylvania and West Virginia coal companies were drawing increasingly upon the bachelor populations of southern and eastern Europe for their labor, the Maryland companies continued to rely upon family men, often drawing upon second and third generation descendants of pioneer miners.

No reform group in Maryland during the nineteenth century was powerful enough to convince the legislature that child labor in the coal mines should be restricted. Nothing in the Maryland law prohibited the employment of children of any age in the coal mines or regulated their hours of labor. An act of 1874 establishing a ten-hour day for all children under sixteen[30] was promptly repealed by the following session of the legislature and reenacted with an amendment which made it applicable only to children employed "in any cotton wollen [*sic*] or other manufacturing establishment in this State."[31] Not until 1902, when the mine ventilation and inspection law was revised, was there any legal restriction on the employment of children in the Maryland mines. In the meantime both Pennsylvania and West Virginia had acted:[32] Pennsylvania law forbade the employment of children under fourteen in the anthracite mines and of children under ten in the bituminous mines; West Virginia prohibited the employment of children under twelve in the coal mines.

The Maryland Act of 1902 said merely:

> No person under the age of twelve years, or female of any age, shall be permitted to enter any mine to work therein; nor shall any boy

under the age of fourteen years, unless he can read and write, be allowed to work in any mine. And the mine boss shall see that this requirement is fully met.[33]

The law did not change the existing situation to any great extent, for by this time half turns were difficult to secure and by custom were not given to boys under fourteen. However, the *Coal Trade Journal* of October 11, 1905, carried a story to the effect that the "State Labor Bureau," aided by the "Consumers' League of Maryland," had been investigating child labor in the coal mines with a view to punishing both employers and parents who violated the 1902 law. The *Journal* added, "Very few boys are employed in the mines of Maryland."

The law of 1902 established a minimum age for mine workers; it in no way limited the kind of work that boys of twelve to sixteen could be required to perform. Consequently, young miners and drivers were involved in a shocking number of accidents of all kinds. Because some of the early mine inspectors did not include the ages of the victims in their reports, it is impossible to determine exactly how many children of sixteen or under were injured or killed while at work. The nature of the accidents is clear; boys were crushed between mine cars, dragged to their death by runaway mules, fatally burned by explosions of powder, and buried under falls of coal.[34]

Coal mining, a dangerous occupation under the best of conditions, offered special hazards to youngsters who lacked the experience and judgment of adult miners. Nevertheless, not until 1912 did it become illegal for coal companies to employ boys of twelve, provided they could read and write, or for fathers to take their sons into the mines as half turns at the same tender age. In that year the Maryland General Assembly passed a child labor act under the provisions of which no child under sixteen was permitted to work "in any tunnel or excavation; nor in, about, or in connection with any mine, coal breaker, coke oven, or quarry."[35]

Chapter 5

Tonnage, Turns, and Off-Takes

Of the various factors which determined coal miners' wages in Maryland, the fundamental element was the amount paid for mining each ton of coal. Other labor in and about the mines was paid a daily rate, but the miner, the man who dug the coal, was on a piece-work basis. This practice was established in the region at least as early as 1839 at the George's Creek Coal and Iron Company mines in Lonaconing.[1] The unit of measurement in the state was not the net ton of 2,000 pounds commonly used in the bituminous mines,[2] but the gross or long ton of 2,240 pounds.[3]

In the sixty years following the opening of the canal trade, the region grew to peak production, but the miners themselves actually gained little or nothing in the way of an increased wage rate. True, the companies paid 30 cents a ton in 1851 and 63 cents a ton in 1910, but they had paid 65 cents a ton in 1880 and 1881, when the Knights of Labor were in the ascendant. The Civil War years, during which wages soared to a dollar a ton, of course presented an abnormal employment situation.[4] The demand for labor was so great that wages rose to several times the prewar level.

Mine operators pointed to the fluctuating fortunes of the coal trade and the competition from other regions as their principal reasons for cutting wages or for refusing to give increases. Differences of opinion between miners and operators concerning the latters' ability to pay in many cases led to strikes and in some cases to acceptance of the situation until times should improve.

In order to get contracts many operators cut prices by cutting the cost of labor at the mines, telling their miners that they could actually make more at the lower rate because they would have steady work.[5]

Competition at first was both local and interregional, but a local operator could not reduce wages unless all other mines in the area agreed to do so. What the rugged individualist did in this case was to cheat his miners by paying them for less coal than

Table 6. Price per ton paid for pick mining in Maryland Coal mines, by month, 1851–1910

ear	Jan.	Feb.	March	April	May	June	July	Aug.	Sept.	Oct.	Nov.	Dec.
351	.30	.30	.30	.30	.30	.30	.30	.30	.30	.30	.30	.30
352	.35	.35	.35	.35	.35	.35	.35	.35	.35	.35	.35	.35
354	.28	strike		.33	.33	.35	.35	.35	.35	.35	.35	.35
355	.35	.35	.35	.35	.35	.35	.35	.30	.30	.30	.30	.30
356	.30	.30	.30	.30	.30	.30	.30	.30	.30	.30	.30	.30
362	.30	.30	.30	.30	.30	.40	.40	.40	.45	.45	.45	.45
363	.50	.50	.50	.50	.50	.50	.50	.50	.50	.50	.50	.50
364	.50	.50	.50	.60	.60	.75	.75	.75	1.00	1.00	1.00	1.00
365	1.00	1.00	1.00	1.00	1.00	.75	.75	.75	.75	.75	.75	.75
366	.75	.75	.75	.75	.65	.65	.65	.65	.65	.65	.65	.65
377	.50	strike	.50	.50	.50	.50	.50	.55	.55	.55	.55	.55
378	.55	.55	.40	.40	.40	.40	.40	.40	.40	.40	.40	.40
879	.40	.40	.40	.40	.40	.40	.40	.40	strike		.50	.50
380	.50	.65	.65	.65	.65	.65	.65	.65	.65	.65	.65	.65
882	.65	.65	.65	.50		strike		.50	.50	.50	.50	.50
384	.50	.50	.50	.50	.50	.50	.50	.50	.50	.50	.40	.40
886	.40	.40	.40	strike	.40	.40	.40	.40	.40	.40	.40	.40
887	.40	.40	.50	.50	.50	.50	.50	.50	.50	.50	.50	.50
892	.50	.50	.50	.50	.50	.50	.50	.50	.50	.50	.50	.50
894	.50	.50	.40	.40	strike		.40	.40	.40	.40	.40	.40
896	.40	.40	.40	.45	.45	.45	.45	.45	.45	.45	.45	.45
900	.45	.55	.55			strike			.55	.55	.55	.55
903	.55	.55	.55	.65	.65	.65	.65	.65	.65	.65	.65	.65
904	.65	.65	.65	.60	.60	.60	.60	.60	.60	.60	.60	.60
910	.60	.60	.60	.63	.63	.63	.63	.63	.63	.63	.63	.63

Sources: For 1851, Scharf, II, 1503; 1852, Roy, p. 52 (*Coal Miners*); 1854, Baltimore *Sun*, Jan. 13–April 6, 1854; 1855, Cumberland *Telegraph*, July 12, 1855; 1856, 1862, 1863, 1864, 1865, 1866, 877, 1878, 1879, 1880, 1882, 1884, 1886, and 1887, Maryland Bureau of Industrial Statistics and nformation, *Report, 1890–91*, p. 226; 1892, Maryand Bureau of Industrial Statistics, *Report, 892*, p. 212; 1894, *Frostburg Mining Journal*, March 24, 1894; 1896, *Ibid.*, March 28, 1896; 1900, *Coal Trade Journal*, Feb. 28, 1900; 1903, Maryland Bureau of Statistics and Information, *Twelfth Annual Report, 1903* (Baltimore, 1904), p. 58; 1904, Maryland Mine Inspector, *Report, 903–04*, p. 3; 1910, Maryland Mine Inspector, *Report, 1915–16*, p. 16.

they dug and sent out. Some companies "by this practice, were enabled to undersell the honest ones, and demoralize the coal trade of this region generally." [6]

All of these developments will be discussed chronologically in Part II. For the purpose of establishing a general picture of wages, Table 6 and the Chart showing wages in George's Creek

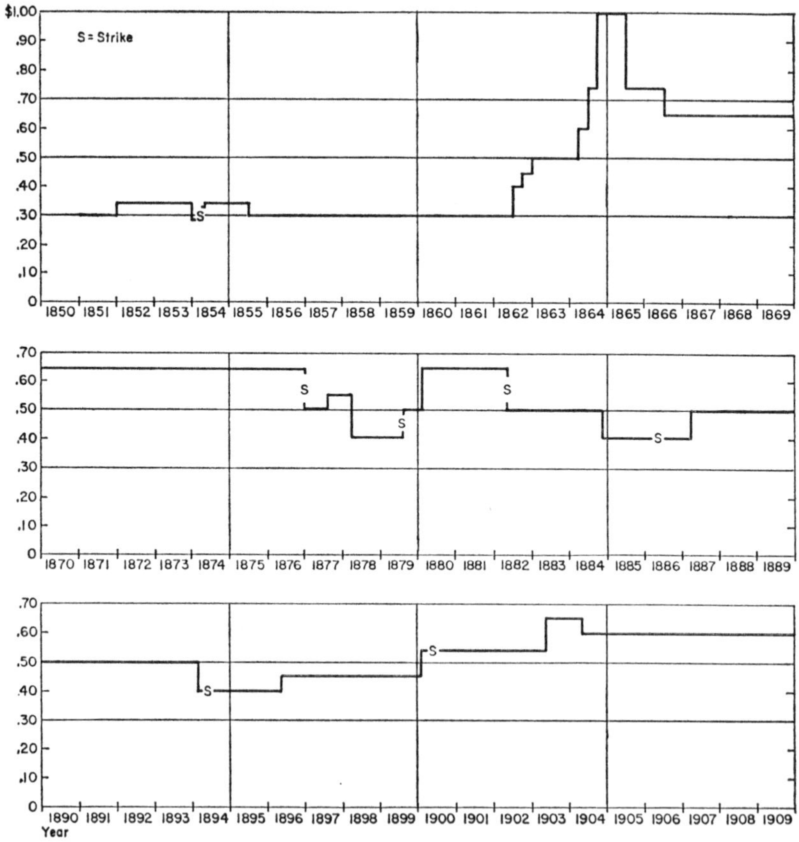

Wages per ton of coal mined, George's Creek, Maryland, 1851–1909

summarize the actual changes in the price paid per ton for mining coal in Maryland from 1851 to 1910 and indicate the major instances in which the miners, by striking, resisted proposed reductions or upheld their demands for increases in wages.

The price paid per ton was only one of a number of factors which determined the amount and character of take-home pay.

The availability of cars for carrying the coal out of the mine, the methods of weighing, the number of days worked—all had to be taken into consideration. Employers made deductions for sharpening tools, for medical care, for rent, and, in the early days, for schools. What was in the pay envelope might be scrip, store orders, or cash, or a combination of any or all of these forms of "money."

No trade publication or state agency routinely reported Maryland coal miners' average annual earnings during the period under discussion. Neither the Bureau of Industrial Statistics nor the mine inspector had a fixed policy of publishing income figures from year to year; therefore Table 7 below is compiled from many sources. In some cases the annual wage is given as stated by the source; in other cases it is computed by multiplying the average number of tons mined per man per year by the average rate per ton paid during that year; in still other cases it is computed by dividing total annual wage payments by total number of miners. Since the statistics on annual income are insufficiently reinforced by data on the cost of living in the mining region, no valid conclusion as to real wages can be drawn.

Table 7 represents the earnings of individuals and does not take into account family incomes, which the sources cite only occasionally. In 1884, for instance, the average income of a family with more than one member working in the mines was $514.88 as against the $445 listed for the average individual worker.[7] In 1889 a boy under seventeen could have added to the family income $5.25 a week, or on the basis of the average days worked, at least $175 a year.[8] In 1891 boys working a half turn averaged $317.50 a year.[9] With even one half turn in the family, a combined income of more than $950 was a very respectable one.

The Bureau of Industrial Statistics and Information report for 1884–85 contained a study of twelve families in the mining district, apparently made with some intention of discovering whether miners' wages were adequate. The rationale behind the presumed sampling is not given. Table 8, adapted from a table

Table 7. Average days worked and average annual income of adult miners, Maryland coal mines, 1850–1906

Year	Average days worked	Average annual income
1850	–	$311.42
1860	–	436.17
1870	–	542.59
1878	210	411.60
1884	254	445.00
1885	254	356.00
1886	–	381.24
1887	–	600.00
1889	203	497.35
1891	254	635.00
1901	–	500.00
1903	219	461.50
1904	226	462.25
1905	252	460.02
1906	250	489.97

Sources: For 1850, United States Census Office, *Seventh Census of the United States: 1850, History and Statistics of the State of Maryland* (Washington, 1852), p. 39; 1860, United States Census Office, *Eighth Census of the United States,* III, 229; 1870, *Ninth Census of the United States,* III, 762 (the wages of 55 "boys" are contained in the average); 1878, one company only, Maryland Mine Commissioners, *Report, 1880,* appendix; 1884, Maryland Bureau of Industrial Statistics and Information, *Report, 1884–85,* p. 81; 1885, *ibid.,* p. 164; 1886, Maryland Bureau of Industrial Statistics and Information, *Report, 1886–87,* p. 136; 1887, Maryland Mine Inspector, *Report, 1887,* p. 5 (*Frostburg Mining Journal* of Sept. 1, 1888, gives a figure of $603.03); 1889, *Eleventh Census of the United States,* XIV, 350; 1891, Maryland Bureau of Industrial Statistics and Information, *Report, 1890–91,* p. 243; 1901, Maryland Bureau of Industrial Statistics, *Tenth Annual Report, 1901* (Baltimore, 1902), p. 47; 1903, 1904, and 1905, *Coal Trade Journal,* Oct. 17, 1906; 1906, *ibid.,* April 29, 1908.

published on page 84 of the report, merely establishes the fact that five of the twelve families lived beyond their income and seven within it; the Bureau made, or at least published, no general conclusions on the mining wage situation.[10]

Table 8. Annual incomes and expenditures of twelve coal miners' families, Maryland, 1884

Family	No. in family	No. of workers	Income	Expenses *	Excess expense over income	Excess income over expense
A	8	1	$470.00	$504.00	$ 34.00	$ —
B	3	1	206.00	344.00	138.00	—
C	6	1	470.00	504.00	34.00	—
D	5	1	414.00	489.32	75.32	—
E	7	3	876.21	937.00	60.79	—
F	1	1	475.00	275.00	—	200.00
G	8	2	900.00	625.00	—	275.00
H	9	1	500.00	324.00	—	176.00
I	4	2	488.00	372.60	—	115.40
J	10	2	701.40	696.27	—	5.13
K	2	1	178.00	157.75	—	20.25
L	6	2	600.00	470.00	—	130.00

* Including oil and tools.

The study summarized in Table 8, although not particularly meaningful in itself, does serve as a reminder not to place too much emphasis on the "average," for here in a year when the average income of an individual miner was $445 we have seven cases of families with one wage earner bringing home anything from a low of $178 to a high of $500. Nevertheless, the incomes tabulated in Tables 7 and 8 presumably approximate what the miners' families had to live on.

Obviously 1870, 1887 and 1891 were the miners' best years in terms of money coming in. But living costs in the mine region were high in 1870 (Table 12, Chapter 7), so that the relatively high income must be discounted in terms of what it would purchase. On the other hand, prices were down in 1891—44 percent less than in 1870. Therefore in 1891, with an average annual income of almost a hundred dollars more than in 1870, and with a dollar which would buy considerably more than in 1870, the miner would seem to have been in a fairly good economic position. Unfortunately he could not maintain this position. Although from 1901 to 1906 he received more per ton than in 1891, his annual income was less than in that banner year. The explanation of this apparent inconsistency may lie in the decrease in the individual miner's productivity.

Miners in some soft coal fields may, in the words of the ballad, have loaded "sixteen tons," but there is certainly no record of that kind of production in Maryland up to 1910. In times of labor strife, partisans of the companies and of the workers hotly disputed the amount of coal a man could dig, but in general the opinion was that five tons was a fair day's work.

Although in 1851 daily production per man was only four tons,[11] by 1869 the Scottish traveler, James M'Killop, found that two men working together in the Maryland mines could fill five two-ton cars in the same number of hours without undue exertion.[12] In 1905 five tons still constituted the normal amount for a man to cut by hand and load into the cars in the course of a day.[13]

Since the miner was on piece-work, he could have stayed underground longer and increased his output to the extent that cars were available, but the five-ton tradition seems to have been established early as a labor organization policy in the Maryland region. The published regulations of the Knights of Labor contained a provision that five cars (ten tons) were to make up a standard day's work for two men.[14] During the great 1882 strike, critics of the Knights protested that a miner should be able to send out eight to nine tons in eight hours,[15] but a work ing miner probably spoke for the majority of his fellows when he wrote, "To dig and load five tons of coal, do the necessary propping and the laying and repairing of tracks, is sufficient work for one day." [16] A miner could not spend all his time underground in remunerative activity. He was not paid for "bailing water from a . . . working place, posting the roof, laying track, cutting clay veins and spars, and removing all obstacles which hindered the mining of coal." [17]

Of course a discussion of what a man *could* dig is academic; the essential bread-and-butter question is what *did* he dig? In 1884 the average miner produced three and a half tons a day, and, as we have seen, made something in the neighborhood of

$445 for the year, during which the companies paid fifty cents a ton.[18] But why, when wages had dropped to forty cents a ton, was the average in 1885 still only three and a half tons? [19] Was there a shortage of cars? Or were the miners perhaps following the Scottish tradition of the "wee darg," the slow-down, "a normal 'instinctive' reaction to any unfair bargain, especially when trade unions were repressed or not recognized"? [20]

In 1903 average daily production per man was 3.4 tons; in 1904, 3.4 tons; and in 1905, 3 tons.[21] Yet on occasion during 1905 Consolidation Coal Company men put out as much as 6 to 7 net tons (roughly 5.4 to 6.2 long tons) each in a day.[22] In 1906 daily production was still at the low 3-ton level, and in 1907 it had gone up merely to 3.2 tons.[23]

More important to the miner than the number of tons he dug in a day was the number of days of the year in which he worked.

Until 1872 coal from this region had only two outlets, the Baltimore and Ohio Railroad and the Chesapeake and Ohio Canal, with the canal carrying roughly a third of the traffic.[24] Even after the completion of the Pennsylvania Railroad connection in 1872, the canal for a number of years continued to carry from a fourth to a third of the coal that went out of the Cumberland area, and as a matter of fact as late as 1888 carried more than a quarter million tons out of a total of a little more than three and a half million.[25] Consequently for at least thirty years mining operations could be and were disrupted by frequent breaks in the canal.[26] At least until well into the 1870's most mine work stopped entirely after the December freeze halted navigation.

Early references to the seasonal nature of Maryland coal mining are frequent.[27] Some persons thought the winter layoff was the reason for the higher wage paid in George's Creek as compared with, for example, the Clearfield region of Pennsylvania. According to Macfarlane, "This irregular work adds to the expense, as miners must have wages sufficient to compensate for those periods when they are without work," [28] and the mine inspector felt that the shortened year "was eagerly seized upon and made use of by [the miners] in insisting upon and enforcing a higher rate of wages than that prevailing elsewhere." [29]

Employment also fluctuated with the demand for and price of coal. In March 1883, for instance, Consolidation and other mines were producing at full capacity; [30] in April Consolidation was back to half time; and by June the *Frostburg Mining Journal* was soliciting aid for needy miners and their families.[31] With this sort of thing happening almost every year—a rush to fill orders followed by a slack period—the coal miner was not employed on all the possible work days in a year (said to number 309).[32] Instead, judging from the scattered reports contained in Table 7, he worked on only two-thirds to five-sixths of them. It was the hope of a longer operating year which occasionally made the miners accept the companies' rationalization that a lower rate per ton would mean a lower coal price, more orders, and "steady work" for the men.

Up to this point in examining miners' wages and income we have considered the rate paid per ton, the number of tons dug per day, and the number of days worked per year. Logically, a simple arithmetic operation should produce annual income "before deductions." The truth of the matter is that the methods of determining what was a ton were so arbitrary and so illogical that complaints about weight, weighing, and weighmasters continually cropped up as a major grievance of labor against management.

To the caprices of weight or measurement must be added the miner's complete dependence on the company for the number of cars he was allowed to fill—the turn.[33] The Bureau of Industrial Statistics explained the intricacies of the turn in its report for 1898 as follows:

For instance, in a cross heading there are located twenty rooms, with two miners working in each. There are two drivers engaged in the heading bringing in and taking out the cars, making twenty-five trips a day, with two cars each. This would average each room or pair of men five cars per day. The rooms are numbered, beginning with one and ending with twenty, so the "turnkeeper," a function performed by one of the drivers, is always conversant with what room is entitled to the first car in the morning by keeping a record where it ended the day previous.

Where the rooms are more numerous than in the illustration, and such is generally the case, the duty of "turnkeeper" is much more difficult, especially when the number of miners employed in the different rooms vary materially. At times there are three, three and a half and sometimes more employed in the rooms, the "half-turn" being held by a boy, and with complications of this character, the "turnkeeper's" lot is at times anything but an easy one.[34]

Unequal turns and discrimination in distributing cars, whether deliberate or caused through carelessness, made for a good deal of hard feeling in the mines.[35]

After the miner had filled a car, he attached a numbered wooden tag or brass check to it so that he would be credited with the coal which went out in that car.[36] Before the 1878 revision of the mining law,[37] the company weighmaster was the final authority on the tonnage allowed each miner. Some companies had scales at the mouth of the mine or at the point where the coal was dumped into the railroad's cars; [38] others had no scales but counted each car as two tons. In order to insure a car's being counted as two tons, the miner had to load considerably more coal than he would be paid for.[39] Complaints about short allowances for weight brought on a whole series of strikes in 1873.[40]

Maryland's first mine inspector found that one of the things the men wanted most was to have the companies put in scales at the mouth of the mine for weighing coal before it was dumped into the railroad hoppers.[41] In 1878 the legislature amended the mine inspection and ventilation bill in two important respects affecting weighing: the operators were required to provide scales,[42] and the inspector was given the power to check them.[43] Furthermore, the company weighmasters were required to appear before a justice of the peace and swear "on the Holy

Evangely of Almighty God" to perform their duties honestly and to keep accurate records.

It should not be imagined that the mere passage of a law immediately took care of all weighing problems. In the first place, the company definition of scales was sometimes rather broad. Take, for instance, the mine inspector's description of the scale at Swanton Mine:

> [It] consists of an iron rod running down through the floor of a house intended for a weigh office; this rod is attached to what is intended to represent the beam; it is about two feet long, with chalk marks denoting, I presume, the age of the world since the Creator of it said "Let there be light and there was light." In the centre of this beam is suspended a large weight; at the other end are seven railroad spikes, nine washers, three pieces of burned-out fire pokers; two heads of old hammers and a number of iron nuts. I was informed that the machine is considerably under the influence of the weather . . . that with the warm, south winds, the manager added junk, before enumerated, but when . . . blizzards came along, he had to subtract. I requested the gentleman who is sworn to weigh the coal to introduce me to the mystery of operating the machine. He informed me it was impossible to do so, as the principal part of it was guess-work.[44]

In the second place, there was the custom of docking. If the weighmaster found any slate or bone coal in a car, no matter how small the quantity, he arbitrarily deducted a thousand pounds from the weight of the coal in that car.[45] In 1877 the amount deducted was a full ton.[46]

In the third place, the miners did not have their own checkweighmen to protect their interests. On May 18, 1893, the *United Mine Workers' Journal* carried a letter from "George's Creek Miner" complaining that the men got whatever allowance the company saw fit to give them. The writer had seen several cars tipped without being weighed at all because the weighmaster had not come back from breakfast. It was not until 1898 that the legislature amended the mining law in such a way as to permit the men to employ checkweighmen at their own expense.[47] The statute allowed the checkweighman to be present whenever coal was being weighed, and also permitted him to examine the scales.

Although the men continued to complain of short weight, they did not avail themselves of the services of checkweighmen. The reasons for this are not apparent. In 1901 there was only one checkweighman in the whole Maryland region, and he resigned to go to work in Ohio.[48] And in 1907, according to the *United Mine Workers' Journal* for May 16, there were no checkweighmen at all in George's Creek mines.

We have followed the miner as he dug the coal and waited for his turn to carry it to the scales. We have watched his coal proceed to the tipple where it was credited more or less accurately and honestly to his account. Now that he is entitled to the reward for his labor, what is it going to be?

The deductions, the "off-takes," have still to be taken care of. In 1869 these were "one dollar for smith, one dollar for doctor, and one dollar for each child that he may have at the colliery school." [49] In 1880 the law pertaining to miners' wages allowed the employer to make deductions from wages for rent, medicine, medical attendance, smithing and fuel, as well as for loans to miners in advance of pay for medicine, medical attendance, smithing, or fuel.[50] The 1880 law was revised in 1900, and "dues for Miners' Mutual Aid Society" was added to the list of acceptable payroll deductions.[51]

Medical fees in 1892 were a dollar a month, which entitled the miner and all members of his family except sons over twenty-one to medical treatment and all the medicines which the doctor might prescribe. Young men over twenty-one paid the same fee as the head of a family.[52] In 1905 the medical fee was still one dollar a month.[53]

After February 1, 1880, as long as the Knights of Labor working regulations were in force, the charge for smithing, that is, sharpening miners' picks and keeping other tools in order, was one-half cent per ton.[54] Thereafter, according to every reference from 1884 to 1905, the charge was one cent a ton.[55]

In addition to the foregoing legal and proper deductions, some employers may, more or less under cover, have held out money for company store accounts. Company stores had been forbidden by law as early as 1868.[56] Yet during the 1882 strike, a letter published in the Baltimore *Sun* of June 26 said, "The 'company store' system, with their exorbitant prices and unrighteous demand on the miners to deal there has been a rod of iron on the backs of the miners for many years."

The Bureau of Industrial Statistics stoutly denied that any company stores were in existence in 1891, although it had received complaints that one store in Barton and one in Frostburg were "virtually company stores." [57] Nevertheless, since the company store was still a subject for legislation in 1898 and 1900, one may reasonably assume that it did affect the miner's pay envelope.[58]

No affirmative evidence suggests that the miners received their pay in any particular medium, but legal provisions forbidding the companies to engage in certain practices lead one to believe that payment may occasionally have been made in something besides legal tender. An act of the legislature in 1880 required that "every corporation engaged in mining or manufacturing, or operating a railroad in Alleghany [*sic*] county, and employing ten or more hands, shall pay its employees the full amount of their wages in legal-tender money of the United States." [59] If the men consented, the companies might also pay "in the notes of any bank payable to the bearer on demand that shall be current at par in this State at the time of such payment." [60] The Maryland Bureau of Industrial Statistics found that wages were being paid in the approved fashion in 1886–87.[61] In 1892 the provisions of the law were extended to Garrett County.[62]

However, it is a piece of legislation in 1900 which arouses most speculation as to the manner in which the miners were paid.[63] This law, which applied only to Allegany County, required the payment of full wages in legal tender and specifically prohibited the issuance of "any scrip or metallic or paper checks in payment of the sums due such employees." Tangible evidence exists that miners in West Virginia were not so fortunate. The George's Creek Coal Company was using metallic checks at Hetzel, West Virginia, as late as 1947. These checks are stamped "Payable in Merchandise Only—Not Transferable."

Until 1896 miners in Maryland were paid once a month. The provisions of Chapter 135 of the laws passed in that year required coal mining companies to pay their employees twice a month.

In concluding this section on wages, it may be interesting, so far as we have the information, to compare the situation of the Maryland miners with that of miners in the competitive Pennsylvania and West Virginia fields (Table 9). Unfortunately we have data only on the price per ton. The figures are from scattered sources, but in every instance the rate quoted for one region is seasonally comparable to the rates quoted for the others.

A rate of forty cents per net ton would be the equivalent of about forty-five cents per gross ton; fifty cents a net ton would equal at least fifty-five cents gross. Therefore in 1884 the Clearfield region had a five-cent advantage over George's Creek, as it did in 1900. But with these exceptions, the George's Creek miners received more than neighboring miners for each ton of coal mined.

Wages and hours are usually studied together. Hours of work were, of course, significant to those mine workers who were employed at daily rates, but to the miners themselves they were not relevant until exceptional circumstances in 1882 made them so. During the early 1880's the Knights of Labor were able to secure a ten-hour day for dumpers and drivers, beginning at seven in the morning and ending at five in the afternoon.[64] Un-

Table 9. Price per ton paid for pick mining, selected bituminous coal fields, 1882–1903

Year	George's Creek, Md.	Meyersdale, Penna.	Clearfield, Penna.	Fairmont, W. Va.
1882 *	.65 [a]	.40 † [a]	.50 † [a]	–
1884	.40 [b]	–	.40 † [c]	–
1885	.40 [b]	.25 [d]	–	–
1887	.50 [b]	–	.50 [e]	–
1893	.50 [f]	–	.40 [g]	.30 [h]
1896	.45 [i]	–	.45 [j]	–
1897	.45 [k]	–	–	.27–.40 [l]
1898	.45 [m]	–	–	.34 [n]
1899	.45 [o]	–	.45 [p]	.375 [q]
1900	.55 [r]	.40 [s]	.60 [t]	–
1903	.65 [u]	.55 [v]	–	–

* Before strike.

† Net ton.

Sources:

[a] *Coal Trade Journal,* Jan. 4, 1882.

[b] Maryland Bureau of Industrial Statistics and Information, *Report, 1890–91,* p. 226.

[c] *Coal Trade Journal,* Dec. 3, 1884.

[d] Pennsylvania Department of Internal Affairs, *Report, 1885,* pp. 169–70.

[e] *Frostburg Mining Journal,* March 12, 1887.

[f] Maryland Bureau of Industrial Statistics, *Report, 1894,* p. 177.

[g] *Coal Trade Journal,* Oct. 11, 1893.

[h] *Ibid.,* July 3, 1893.

[i] *Frostburg Mining Journal,* March 28, 1896.

[j] *Coal Trade Journal,* March 25, 1896.

[k] Maryland Bureau of Industrial Statistics, *Report, 1897,* p. 222.

[l] West Virginia Commissioner of Labor, *Report, 1897–1898,* pp. 93–94.

[m] *Coal Trade Journal,* June 29, 1898.

[n] *Ibid.,* May 18, 1898.

[o] *Ibid.,* April 19, 1899.

[p] *Frostburg Mining Journal,* May 27, 1899.

[q] *Coal Trade Journal,* Aug. 30, 1899.

[r] *Frostburg Mining Journal,* Feb. 24, 1900.

[s] *Ibid.,* March 10, 1900.

[t] Baltimore *Sun,* March 23, 1900.

[u] Maryland Bureau of Statistics and Information, *Report, 1903,* p. 58.

[v] *Coal Trade Journal,* Dec. 2, 1903.

der terms of the settlement of the 1882 strike, mine workers of all kinds were forced to accept a twelve-hour day with an hour for dinner.[65] Chapter 427 of the Maryland *Laws* of 1884, however, established a ten-hour day, and this was still in force in 1912.[66]

Chapter 6

One Continuous Street and Town

Just as the men who mined coal on George's Creek may have been different from the usual run of miners, so, too, did their environment differ from the usual run of mining areas. Engineers reporting on the resources of various coal and iron companies in Allegany County in the 1830's resorted to extravagant, almost poetic, language in praise of their clients' properties: "It seems as if nature, in a freak of prodigality, had bestowed, with a lavish hand, the choicest blessings on the county of Allegany. She has given her a magnificent river for a boundary; a climate mild, equable, and healthy; a soil rich in fruits and agricultural productions; luxuriant ranges for vast herds of cattle; water wholesome and abundant, and mineral treasures almost immeasurable in extent and incalculable in value." [1]

One can understand why miners who came back after trying their luck in the West preferred this country to the flatness of the prairies. "The extremely varied surface of this beautiful region produces great richness of scenery, presenting bold and picturesque views that change with every movement of the observer. High mountain ridges form the outline, finished by dense slopes and crests of forest trees, and the vast swelling hills often cultivated in meadows and fields of corn and grain, delight the eye by the beauty of the surface, while they rarely give any distinct information of the mineral treasures that lie beneath the soil." [2]

How did exploitation of mineral resources affect this almost virgin territory? William Cullen Bryant found at Mount Savage in 1860 "smoking furnaces and forges and vast heaps of cinders," but the region around the village was still notably beautiful, with its "deep winding valleys and broad mountain sides, forests of grand old trees, and grassy fields." [3] Scharf reported in 1882, "Along [the] slopes and swells the miners' houses crop out at frequent intervals, where the rugged surface has been denuded of the forests which once gave shade and moisture to the earth." [4] Destruction of this sort was characteristic of the ruthless exploitation of American forests and mineral deposits after the Civil War. What interests us is that in western Maryland the damage appears to have been less than in other coal regions. Sixty years after the original exploratory reports, the *Coal Trade Journal*, which dealt with all the coal regions in the United States, provided a detailed description of the Maryland mining district, spoke of the pleasing scenery, the alternation of cultivated fields, meadows and woods, and concluded, "It surely looked very unlike the typical coal property." [5] As late as 1930, two historians of the Maryland coal industry found it possible to say, "There is a marked absence of the unattractive type of mining village which is frequently found in certain other mining States." [6]

Before 1837 the whole district from Frostburg southwest to Westernport was a single forest through which wound a roadway scarcely wide enough to allow a wagon to pass.[7] By 1881 the George's Creek valley from Mount Savage to the same southwestern terminus was "one continuous street and town, twenty-four miles in length, inhabited by miners and their families." [8]

The first of the Maryland mining villages were unmistakably company towns, for the earliest operators, developing in relatively isolated areas, had to provide everything for their workmen. Thus the George's Creek Coal and Iron Company built houses for its workers, a company store, a mill race and dam, a lumber mill, smith's shops, lime and brick kilns, an administration building, superintendent's house, church, and schoolhouse.[9] The directors of this company were able to report in 1839, "A flourishing village has grown up . . . which has at least this merit that there is neither drunkard nor beggar in its population of 700 souls."[10] This was the settlement of Lonaconing. The directors' report stated further, "Hitherto all the buildings on the company's land have been put up by or are the property of the Company. At a future time we may find occasion to alter our policy in this regard."

In 1842 the Maryland and New York Iron and Coal Company had constructed at Mount Savage "a village consisting of 22 dwelling houses, a school house, [and] a store, together with vast buildings for the manufacture of iron, one of which, the rolling mill, is 224 feet in breadth."[11]

Similarly the Maryland Mining Company developed the village of Eckhart Mines, which in 1852 contained "about 100 buildings of wood, brick and stone, . . . among them . . . 70 Dwellings . . . all new and in the best repair." The village was located near the center of the company's property on the National Road about a mile and a half east of Frostburg, and had a population of "7 or 800 souls."[12]

The stockholders of the Cumberland Coal and Iron Company were told in 1853: "There have been erected on the lands of the Company 152 dwelling-houses for miners; two churches, one of stone; one large brick schoolhouse; a handsome stone dwelling and outhouses, for the residence of the engineer; 6 large stables, 3 granaries, 1 large machine shop, 5 blacksmith shops, 6 offices, 1 large store house, 1 large lumber house, 2 engine houses, 5 weigh houses,—sand, wood, and lime houses . . . in short, all the fixtures, tools, and appurtenances requisite for the operations of the Company."[13] In addition the company owned sixteen large and productive farms well fenced and in a high state of cultivation. "The farms and miners' dwelling houses, store &c., will yield a very handsome annual rental."[14]

The preceding four examples are typical of mining estates in the English tradition. Company control of such estates, as demonstrated by the management of the George's Creek Coal and Iron Company during its formative years, could be absolute.[15]

Although not in the paternalistic mold of Lonaconing, Mount

Savage, and Eckhart, the villages of Barton, Moscow, and Pekin were also laid out specifically to accommodate the employees of new mines opening nearby. They were originally part of a twelve-hundred-acre estate owned by one of the early settlers of Allegany County.[16] Between 1825 and 1857 the villages of Phoenix, Franklin, Jackson, and Detmold grew up around the mines bearing those names.[17] Up toward the northern end of the valley was the Parker Vein Coal Company, with its mining village of Barrelville.[18]

In these growing communities at least until the Civil War period the coal companies supplied the bulk of housing for miners and mine employees, following the practice begun by the George's Creek Coal and Iron Company.[19] The directors of that corporation included in the estimated cost of establishment accommodations for 400 workmen and families at a total expense of $25,000.[20] Before going ahead with new dwellings, the company had first to provide for the workmen who would construct these accommodations, along with the rest of the plant, and so bought up all available houses in the neighborhood of its works. It rented the houses to certain of its employees, who in turn operated them as boardinghouses at a set charge of three dollars a week for food and lodging. Simultaneously, carpenters rushed the construction of a number of log cabins.

There seems to be no substantiation for Scharf's comment that "pole houses" were "thrown together" for the George's Creek Coal and Iron Company workers.[21] The company's directors stated in their progress report of 1839, "Residences for the workmen had to be built and neither the climate of the Region nor its industrial products allowed of their being erected of the same temporary character and materials which on the public works in milder latitudes are allowable and usual." [22] In his journal of day-to-day occurrences the superintendent noted on October 5, 1837, "As the nights are so cold the hands complained & the masons agreed to do the lining of they^{r} boarding house in two nights." [23] From later references in the journal we know that the company had some of these old buildings, presumably farmhouses, lined with brick to make them more comfortable.[24]

All companies furnishing quarters for their workers anticipated that rents would soon wipe out the initial cost. In 1847 the Mount Savage Iron Works had 280 houses for the accommodation of its officers and men.[25] In the same year the directors of the Maryland Mining Company reported to their stockholders that they had built twenty-three dwellings for miners in addition to a store and office. These buildings had cost $10,000 and were expected to bring in rent of $1,000.[26] The following year this company made an "additional expenditure of $9000 for miners' houses, which pay a rent equal to 15 per cent." [27]

As the Chesapeake and Ohio Canal approached Cumberland, a number of companies announced plans for housing the additional miners who would be needed with expanding production. Among these were the Maryland Mining Company, the Frostburg Coal Company, the Allegany Mining Company, the Cumberland Coal and Iron Company, the Parker Vein Coal Company, and the Franklin Coal Company.[28]

Even during the Civil War, some mining companies were able to construct new houses; the Hampshire and Baltimore Coal Company, for example, was building six at its Hampshire Mines in 1864.[29] But the real boom came after the war, as mining operations expanded with the industrial expansion of the nation. In the Maryland coal region the process of providing company houses continued through the end of the period covered in this study. The Consolidation Coal Company gradually assumed leadership, absorbing older establishments and their housing facilities, and providing new facilities of its own at existing and new mines. In 1871 this company owned 311 dwelling houses,[30] and in 1873 it listed among its assets dwellings at the mines worth $115,045.[31] Consolidation's attitude toward housing was probably typical for Maryland coal operators in the 1870's. "In order to secure and retain good miners (men of family) it will be necessary to erect fifteen or twenty small tenements at one of our principal mines, the total cost of which will be from twelve to fifteen thousand dollars. As these houses will yield a fair return in rent, the company will be fully reimbursed in interest on the outlay. By furnishing such tenements, skilled miners are secured, who become permanent residents of the region, while their sons as they grow up, become the faithful employees of the company which has given steady work and comfortable homes to their parents." [32] Fifteen new houses were built at Consolidation's Ocean Mines in 1874.[33] Still expanding in 1897, the company began a whole new town near its Ocean No. 7 mine, and by early 1898 had built forty-two houses there.[34]

Other companies were equally active. The Maryland Coal Company put up eighty new houses for its employees in 1870,[35] and in 1889 a new company, the Barton and Georges Creek Valley Coal Company, began building "quite a neat little village" near its mine. At the same time, this company also laid out large lots and sold them to its employees.[36]

A preponderance of company housing was common in all coal regions. The conspicuous difference between Maryland and other coal fields was the amount of real estate owned by miners. For example, it could not be said of any Maryland mining village, as it was said of the town of Morris Run in Tioga County, Pennsylvania, "There was not so much as a pig pen or the paling of a garden fence that did not belong to the company. . . . There was not a family that held its home on any other tenure than that of tenant at will subject to compulsory removal at ten days' notice." [37]

During the time that the early coal companies were dividing their rather slender resources between production costs and the costs of housing their labor, miners were themselves buying land and putting up dwellings. When developers planned some of the new villages, they apparently intended that lots should be available to the general public. For instance, William Shaw, Jr., in 1853 set aside 66 lots for a mining town at Barton,[38] but there is no indication that the houses or other buildings erected on these lots were exclusively a company production. Pompey Smash, unlike many of the other early mining villages, did not consist mainly of company houses. "It was laid out in 1851, and was owned by miners, who bought the lots and built the town. . . . Some of the coal was hauled in wagons and sold to the neighboring blacksmiths by a negro [*sic*] slave named Pompey, who, on one occasion, had the misfortune to smash his wagon on the spot where the village was afterwards laid out; and it was named Pompey's Smash. Years afterward the miners became ashamed of the name and had it changed to Vale Summit."[39]

Engineers reporting on the assets of Midland Mines suggested in 1856 that sixty acres of level land owned by the company could be sold for miners' houses and stores, and would, if laid out in lots, "bring at least $300 per acre."[40] Some of this land was undoubtedly sold to miners, since by 1864 only twenty-two company houses had been built on it.[41]

After the Civil War, the George's Creek Coal and Iron Company relinquished its grip on some of the Lonaconing land and "sold lots to the amount of over fifty thousand dollars" (some of it presumably to its employees),[42] and Scharf tells us that in 1882 many of the company's workmen owned the homes in which they lived.[43]

The editor of the *Frostburg Mining Journal* reported that he had been secretary of two building associations in Frostburg between 1873 and 1897. During that period the associations had advanced money for the construction of 247 houses. Of these loans, 222 were made to miners who eventually paid them in full and took title to their property.[44]

The most significant information on home ownership by miners comes from the manuscript population census schedules for 1860 and 1870. The schedules for 1860 show that of 928 miners, 49 (or 5 percent) owned real estate totaling $25,275 in value. Individual miners' holdings varied from $100 to $4,200, the median being $300. The schedules for 1870 show that of the 1,160 miners who can reasonably be identified as heads of families in the districts of Westernport, Frostburg, and Lonaconing, 231 (or roughly 20 percent) reported that they owned real estate. Individual holdings ranged from $100 to $8,000, with the median $800. The total reported value of miners' real property came to more than $217,000. Comparing the figures, we are struck with the economic advancement made by the miners in a ten-year period. Only 5 percent reported owning real estate in 1860, as against 20 percent in 1870, and the median amount of such holdings had risen from $300 to $800. It is impossible to determine how much, if any, of the increase in the average holding was due to Civil War inflation. Generally speaking, the postwar drop in prices had set in by 1870, but we have no way of knowing on what basis any individual in fact readjusted the value of his property for census purposes. Actually wages remained high in the George's Creek region for a long time after the war, and the Maryland miners had also benefited considerably by prolonged labor trouble in the Pennsylvania mines in 1869. Many of the real estate holdings reported in 1860 may have been building lots which by 1870 had been improved with dwellings and other structures.

All through the 1870's Cumberland newspaper reports of real estate transfers show frequent sales of lots by mining companies to individuals, sometimes with specific reference to the reservation of mineral rights. During the early 1880's various outsiders expressed a belief that many Maryland miners owned their homes. For example, a Clearfield, Pennsylvania, operator was quoted as saying during the strike which affected both regions in 1882, "In the Cumberland region the miners own their own houses and hence they have held out as long as they have." [45]

Their defeat in 1882 resulted in a serious economic setback for the miners. The chief of the Bureau of Industrial Statistics and Information reported in 1885, "Some years back, when miners were receiving 65 cents per ton and there was abundant work, some of the old hands bought their own houses. These do not, however, constitute more than one third of the whole number now in the district, and many of them find their property encumbered with debt." [46] As a matter of fact, home ownership by one-third of the Maryland miners was statistically significant. Even in 1902 only one-eighth of Pennsylvania bituminous miners owned their homes.[47]

At any rate, the setback was temporary, and references to

home ownership continued throughout the 1890's. Of these, probably the most impartial source is the *Coal Trade Journal*, which observed on October 6, 1897, "This is one of the very few mining districts where any number of miners own their own homes and here it is quite the general thing." [48]

It must be emphasized that houses in this region were not shacks. For the most part, whether built by companies or by individuals, they were substantial and comfortable for the times, and they were infinitely better than the general run of houses in competitive bituminous coal fields, such as those described by the Pennsylvania Department of Internal Affairs in 1886.[49]

The available supply of timber made log cabins a logical answer to the initial housing shortage. The George's Creek Coal and Iron Company provided them for its men in the late 1830's, at a cost of $40 to $50 per cabin without chimneys,[50] which were supplied for an extra $18 or built by the tenants with a six-dollar credit for their labor. At Mount Savage in 1845 one company superintendent wrote, "We have generally built log houses for our miners, they contain one room and garret and cost from $70 to $80. The better class of houses contain 2 rooms & hall and kitchen on the ground floor, and two Rooms up stairs the price paid for them by contract was 400 Dollars. They would cost more now in this season of high wages." [51]

The actual specifications for another type of miner's house built in this area in the 1840's are contained in a letter addressed to the superintendent of the George's Creek Coal and Iron Company and dated July 15, 1845.[52] These called for a one-and-a-half story house, 32 feet square, with a roofed porch across the entire front. There were no cellars, but the houses were to be set on stone foundations. The frame was to consist of "good substantial timbers well put together with mortices and tenons, to be weather boarded outside with ¾ or 1 inch weather boards nailed horizontally." The inside was to be plastered and painted; the outside, whitewashed. These were double houses, each unit consisting of three rooms on the first floor and one on the second, with a brick chimney running up through the middle and providing each tenant with two fireplaces. Ceilings were to be nine feet high on the first floor and seven feet on the second. Twelve windows, including dormers upstairs, would provide light and ventilation. The cost of these houses, constructed in a "neat, plain and substantial manner" with "good Carpenter locks and serviceable hardware" was about $400. Double houses built for miners by the Midland Coal and Iron Company in 1856 also cost about $400 each.[53]

Since building stone, like timber, was in good supply, it was sometimes used. By 1855 Mount Savage had "several hundred houses for the operatives [of the ironworks], built principally of stone, and situated in parallel rows on the hill-sides; nearly every house has a small yard in front, enclosed with a thick stone wall, which gives to the village a peculiarly picturesque effect."[54] However, in fairness it must be reported that William Cullen Bryant in 1860 described these same picturesque houses as "dirty cottages."[55] Some Eckhart houses also were of stone.

The double house was typical of this region. These two-family dwellings known as "blocks" were usually of frame construction, two stories high, providing three or four rooms and sometimes a shed kitchen for each occupant.[56] A small garden plot ran along each side of the house. At Eckhart, the Consolidation Coal Company had inherited some of the Maryland Mining Company dwellings built before 1852—single-family affairs of stone, frame or logs—which it rented to the miners at about three dollars a month in 1882.[57] A few of these old houses at Eckhart still survive in relatively good condition.

Farther down George's Creek, along the railroad south of Frostburg, were rows of "pretty reddish frame cottages, nearly all with a small garden, surrounded by a white railing in front."[58] Front yards were neatly sodded and ornamented with flowerbeds. In winter potted plants filled the sunny windows.[59]

Not all Maryland coal company employees were idyllically sheltered. One writer, the descendant of an 1882 striker, says that most of the houses at Eckhart in 1882 were "mere shells" and that the occupants had no money to make them comfortable.[60] In Mount Savage some tenements or rows purportedly accommodated from three to six hundred people.[61] Tenements of this size seem to have been limited to Mount Savage and may have been provided originally for the employees of the iron works. Scharf refers to rows in Lonaconing: the "Log Row" built in 1852, the "Pink Row" built in 1854, and the "Detmold Row" in 1866.[62] Nowhere, however, does he indicate their size. In any case, housing was apparently not an issue in attempted labor organization of the region. A union miner writing from Eckhart Mines to the editor of the *United Mine Workers' Journal* and complaining generally about working conditions in the George's Creek area in 1909, felt impelled to add to his letter, "I may say of this place there are no company stores and none of which we understand as 'company blocks.' "[63]

The result of the building activities of companies and individuals was the establishment of a loosely-knit community of two relatively large towns—Frostburg and Lonaconing—with small satellite villages near each of the mine sites. The pattern of settlement was evident by 1880, when the tenth census reported the population of the minor geographical units as follows: [64] Barton, 1,112; Borden Shaft, 289; Eckhart Mines, 822; Hoffman's Hollow, 203; Midland Mine, 276; Midlothian Mines, 398; Miller Mine, 100; Moscow, 264; Ocean Mine, 126; Pekin, 505; and Pompey Smash, 535—a total of 4,630 souls.

Besides miners' houses, each small village usually had a church, a school, and at least one store. Ocean Mines, for example, in 1880 had a Methodist church, a public school, a post office, a saloon, and a general store.[65] Eckhart Mines had three churches, two public schools, a post office, four general stores, a grocery, and a saloon; it also had a physician.[66] Barton, which was the home of a mining company superintendent, had a school and a post office and supported five churches. In addition to two general stores, it offered the services of two hotels, a restaurant, two saloons, a druggist, a barber, a blacksmith, and a saddler.[67]

But the rival centers of civilization, to which miners and their wives flocked on pay nights, were Frostburg and Lonaconing.

They were easily accessible, because no mine in the area was beyond walking distance of the Cumberland and Pennsylvania Railroad, which ran from end to end of the George's Creek valley and provided passenger as well as freight service.

Frostburg in 1850 was a "small and straggling village." [68] Incorporated in 1870, it had within three years a population of more than 3,400, and its merchants were doing a thriving business with at least 5,000 other persons residing outside the corporate limits and at neighboring mines.[69] Just after the turn of the century, Frostburg's mayor, in a letter to the Baltimore *Sun*, was able to boast of a population of 6,000, served by two lighting plants (gas and electric), large modern stores, three banks, a town hall, a municipal water system, fire protection, fine hotels and boarding houses, thirteen churches representing eight denominations, one high school and a lower school system of over a thousand pupils, a state normal school "with a collegiate faculty," two large opera houses and five public halls for assembly and lodge purposes, theatrical and other entertainment, and electric railway service to Cumberland.[70]

Lonaconing, the coal region's second metropolis, had by 1871 become Allegany County's third town in terms of population and importance.[71] It was incorporated in 1890, and in 1898 had about 7,000 population, with churches, schools, hotels, a savings institution, a state bank, and two newspapers.[72]

Even before 1880, former miners owned a number of retail enterprises. Sons of Matthew Sloan, a pioneer coal worker in the George's Creek region, established general merchandise businesses at Lonaconing, Pekin, and Ocean in the 1870's.[73] In Frostburg, Owen E. Hitchins, a Welsh immigrant, after "varied experiences as a coal miner, butcher and contractor for the United States Government during the Civil War," in 1864 bought a mercantile business which he operated with his brother.[74] Despite its origin, Hitchins Bros. was the store most frequently complained of in the 1890's as being a company store.

By 1900 former miners or miners' sons were engaged in selling a great variety of goods, not only in the all-inclusive general stores, but also in such specialties as stationery, jewelry, boots and shoes, furniture, and building supplies. Among those offering services as distinct from goods were restaurant, saloon and hotel keepers, barbers, painters, professional photographers, and undertakers. Several former miners became presidents of the banks which they had helped to organize. Others were active in promoting public utilities.[75]

A professional class was beginning to appear. Before 1900 mining families had produced at least one physician, one dentist, a school principal and several teachers, and a number of practicing attorneys. Clergymen for the most part came from outside the region, all the larger Protestant congregations being served by trained theologians, and the Catholic parishes by priests assigned from Baltimore. By 1880 the lay preacher had practically disappeared except in the smaller, more primitive sects.

Apart from the incorporated towns, there was a minimum of government. The county commissioners provided money for schools and maintained the principal roads. A state's attorney and a sheriff were generally responsible for law and order. These elected officers were aided by constables appointed by the commissioners. Justices of the peace appointed by the governor acted as magistrates and disposed of minor cases involving disorderly conduct, drunkenness, disturbance of the peace, and the like. More serious cases went to the circuit court at Cumberland.

Frostburg, the only sizable incorporated town for most of this period,* had a mayor and council, clerk, treasurer, bailiff, and its own police in addition to the county constable. The town officers had tax powers and the authority to maintain local roads and to establish standards of cleanliness and sanitation.[76]

This latter authority, though rudimentary, helped to make Frostburg a cleaner and possibly healthier town than its neighbors. Advancing civilization had early brought cholera, typhoid and dysentery to Lonaconing and Frostburg.[77] Typhoid epidemics occurred at Lonaconing in 1877 [78] and at Mount Savage

* Westernport had been incorporated in 1859 and again in 1868, but was more remote from the largest mines in the valley.

in 1904.[79] In the mining towns and villages, as elsewhere in Maryland during the same period, sanitary conditions were primitive and epidemics were frequent. In most small towns of the 1880's or 1890's, in or out of the mining region, hogs and cattle roaming in the streets inspired indignant letters to the editor of the local newspaper. State officials were aware of deficiencies. As early as 1885 the Bureau of Industrial Statistics reported a "great want of proper sanitary regulation in the mining districts." [80] Of the mining towns, only Frostburg had "fine water . . . supplied by a reservoir on the mountain." [81] The rest were dependent on streams and wells.

Lack of public sanitation services that we take for granted today created a disposal problem. The Borden Mining Company in 1892 found it necessary to publish an advertisement warning persons "not to trespass upon the property of the Borden Mining Company . . . by depositing upon the same Dead Animals or Garbage of any kind or character whatever." [82]

Lonaconing, not incorporated until 1890, was a prime target of state health officers, who described it as "a constant menace to the health and life of its inhabitants," and a potential breeder of "severe and fatal disease." Its streams (which also supplied drinking water) were polluted with slaughter-house and stable refuse and the waste from privies.[83]

However, these conditions were not peculiar to the mining district. Cumberland, county seat of Allegany County, had no sewers in 1887, the same year in which the report on Lonaconing was made, and its water supply was menaced by pollution from domestic and industrial sources.[84] Even in the vast city of Baltimore, the sewers could "scarcely be called a 'system' " and discharged their contents into the water courses surrounding the city.[85]

In view of the minimum nature of the local government, there was a surprising amount of political activity in which miners were prominent participants.* On the county level, where politics were partisan, miners were elected as county commissioners,

* The election of delegates to the Maryland General Assembly will be discussed later in connection with mining legislation.

county clerks, judges of the Orphan's Court, and as sheriff, and were appointed constables, school trustees, and roads commissioners. Frostburg's charter provided property qualifications for the offices of mayor and councilman (real and personal property to the extent of $500 for the former). In the Frostburg elections, where the tickets were labeled with such nonpartisan tags as "People's," "Tax Reduction," and "Citizens,"[86] the office of mayor often went to a prominent professional man or to a merchant, in the latter case frequently a former miner. John J. Price, for example, served four terms as mayor and three as county commissioner; a miner for 37 years, he had become a grocer and a substantial property owner.[87] After 1880 the Frostburg council almost routinely had at least two members who were miners.

As later chapters dealing with strikes will disclose, the mining population of this region was an orderly one. It was therefore unnecessary for the community to provide an elaborate public protection apparatus. Frostburg as late as 1903 had only two policemen.[88] Lonaconing, although for many years without a municipal government, was as "quiet as if it had been policed night and day."[89] In 1880 there were only seven constables in the entire mining district. At least one of these had been listed in the 1870 census as a miner. Such magistrates as "Squire" John Douglas and "Squire" John Ryan in Lonaconing, up from mining ranks themselves, administered stern but not company-directed justice. The sheriffs and their deputies, not infrequently former miners or merchants depending on mining trade, were generally sympathetic with strikers and not inclined to make wholesale arrests. On occasion, when trouble threatened, they even drew their posses from among the conservative element of striking miners (see Chapter 14). It was ostensibly this absence of a large law-enforcement body (as well as lack of zeal on the part of the existing officers) that led the coal companies to argue the need for "special police" in 1882 and to summon the state militia in 1894.

Chapter 7

Some Keep a Cow

"When the miner comes home from his work, the first thing he does is to strip off his working clothes and get into the big tub of warm water the good wife has provided for him. Clean clothes and a hearty supper follow, after which a smoke, for the miners all smoke; a chat with the children or a visit to a neighbor or the store at which he deals; then home, and early to bed." This was the miner's weekday life as the Baltimore *Sun* described it in 1891.[1]

How much did it cost him to live in his home? What were its furnishings like? How well did he eat?

We know very little about how miners lived along this "one continuous street and town" which made up the George's Creek area. However, the bits of information on rents, land values, and food prices, together with descriptions of miners at home, can be combined to give an inkling of what life was like in a Maryland mining town during the last quarter of the nineteenth century.

Those who did not own homes could rent accommodations from the coal companies or from fellow miners. Families which occupied company houses were particularly vulnerable in times of strikes; on the other hand, the man who rented a company house might be the one retained in slack times. In 1875, when the Borden Company discharged ninety men, they were "those who do not occupy or board in the Company's houses." [2]

Rentals did not vary a great deal over a long span of years. There is no precise information on pre-Civil War rents, but a report of the Maryland Mining Company in 1848 contains a rule-of-thumb formula: "Miners' houses . . . pay a rent of 12 to 15 per. cent. on their cost."[3] Applied to a double house costing $400, the price of a "better class" house in 1845, this formula would result in an annual rent of $48 to $60, or $2 to $2.50 per month for each of the two families occupying the block. Assuming an unchanged rate of return, rents in 1856 would have been the same as in 1848, for we know that in 1856 "your miners' houses (double) will cost each about $400."[4]

During his 1869 visit, the Scotsman James M'Killop observed that in the Maryland mining region "the house-rent for masters' [company] houses range from 2 to 5 dollars per month according to their conveniences. The highest rents paid in the valleys are for those homes that are owned by workmen themselves, who sometimes exact even as much as 6 or 7 dollars a month for houses that do not possess the accommodation of the operators' houses; and it is so far fortunate that masters are not guided by the exorbitance of workmen in this respect."[5] Since the number of Maryland coal miners doubled in the decade 1860–1870, the high rents charged by miner-landlords probably reflected a housing shortage. Obviously the companies could not join in this gouging; in order to attract a labor force they would have to expand their living accommodations and at the same time keep the price to a reasonable level. The quotation demonstrates that miners in Maryland could and did acquire real property for investment as well as for shelter. It also confirms Scharf's statement that after the Civil War an unspecified number of miners bought lots from the George's Creek Coal and Iron Company and "built double houses . . . , living in one and renting the other."[6]

No record remains of the actual amounts charged by the Consolidation Coal Company in 1877, but in that year, after reducing wages roughly 23 percent, the corporation reduced rents by 20 to 25 percent.[7] In the early 1880's, company houses rented for $3.50 to $4 per month; other houses rented for $4 to $6, according to location. "When a mine is opened houses spring up in the immediate vicinity and rents rise. After some years the mine runs out and the men seek work a greater distance and rents fall.

Very many miners living under cheap rents walk long distances to their work, four and five miles not being unusual . . . the coal is constantly receding in the mines, some of the workings penetrating the hills as far as one and three-quarter miles, therefore houses in the immediate neighborhood command the highest rents." [8] In Pennsylvania during the same period it was reported, "In many, if not in most, of the mining districts of this State, the wage-workers live in houses owned by the companies, for which they pay an average rent of $5 per month." [9]

In 1897, referring to the houses being built by the Consolidation Coal Company at its new town of Lordolph (Lord), the *Coal Trade Journal* quoted rentals of four dollars a month on houses costing five hundred dollars apiece.[10] The *Journal* stressed the fact that no Consolidation Coal Company miner was compelled to rent a company house, and said, furthermore, that these new houses were built solely as a convenience to the miners, not as a source of profit.

The average yearly income for a Maryland miner in 1884 was $445 (Table 7). If he paid rent of $4.50 a month, then a little over 12 percent of his earnings went for shelter. In 1901, not quite two decades later, his average annual income was $500. Of this he might pay a little more than 9.5 percent for the same item, assuming he occupied one of the Consolidation Coal Company's new houses at four dollars a month.

If a miner wished to buy land and build his own house, what would that land cost him? In the 1850's the Midland Mining Company expected to sell building lots at $300 an acre.[11] Whether it succeeded in doing so, we have no way of knowing, but we do know that in 1880 land prices were far below this early estimate. According to the *Maryland Directory* for that year, agricultural land at Barton was for sale at $11 per acre; at Eckhart Mines, $20 per acre; at Mount Savage, "exclusive * of mineral rights," from $200 to $400 per acre; at Pekin, from $5 to $25 per acre.[12] Coal lands sold at $500 to $1,000 an acre. Prices for agricultural land in 1887 are available for a few of

* In view of the price, undoubtedly "inclusive" was intended.

these areas: Barton, $3 to $15 per acre; Lonaconing, $20 to $45; Mount Savage, $50 to $75 per acre.[13]

At Lonaconing in 1880, "All the land in the vicinity [was] owned by the coal companies," which therefore could have established a "company town" in the Pennsylvania tradition had they been so disposed.

In addition to owning his house or renting from a company or from a private individual, the Maryland miner had another option. "Many of the houses are built on 5-year leaseholds from the company, the understanding being that the lease will be renewed unless the land is required for mining purposes in which event improvements will be paid for [by the company]."[14] In developing living accommodations near its new Ocean No. 7 mine, four miles south of Frostburg, the Consolidation Coal Company in 1898, besides building company houses, also laid out lots 50 by 100 feet and 75 by 200 feet which it leased to miners at $3.00 a year, payable semiannually.[15]

Coal Age reported on October 14, 1911: "The Consolidation Coal Company . . . owns nearly all the land on which the employees live. Any employee can lease for $2 to $5 a year enough land for a home with a liberal curtilege for truck farming. This lease lasts for five years. If the company does not desire to renew the contract, it pays for all improvements at its own price or at a price to be determined by arbitrators chosen by the company and the tenant and by a third arbitrator chosen by the other two. If the tenant is still not satisfied, the terms of the lease permit him to appeal to the county courts. The houses built on this leasing system are not lined up simply in serried rows of uniform plan and color, but are designed and placed to suit the whims of the many owners; they are kept clean and orderly, the tenants delighting in cultivating the pretty flower gardens with which, in nearly every case, the fronts of the houses are graced."

So far we have considered only provisions for families. Boardinghouses, of course, dated back to the boisterous early days of the George's Creek Coal and Iron Company, when boardinghouse keepers on the company's lands, limited to charging three dollars a week, sometimes supplemented their income by selling strictly forbidden "spiritous liquor." [16] The observant M'Killop described boardinghouse fare for Maryland miners in mouth-watering prose: "The meals are always taken at fixed hours, breakfast at six, dinner at twelve, and tea at six. In boarding houses these are generally announced by the ringing of a hand bell, and those who absent themselves from the table have no resource but wait on the next meal, as it is seldom they will put themselves to the trouble of preparing separate meals. The cost of boarding for a single man is $25 a month. Animal food [meat] is placed on every table three times a day; and there is no such things as meals without apple or pumpkin pies, with a probable assortment of other dishes, containing pudding, tomatoes, squash, beet root, pickled cabbage, pickled cucumbers, preserves, &c., all of which the fertility of the district is capable of producing." [17]

In 1893 board was sixteen dollars to eighteen dollars a month.[18] At this time the mining rate was fifty cents a ton and production about four tons a day; [19] thus, out of a theoretical daily wage of two dollars, the Maryland miner paid between fifty-three and sixty cents a day for board.

Since the authorities quoted did not define the term "board," they may have referred only to meals, in which case the miner would have had an additional expense for lodging. This is a distinct possibility, for an unmarried miner in the Pennsylvania bituminous mines, with a yearly income of $563 in 1872 reported expenses for "rent," ten dollars a month; for "board," two hundred sixty dollars a year.[20]

Compared with the British workmen, and for that matter with the miners of the rest of the United States, the Maryland coal miner had a high standard of living. M'Killop remarked, "The working classes [in western Maryland] live far more luxuriantly than they do in any part of England, which ranks far ahead of the plain vegetable diet of Scotland." [21] Reading the letterbook of the George's Creek Coal and Iron Company, one must conclude from the nature of the orders placed by the storekeeper that from the beginning the Maryland miners had a taste for small luxuries. For example, take the following complaint to a Baltimore merchant:

> With respect to the rejected Cask of Wine called Sauterne you say "To insure its not turning Sour I put in some Marseilles Madeira —I think it far more palatable, & decidedly more healthy than the french port."
>
> The germans never drink french port—they shewed the purity of their taste by immediately detecting that this Cask was not genuine Sauterne. I will not occupy your time by arguing the question of your own peculiar taste, it is sufficient to say that our german customers who have to pay for the Wine choose to drink it pure. . . . The Germans will not touch a drop of it if adulterated with a Strong Wine, or with Spirits, & we cannot contend against the rooted antipathies of a national Taste.[22]

State bureaus of labor or industrial statistics in the latter part of the nineteenth century were fond of comparing living conditions in their states with those abroad, particularly in England. The chief of the Bureau of Industrial Statistics and Information in Maryland concluded in his first report that Maryland workmen had a more varied diet and food of better quality than those similarly employed in Europe: "Seeing the class of goods disposed of in the mining regions and other parts of the State, I am satisfied that our people are fed at a higher standard, and from the best facts in my possession, at no higher rate, generally, than the European cost." [23] But the Pennsylvania industrial statistician's report for the same period cautioned that diet was a matter of choice: "The question of what a wage-earner eats may be governed by considerations other than the supposed one of mere ability or inability to purchase a particular kind of food." [24]

Considering the nature of the country surrounding the mines, we should expect the miners to be well supplied with foodstuffs. Allegany and Garrett counties were farm land long before the mines were opened, and from the beginning of operations the entrepreneurs counted among their assets the general fertility of the region. J. H. Alexander and P. T. Tyson, among the original incorporators of the George's Creek Coal and Iron Company, wrote in 1836, "The fertility of the soil, too, for agricultural purposes, is another feature of moment in considering any scheme which would involve the assistance of many working hands. The rich returns of the earth, and the low price of provisions, would render labor likewise low." [25]

Both soil and climate were suitable for cultivation of the usual staples. "This country produces abundantly all the crops common to the northern and middle States, although from the elevation of the land the summer is sometimes rather short for ripening Indian corn; still excellent crops are raised, wheat, rye, oats and potatoes are sure and abundant,—and it cannot be doubted that the hardier varieties of New England corn would ripen here." [26] William Cullen Bryant observed good crops of wheat and other grains.[27]

Many miners themselves had gardens, and the Maryland mine commissioners were of the opinion that "with proper economy of time on the part of the miner all the comforts of life could be had in the cultivation of vegetables, grapes, pears and small fruits." [28] During the strike of 1882 the correspondent for the Baltimore *American* remarked that the strikers were "nearly all engaged in repairing their pretty little cottages or cultivating their equally attractive gardens. The latter appears to be the standard means of passing away the time, and, as a consequence, the market garden produce of this region will be phenomenally large this season." [29] The Maryland Bureau of Industrial Statistics and Information estimated, "The garden, cultivated with vegetables, may be considered worth $10 per annum to the miner," and added, "Some keep a cow, a few pigs and chickens." [30]

This Baltimore journalist found that "the miners live well, some of them even luxuriously for their class, and spend their money freely, thus making an active demand for all classes of goods":

In all the larger towns, such as Frostburg, Mt. Savage, Borden Shaft, Barton and Pompey Smash, the number of stores is remarkably large, and, of course, the entire valley is overrun with drummers, mostly from Baltimore. . . . Those who have never been in this valley can form no idea of the manner of living common among the miners. In several of the cottages which I entered I saw pianos and organs, lace curtains, Brussels carpets, and other luxuries, which many an educated employe of Baltimore would regard with envy, as beyond his means. The wives and daughters are all comfortably clad, and none, apparently, consider their wardrobe complete without a shining black silk dress. The little rooms are all prettily, and some of them artistically, decorated with cheap ornaments, pictures, &c., and one of the girls who played a few selections on a small piano for me, displayed a remarkable degree of skill.[31]

The wide variety of shops in Frostburg in the 1880's may be indicative of the demands of consumers in the mining area.[32] Twenty-seven stores sold groceries and provisions; thirteen, general merchandise. In addition one could patronize bakers, booksellers, butchers, confectioners, furniture stores, haberdashers, hardware shops, milliners, a photographer, saddlers, shoemakers, stationers, tailors, a jeweler and watchmaker, or a wine and liquor store, to name only a few. Attorneys, dentists, engineers, physicians, and pharmacists provided professional advice and services.

The high standard of living and the desire of the workers to maintain it provoked the mine inspector to an extraordinary outburst against the miners in 1883. In his report on the strike of the previous year he wrote: "I was not furnished with any argument in support of their cause [maintenance of the existing wage rate] less ridiculous than the claim, actually asserted by prominent members of the strike, that the George's Creek miners, as a body, laid claim to a higher degree of culture and refinement than other classes of workingmen, hence their physical and intellectual wants were greater, and that a higher rate of pay was required to sustain the dignity of a roast beef, silk dress and broadcloth aristocracy of labor, as against the corn bread, pork and beans and linsey-woolsey common places of competing neighborhoods. They seem to contend for a sort of civilization of reed-organs and concertinas as against the hand-organs and Jew's harps of the *imported labor* of Clearfield." [33] (Italics supplied.)

Standards were still high at the end of the century. The *Coal Trade Journal* remarked, "The Cumberland miners are much better off in a social way than those in any other district, the most casual observer of their houses, their children, their gardens and livestock, can see that at a glance." [34] This writer introduced an additional measure of the miner's prosperity—his means of transportation: "The men go to and from their work on horse-back or in little road carts. It is surprising to see the number of very presentable horses which are thus employed. The expense of keeping the horses is of course far below the city standard, such an important charge as shoeing, for instance, being paid for with a bushel of potatoes from the home garden patch." At the Old Bowery mine of Borden Mining Company, he continued, "there are a lot of small stables built by the miners to hold their horses. Some appear to fit the horse as closely as a blanket, but others are a fair size. The carts and harness are left about on the grass without any fear of loss apparently."

The miners' way of life in Lonaconing in 1902 was described in the State Industrial Edition of the Cumberland *Evening Times* on August 30: "The suburbs of the city are largely built up with the miners' homes, generally 1½ story or 2 story residences, costing from $1,000 to $2,500 apiece, each with its plot of ground, giving plenty of yard and garden room. These homes are a credit to the men and to the city. They are generally neat, well painted in pleasing colors, with gravelled walk, frequently with flower garden and porch, vines in front and a well tilled kitchen garden in the rear. They are tastefully kept, well furnished, the walls hung with pictures and the nucleus of a library is not uncommon, while many of them show refined taste and cultivation, an organ or piano adding a touch of artistic comfort and homelikeness to them, carpets floor the best rooms, and the kitchen, dining room and pantry show a love of good living that is decidedly appetizing and the miner lives well. He sets a good table, has the comforts of life about him for himself and loved ones and in dress, when he takes holiday from work, on Sunday or legal holiday, few if any laboring men and their families in the country dress better or in more excellent taste."

Without adequate data, one can merely speculate on the cost of living in the mining areas. Occasional comments suggest that until the 1880's prices in the western counties were high. The Cumberland *Civilian* of February 8, 1850, expressed surprise that the farmers of the area had not more fully exploited the possibilities of their market:

> Allegany County is not remarkable for the abundance of its Agricultural produce. The large number of workmen engaged upon the Baltimore and Ohio Railroad, the Chesapeake and Ohio Canal, the mines, furnaces, &c., in the county, not only consume what is raised in it, but create an urgent demand for more. The demand for any article raises its price. It is hardly probable that their is any other place in the country where higher prices are paid for the produce of the farmer than in Cumberland—the store house of this county. We will annex a few quotations:—For Beef, per cwt. \$5.50 to \$6.50; Pork do. \$5.00 to \$5.50; Potatoes per bushel, 87½ to \$1, and sometimes as high as \$1.25. (Remember the number of the sons of the Emerald Isle we have in our midst.) Wheat per bushel \$1.00 to \$1.05. . . .
>
> As "Goods" can be got almost as cheap here as in Baltimore, it is a wonder that the farmers of Western Virginia, and the adjoining counties of Pennsylvania, do not bring their produce here altogether, and lay in their stocks of goods; or that the country merchants to a man do not make this their farthest market. As it is, the trade with these sections is thriving.

This quotation reflects a mixed situation, with high prices for agricultural products, but apparently with no reflection of freight costs on the prices of nonagricultural "store goods." In 1850 the Maryland miner received twenty-eight cents a ton and was making a daily wage of \$1.50 to \$2.00.[35] Therefore, although a half day's work would buy him a good-sized roast of beef or pork, if he insisted on having potatoes he would have to dig three or four tons of coal to pay for a bushel of this staple vegetable at winter prices.

After the Civil War had brought about high wages and prices, M'Killop observed, "A dollar is looked on with as much indifference, and is as easily spent, as a shilling would be at home. Nothing can be done without the mention of it. For drawing a tooth a dollar is charged, and sewing a button on a man's coat would cost a quarter of a dollar." He added, "The following prices of some of the various necessaries occur in our notebook: Flour selling at $9 per barrel; flesh meat from 10 to 15 cents per lb.; sugar-cured ham, 25 cents; butter, 35 cents; cheese, 10 to 13 cents per lb.; tobacco, 40 cents per lb.; eggs, 30 cents per dozen." At this time, according to the same source, Maryland miners were paid 65 cents a ton.[36]

When the coal companies tried to cut wages in 1874, the editor of the *Frostburg Mining Journal* pointed out that a reduction was unreasonable in view of the fact that the cost of living had not gone down, and that rents and food prices were high.[37] In 1876 the companies again tried to bring their labor costs down, and the same newspaper remarked, "There is probably not a coal region in the country that can less afford a reduction of wages than this," arguing that prices were high because of high transportation costs.[38]

Ten years later, when additional rail facilities had introduced competition in freight charges, the Maryland Bureau of Industrial Statistics and Information published a "List of Prices of Miner's Supplies, as charged in the Neighborhood of the Mines." The prices of certain items can be directly compared with and are not significantly different from prices in Baltimore and at the Chesapeake Bay oyster fisheries published elsewhere in the same report (Table 10).

The miner paid $2.50 a ton for his coal, while the price throughout the rest of Maryland ranged from $5.50 to $7.50. He paid from $1.50 to $2.25 for a pair of shoes which might cost from $3.00 to $5.00 elsewhere.[39] The mining rate in Maryland at this time was 50 cents a ton in 1884 and 40 cents a ton in 1885.

Table 10. Comparison of the prices of selected commodities in the Maryland mining region, the city of Baltimore, and the Chesapeake Bay oyster region, 1884–1885

Commodity	Unit	Mining region	Baltimore	Oyster region
Bacon	lb.	$0.09	$—	$0.08
Butter	lb.	.20–.30	.25	—
Coffee	lb.	.17	.18	.16
Coal oil	gal.	.15–.20	.14	.12
Corn meal	lb.	.015	.02	—
Eggs	doz.	.28–.30	.25	—
Flour	—	4.25–6.00 bbl.	.035 lb.	—
Ham	lb.	.125–.14	.14	—
Molasses	gal.	.30–.40	—	.35
Sugar	lb.	.06	.06	.07
Tea	lb.	.30–.80	.60	.60

Source: Maryland Bureau of Industrial Statistics and Information, *Report, 1884–85,* pp. 191, 198.

Food prices in Clearfield, Pennsylvania, a competitive region, were roughly comparable to those in the Maryland region in 1885 (see Table 11). Clearfield miners in 1885 were paid 40 cents a ton, but this was a net ton of 2,000 pounds as against the Maryland miner's gross ton of 2,240 pounds.

We have no evidence of real wages except for the suggestion in Table 12, taken from the 1890–91 report of the Bureau of Industrial Statistics and Information. The heading is that given the original table.

From its statistics the Bureau concluded that the cost of living had been greatly reduced in these two decades. It pointed out that the lowest prices were those of 1885, at a time when wages also had been considerably cut below previous levels. However, the Bureau's overall conclusions must have been drawn from a greater number of commodity prices than those listed in Table 12, because only three items—flour, coffee and corn meal—cost less in 1885 than in 1890. The Bureau also observed, although it did not publish the figures in this particular report, that the next lowest prices were for the year 1891. Costs for that year were quoted as being 44 percent less than in 1870; 32 percent less than in 1875; 18 percent less than in 1880; and 2 percent *more* than in 1885.[40]

Presumably the prices quoted in Table 10 and Table 12 were those charged by independent merchants. Truck stores could

Table 11. Prices of selected commodities in the Clearfield, Pennsylvania, mining region, 1885

Commodity	Unit	Price
Flour	24 lbs.	$1.00
Rice	lb.	.10
Potatoes	bu.	.50
Cabbage	head	.20
Pork	lb.	.16
Sausage	lb.	.16
Beef	lb.	.16
Mutton	lb.	.16
Tea	lb.	.50
Coffee	lb.	.25
Milk	qt.	.10
Sugar	lb.	.10
Butter	lb.	.25
Eggs	doz.	.25
Lard	lb.	.14
Cheese	lb.	.18
Ham	lb.	.18
Boots	pr.	5.00
Coal	ton	2.25

Source: Pennsylvania Department of Internal Affairs, *Report, 1885, Part III, Industrial Statistics* (Harrisburg, 1886), p. 145.

have a bearing on the cost of living, as demonstrated in the testimony of a Clearfield County, Pennsylvania, miner: "This town has four company stores, and all of them charge from three to five cents a pound on each and everything bought more than the other stores have; dry goods, from five to ten cents a yard more; flour, from one dollar to two dollars a barrel more, and the same in proportion for all other articles. The miners have to deal with these company stores, or consider themselves discharged." [41]

Table 12. Comparative cost of living and prices paid for [coal] mining [in Maryland] at periods named below

Commodity	Unit	1870	1875	1880	1885	1890
Flour	bbl.	$10.00	$8.75	$7.50	$5.75	$6.25
Shoulder	lb.	.20	.125	.10	.09	.08
Ham	lb.	.30	.20	.15	.125	.125
Gran. sugar	lb.	.18	.14	.125	.083	.05
Potatoes	bu.	1.00	1.00	.75	.70	.60
Coffee	lb.	.35	.35	.25	.17	.25
Syrup	gal.	1.00	.80	.60	.60	.40
E. B. tea	lb.	1.10	.80	.75	.60	.40
Rice	lb.	.125	.12	.10	.09	.08
Best cheese	lb.	.20	.18	.18	.16	.15
Butter	lb.	.40	.30	.25	.20	.20
Corn meal	cwt.	2.80	2.00	1.50	1.30	1.75
Lard oil	gal.	1.75	.95	.80	.60	.50
Coal oil	gal.	.40	.40	.30	.20	.12
Lard	lb.	.20	.18	.125	.10	.10
Salt	10 lbs.	.20	.20	.15	.15	.125
Price of mining, per ton		.65	.65	.65	.40	.50

Source: Maryland Bureau of Industrial Statistics and Information, *Report, 1890–91*, p. 229.

Unquestionably the early mining companies in Maryland maintained stores, which were in fact necessary while the area was thinly populated. Before the Civil War, according to the *United Mine Workers' Journal* of April 15, 1909, "Every mine had a store attached to the work, and a species of mild coercion was exercised to get the miners' trade. The better class of the underground workmen dealt with the independent stores because goods were cheaper, and when they patronized the company store they would not permit the company to deduct the store account. They had worked for money, not for dry goods and groceries, and paid the store account from the money they had earned, after receiving their monthly pay." It was the thriftless miners who made use of company-store credit and consequently found themselves with little cash in hand on payday, says the same writer.

In its early days the George's Creek Coal and Iron Company paid by checks drawn on Baltimore or Cumberland banks, and one of the superintendent's complaints about his workers was that they preferred to cash their checks and spend their money elsewhere than in the company store. As he wrote to one of his correspondents on October 14, 1839: "It is a mistake to suppose that our hands ever pay into the Store Checks. No sooner are checks out but they are in a fever 'till cash is obtained, and as they all have credit with the store, all their purchases are charged. Indeed in order to secure as long a credit as possible, they all crowd in on the 1st of the month, day after accts. are made up—& towards middle of the month the store is empty—Thus putting off the day of reckoning as long as possible. Not a check is presented at the Store in payt. of Goods." [42]

The George's Creek Coal and Iron Company did not operate a store after 1858,[43] but at least two other Lonaconing coal companies were imposing the truck system on their employees as late as 1867. When Alexander McDonald, the British miners' leader, visited George's Creek, he urged the miners to wipe out the "social curse" of "base" and "foul" truck at the New Central and American Coal companies.[44] In a farewell letter sent to the *Workingman's Advocate* at the close of his first American tour, McDonald wrote that in Maryland the system had "hatched in the most villainous form, which is the Manager or the Superintendent of the mine keeping a store or truck shop," and again urged a fight against it.[45]

What pressures were put upon the legislature by miners, we cannot now determine. However, in 1868 the Maryland General Assembly passed an act forbidding railroad and mining companies to "own, conduct or carry on any store, or have any interest in any store, or receive any portion of the profits thereof." [46] This act evidently did not accomplish its purpose. "All that it was necessary to do to still keep the men in a state of subjection was to have the store run in the name of the boss miner or other officer."[47] The *Frostburg Mining Journal* of March 2, 1872, printed the following effort entitled " 'Foolish Joe,' the Miner," signed by "Blackbird":

> Consolidation says they keep
> No stores of any kind;
> Now that's a piece of "flabbergash,"
> And nothing but a blind.
>
> Six thousand dollar licenses
> To skin the working man,
> Are issued by the Company,
> So now it's worse than when
>
> They kept the stores and had their clerks
> To peddle soap and tea—
> To 'tend the books and keep accounts
> Of men like you and me.
>
> These clerks they were a jolly set
> And did not care a "brown," *
> If men did buy their pepper at
> Some other place in town.
>
> But now those licensed skinners look
> Around with watchful eye,
> And guard our little income so
> No copper can go by.

Certainly strikers in 1882 complained about company stores, and an observer of the period wrote that in Eckhart almost everyone traded at the commissaries and was in debt.[48]

The Bureau of Industrial Statistics and Information stated in its report for 1884–85, "There are some stores owned by the companies, and employees find it advisable to deal with them, but the majority of the men deal at private stores. The miner is generally in debt to the storekeeper."[49] In 1891 the Bureau

* A copper coin; a penny or a halfpenny. (English slang.)

averred, "All such stores have long been abolished, and for many years there has been no complaint of the 'pluck me' system being in vogue in the district." However, the Bureau continued, "It has recently been reported that one firm in Barton and one in Frostburg were virtually company-stores, although no specific complaint has been made to the knowledge of this bureau. It is certain that no outward compulsion is used to secure the patronage of the miners, and if those referred to are in fact 'company' stores they are not conducted in a manner to arouse opposition." [50]

Nevertheless, by 1898 there had been so much complaint about the store in Frostburg that the legislature finally conceded that the existing law was inadequate, and passed "An Act to prohibit railroad and mining corporations, their officers and agents, from selling or bartering goods, wares, or merchandise in Allegany County to their employes." [51] This act, the *Frostburg Mining Journal* stated, was aimed at the Hitchins Brothers Company, one of whose stockholders was a director in the Barton and Georges Creek Coal Company.[52]

Chapter 8

Polite Culture and General Refinement

Maryland miners and their families were able to engage in a wide range of religious, educational, and social activities. An organizer for the United Mine Workers of America, coming to the George's Creek area in 1905, reported:

> Here is a community of men, intelligent and educated to a very high degree. If you talk politics they can certainly hold their own, for they all take great interest in political matters from a county, state and national standpoint. In all kinds of sport they are well to the front; in dancing and twirling they are certainly past masters, for they have educated the opposite parts of their anatomy to their heads to such an extent that in each of the dozen or so of dancing halls along George Creek you will find fifty or a hundred young couples tripping the fantastic toe every night until the small hours of morning. Sobriety is a very marked feature and drunkenness is discountenanced and is not tolerated, especially on the streets. Music is thoroughly understood and practiced, both vocal and instrumental (especially the former), and to a Welshman like myself this feature is a very pleasing one. Religion is also well represented, for the churches are full and prospering.[1]

One of the miner's salient qualities was his awareness of current events, especially of politics. The various beneficial societies furnished meeting places where the men customarily gathered to discuss the news of the day.[2] "Much of the leisure time . . . is spent in the lodge rooms, and here the men have many opportunities to argue questions and practice the art of speech making," [3] a training particularly valuable to union leaders and to those who represented their fellow-workmen in the Maryland House of Delegates. A correspondent of the Baltimore *Sun*, in commenting on the miner's ability in debate, wrote, "It is never safe to undertake to joke with a miner. He is good at repartee, and in discussing current questions, especially politics, is quite an adept." [4]

We have no way of knowing how much formal education was possessed by the generation of miners which arrived in western Maryland between 1860 and 1870, but we do know that more than 92 percent of those listed in the 1860 census could read and write.[5] Numerous sources comment on the literacy of the coal workers. They were described as "a reading people," many of whom had accumulated good private libraries.[6] It was said, also, "More literature finds its way into the hands of the miners of this region than in any other coal field in the country." [7] Newspapers were the most popular reading matter.[8] By 1890 two daily and eight weekly papers were being published in Allegany County, and two weekly papers in Garrett County.[9] After 1891 a German weekly newspaper, the Cumberland *Freie Presse*, circulated in all the mining towns in the vicinity.[10]

In his Civil War reminiscences Andrew Roy described the informal education of perhaps rather exceptional miners: "Lieutenant Spiers and I had worked in the same mine together before the war. He was the most intelligent and scholarly miner I ever knew. He was a great reader of books. We used to meet after our day's work in the mine and read aloud to each other, 15 minutes alternatively. In this manner we read Allison's History of Europe, Bancroft's history of the United States and a number of Shakespeare's plays, Burns' Poetry, Childe Harold's Pilgrimage, Paradise Lost, Longfellow and Bryant." [11] Lieutenant "Spiers" was Andrew Spier, later president of the Blaen Avon Coal Company.[12]

It was fortunate that miners' homes provided educational materials, because a boy's schooling, as we saw earlier, often ended when he became fourteen and might be entitled to a half turn. After that time his education was mainly what he acquired from books, periodicals, newspapers, and "his contact with men of the world." [13] In spite of hardships, Maryland miners were "good parents and good citizens," and wished their children to advance in the world and to have advantages which they themselves had lacked.[14] "They earnestly endeavor to comfortably maintain their families and give to their children the benefit of that system of education which the State so liberally provides." [15]

From very early in the coal-mining history of Allegany County, schools were available for the children of miners who were willing to pay for the teaching. The George's Creek Coal and Iron Company built a schoolhouse and arranged for the company's Episcopal clergyman to act as the teacher.[16] An inventory of the Lonaconing children made by company officials revealed sixty-four of school age and ninety-four "too small." [17] School opened on March 6, 1839. The "Rules of the Lonaconing Residency" provided for the deduction of "contributions for the Doctor and School Fund," but the amount of the school payment is not specified.[18] However, it would not be likely to exceed the charge for medical service, which was fifty cents a month. This company's school was still in operation during the Civil War, taught by a Scotsman who had "but one arm, a cowhide and a violent temper, and many a boy's back felt all three of them." [19]

Until 1860 education in Allegany County remained a matter of private schooling at varying rates of tuition. A "Free School Fund" was available, but, said the editor of a Cumberland newspaper, "There is a large class of our citizens in moderate circumstances who cannot afford to educate their children at the present rate of tuition, and who, rather than receive anything from the Free School Fund, retain their children at home during the greater portion of the year." [20] This was not an unusual situation. One of labor's chief goals in the antebellum period was the establishment of free schools divorced from the stigma of pauperism. The 1860 census schedules showed that Maryland miners were making a real effort to educate their children, for more than half of those with children aged six to sixteen were sending one or more youngsters to school; [21] some families were paying for the education of older or younger children.

Urging the passage of a public school law, the Cumberland editor had earlier written, "We need relief for our Mechanics and working men whose earnings do not permit them to send their children to school at present. We do not want State charity for them, but a system of education securing good teachers and tuition for those who are able to pay at the rate of about four dollars a year, and gratuitous instruction for those not able to pay that amount." [22]

In 1860 the Maryland legislature authorized the establishment of a uniform system of schools throughout Allegany County.[23] Tuition was to be one dollar for each quarter of twelve weeks, but free instruction was to be furnished those who could not pay.[24] The curriculum was to comprise "the usual branches of a common English education, embracing at least orthography, reading, writing, arithmetic, English grammar and geography." Funds were to be provided by county taxes, with some contribution by the state.

In the fall of 1861 the board of commissioners of public schools had to explain to the public in a letter to the editor of the Cumberland *Civilian and Telegraph* that the schools would be discontinued that winter because the county commissioners refused to levy the tax "in the pretense that the school law is unconstitutional and void." [25] This left the school situation pretty much as it had been before the passage of the 1860 law, and the exigencies of wartime prevented further reform until 1864, when school legislation was passed for the state as a whole.[26] Allegany County's public school system, established the next year,[27] provided for two districts in the mining area: Frostburg, including the upper George's Creek valley; and Westernport, covering the valley below Frostburg. Their commissioners were the Reverend O. Perinchief of Mount Savage and Major Alex. Shaw of Lonaconing.[28] Mr. Perinchief reported for his district fifteen schools and 1,300 children; Mr. Shaw, sixteen schools and 1,129 children.[29]

The Allegany school board ordered from the state superintendent of schools books for teaching reading, writing, spelling, grammar, geography, history, and "primary, elementary, intellectual and practical arithmetic," [30] at the same time deciding that "the school books must be *sold*, at actual cost, and not *hired*—except to the poor who are to be supplied gratis." [31] In the autumn of 1865, the Cumberland *Civilian and Telegraph* informed the public, "The Public Schools now opening are literally FREE to all, without money and without price. Books only to be purchased, and these furnished at very low rates." [32] However, board members had agreed that "scholars over age" might be admitted "by paying $1½ per term." [33] The phrase "over age" presumably means over nineteen, the upper limit mentioned in the earlier school population report.

The initial difficulty in financing the school system in Allegany County may explain why some coal company schools were still holding sessions in 1869. M'Killop commented in that year that among the "off-takes" on the miner's wage was a dollar a month "for each child that he may have at the colliery school." [34]

In his first official report under the new state law, the president of the Allegany County board of school commissioners elaborated on some of his problems.[35] He complained especially that under the local law little money had been spent on buildings. At the time of his report, he said, the county had money enough to keep the schools going for only six months. About two thousand boys and girls were then attending school, although there was "literally, not one School-House in the County." Of the half-dozen "approximations," the best was at Mount Savage, but even that was so small that it would not accommodate a fifth of the children.

Building lots were soon donated (presumably by the coal companies) at Lonaconing, Mount Savage, and Ocean Mines.[36]

The citizens of Frostburg contributed $2,000 for a school lot in their town in 1866,[37] and in 1867 the people of the Lonaconing-Barton area held a "pic-nic" to raise funds for building a church and schoolhouse at Barton.[38] In 1869 a schoolhouse was being erected halfway between Midland and Ocean Mines in the Lonaconing district.[39] By 1869 the county was able to furnish books and stationery free to all its students, although in most counties in Maryland at this time children paid for these materials.[40]

The school tax of two mills on the dollar could by no means build and furnish all the necessary schools. The Cumberland *Civilian and Telegraph* for November 21, 1872, pointed out the need for still more schools in the mining villages: one at Westernport for 400 children; one at Midlothian for 100; one at Pompey Smash for 140; one at Hoffman Hollow for 100; one at Eckhart Mines for 200. In the same year a petition was being circulated in Frostburg asking the General Assembly for a law permitting the Allegany County school authorities to appoint at least one teacher capable of giving instruction in German and English.[41] This would aid "a large class of our citizens, who are now debarred advantages accorded to others, in the expense of which they share."

Once the school building program really began, the mining areas were well looked after. In 1873 the school commissioners announced that new schools were to be built at Barton, Eckhart, and Pekin.[42] In 1875 plans for the school at Pompey Smash were made public.[43] By 1876 the five schools of the George's Creek district had 900 pupils.[44] In 1877 the five Lonaconing schools had over 1,100 pupils and 16 teachers during the winter term.[45] From Scharf's list of school locations in 1881, we are able to identify the following definitely as mining communities: Westernport, Franklin Mines, Barton, Moscow, Castle and Detmold, Pekin, Ocean Mines, Frostburg, Borden Mines, Allegany Mines, Borden Shaft, Midlothian Mines, Eckhart Mines (2 schools), Hoffman Hollow, Pompey Smash, Mount Savage (2 schools), Barrelville, Miller Mine, Lonaconing, Midland, and

Jackson Mines.[46] The number of schools would seem to support the observation of the mine commissioners in 1880, "The State has made ample provision for the education of their [the miners'] children." [47]

Attendance varied with prevailing circumstances. In 1881, for example, the Allegany County school commissioners reported, "The enrollment of pupils during the last year was larger than that of the year previous, but the average attendance was smaller, owing to the unusually severe winter and epidemic diseases which broke out among the children of the mining region, for a time nearly depleting several of our larger schools." [48]

Some boys in the mining area, as in farm country, went to school only when they could not be at work. The Bureau of Industrial Statistics and Information reported in 1891: "Superior opportunities for acquiring a sound common school education are at hand, though the young boys cannot attend school as easily as the boys did in the days when shipments were made almost entirely over the canal. Then work was often entirely suspended during the winter months, owing to the freezing up of the water, and the boys all attended school. Now work is carried on all the year, and when a boy enters a mine his education has to be acquired outside the school. There are few, if any, night schools. They do not appear to flourish, as the work of the boys is so laborious that they come home from the mines too weary to attend such schools, yet whenever and wherever possible the advantage of an education is gladly availed of. At Lonaconing, the centre of the mining region, over 1,200 children were enrolled as pupils during the past year." [49]

Actually the situation was not as bad as implied in the preceding quotation, for only 300 boys under 16 were employed in the Maryland mines in 1890,[50] and the employment of boys decreased sharply after that date.[51] By 1898 a boy had to be "near the age of manhood" before he could get a regular job in the mines. However, even assuming that at fourteen a boy were to obtain one of the rare half turns, "a studious boy can be well grounded in the fundamental principles of the English language, and have a fair enough knowledge of mathematics to carry him successfully through life before he enters the mines."[52]

As it became the custom not to take boys younger than fourteen into the mines, there was some feeling outside the mining area that the state ought to adopt a compulsory school law to keep the nonworking boys out of mischief. The editor of the *Frostburg Mining Journal* at once spoke up: "For the vast majority . . . the State need make no such compulsory school law. The miners, like other people, are looking to estates for their children higher than their own. They are moving upward pretty rapidly, too. Education, polite culture, general refinement, and the *means* of advancement along these lines are far ahead of the social and ethical conditions of twenty years ago."[53]

When the employment of children in the mines was at its highest level, some sort of night school would have been valuable. The General Assembly in 1884 passed "An Act to establish a Mining School in the County of Allegany, in connection with the Public Free School System of the State of Maryland."[54] Not more than $3,000 a year was to be spent, the money to come from the county's public school fund, supplemented by a tax levy if necessary. In spite of its title, the statute did not provide for a "mining school" but merely for free evening classes, open to male students of fourteen years or over, "for the studies of mining and mechanical knowledge, and all necessary studies preparatory thereto." Instruction was supposed to begin in November 1884. In October the board of school commissioners waited upon the county commissioners and asked for an appropriation in order to comply with the law. This request was not granted.[55] Eighteen months later the county commissioners were still unwilling to increase the school budget, which already represented "more than ¼ of the entire expense of the County."[56] After futile attempts to get state financial aid in 1886, the champions of the "mining school" apparently abandoned it.[57]

Not until 1904 was another legislative attempt made to provide free evening schools in Allegany County for the teaching of reading, writing, arithmetic, and "such advanced studies as may be necessary." Again the night schools were to be for males of fourteen or over. Teachers were to be paid out of the public school fund at rates determined by the school commissioners.[58]

On November 17, 1904, the school commission voted to open night schools at Mount Savage and Lonaconing under specific conditions, most of them matters of administration and not relevant here. Classes were to meet three nights a week for two and a half hours each evening, and books were to be supplied at cost. Significantly, one of the regulations provided for discontinuance "when funds are exhausted." No further mention of night school appears in the school commission journal until April 8, 1905, when there is an item in the financial report for March reading, "March 15—Received from Co. Treas. Night School 300.00." State board of education reports for 1904 and 1905 contained no mention of the Allegany County night school.

Religion, like education, was of vital concern to the people of the mining towns. Allegany was described as "a great county for churches," every small town having two or three. "The denominations are mixed, the Protestants predominating, although there are a large number of Catholics, as is found in every mining region. They are all in perfect harmony." [59]

Lonaconing was well served religiously from its very beginning.[60] Visiting ministers—Episcopal, Baptist, and Methodist—attended to the spiritual needs of the George's Creek Coal and Iron Company workers and officials while the company was establishing its own (Episcopal) church.[61] Early in 1839 the company considered the petition of twenty-five Presbyterians who wished to lease a lot on which to build a church.[62] Although this request was not granted, permission was given to twenty other petitioners—and presumably also to the Presbyterians—"to use the permanent carpenters shop for divine services." [63] Among the company's employees were several Welsh lay preachers, at least one of whom could conduct services in both English and Welsh.[64] We gather that the Welsh miners belonged to various dissenting denominations, for a company diary entry refers to an act of conscience on the part of one of the preachers who "gives up the Method church to D. Jones the Independent preacher." [65]

The management was particularly interested in organizing an Episcopal church, which would be, as it were, the official church of the community. On February 15, 1839, the company diarist wrote of his activities in selecting and staking out a site and estimating the cost "of a log building . . . having 16 pews and capable of containing 100 persons." The contract for constructing the shell of the church was awarded a few days later.[66]

On October 27, 1839, the diarist commented, "This Sabbath it is to be hoped will be the last that will find the place without a preacher"; and on the following Sunday an Episcopal clergyman was in residence.[67] Although the congregation was not very numerous, the resident officials proceeded with the organization of the "Church of St. Paul, Lonaconing," [68] only to be taken up rather sharply by a letter from a major stockholder in Baltimore "advising that the organization of the parish should be annulled the name bestowed upon it not having been that intended." [69] Accordingly the necessary steps were taken, and on November 23, 1839, "the proceedings of the last Saturday were annulled and the church again organized as 'St. Peters Parish.' "

The company also organized a Bible Society, presumably nondenominational.[70] Attendance was "not as good as might be wished," but thirty-seven members agreed to pay dues of two dollars a year, and orders were taken for the purchase of Bibles. The company store had stocks of English and German Bibles and Testaments and a good number of "Common Prayers." [71]

Before 1865 all other denominations in Lonaconing worshiped in the old schoolhouse erected by the Lonaconing Coal Company.[72] Scharf says that the George's Creek Coal and Iron Company gave building lots for all the churches,[73] and lists among the denominations: Presbyterian, Lutheran, Catholic, Protestant Episcopal, German Evangelical, Independent Primitive Baptist, and Primitive Methodist.[74]

Mount Savage's first church was built in 1825 by a few of its early settlers, who were all Catholic farmers.[75] "When iron-works and coal-mines were started [in 1840] there was a sudden influx of iron-workers, miners, mechanics and laborers to this place. Three-fourths, at least, of these were Catholics, which necessitated an addition to the old church." When the priest in charge of this parish decided to build a new church, the Mount Savage Iron Company gave half an acre of ground.[76]

Camp meetings and revivals were popular, sometimes large enough to warrant the running of special trains.[77] Newspaper accounts throughout the 1860's and 1870's told of the building of more and more churches, both Catholic and Protestant, in the various mining towns. Attendance was large and the congregations generous: when the Presbyterian church was dedicated at Lonaconing in 1867, 500 persons worshiped at the morning service and the day's collection totaled almost $800;[78] at Pekin Mines in 1872 the collection of several hundred dollars taken at the dedication of the Methodist Episcopal church exceeded the cost of the building.[79]

By 1880 Frostburg congregations included Methodist Episcopal, Welsh Baptist, English Baptist, Methodist Episcopal South, Reformed (German and English services and Sunday schools), Presbyterian, English Lutheran, German Lutheran, Protestant Episcopal, Catholic, and Evangelical German Lutheran.[80]

Methodism had spread rapidly among English miners during the nineteenth century, and to it they owed much: free education for their children, improved living and working conditions, and training in leadership.[81] Consequently, the Maryland mining towns supported many Methodist and Methodist Episcopal churches.

Sunday schools may have helped to educate boys who had dropped out of the public schools to work in the mines, particularly in the 1870's and 1880's, when half turns were given to youngsters of twelve to fourteen. The report of a Sunday school convention in Allegany County in 1869 gives a good idea of the important part played by religion in the lives and education of mining families.[82] According to that report, the Lutheran church of Frostburg had 150 scholars, 19 teachers, 4 or 5 Bible classes (one of young men), and a Sunday-school library of 300 volumes. The Welsh Baptist church had 80 scholars, and 93 books in its library. The Presbyterian church had 75 scholars, 10 teachers, and 250 volumes. On the "Lonaconing Circuit" there were Sunday schools at Lonaconing, Barton, and "on the mountain." The school at Lonaconing had 135 scholars, 22 teachers, and a 600-volume library. At Barton average attendance was 120, with a total of 160 scholars, 24 teachers, and 550 volumes. On the mountain, at the Laurel Run school, there were 35 scholars, 11 teachers, and 100 volumes. The Presbyterian church at Lonaconing had 313 scholars, with an average attendance of 225, 30 teachers, 857 volumes, and a missionary society. In a new Sunday school at Midland there were 80 scholars, 13 teachers, and 300 volumes. At Allegany Mine the Methodist Episcopal school had 80 scholars and a small library.

The mine inspector was probably not exaggerating when he reported in 1878 that the great majority of the miners were "worshipers in the churches of the several denominations of Christians amongst us." [83] Nor did ensuing generations desert their religion; in 1891 a correspondent of the Baltimore *Sun*, in describing the miners' daily life, wrote, "Sunday they attend with their families their respective churches." [84]

Closely tied to the churches was the temperance movement, which appears to have been strong in this area. As we have seen earlier in this chapter, the sobriety of the mining population was remarked upon by a United Mine Workers' organizer, and his observation was only one of many made over a period of years. The Bureau of Industrial Statistics and Information reported that "while there are no restrictions on the use of intoxicating drinks, intemperance is the exception," [85] and the compilers of the *Maryland Directory and State Gazeteer* in 1887 noted a "tolerable abstemiousness as to strong drinks." [86] During the strike of 1882 a newspaper hostile to the miners conceded, "The fact that there is little, if any, drunkenness is one of the surprising features of the strike, and cannot be too highly commended."[87]

Possibly the earliest record of the temperance movement in the George's Creek region is the entry in the George's Creek Coal and Iron Company journal dated January 21, 1839: "Temperance meeting at 7 P.M. Society organized. . . . 7 members on principle of Total Abstinence and 26 on that of Abstinence from Ardent Spirits—Monthly meetings to take place on the 3d Monday of each month." An indication of the extent of the crusade carried on by many of the churches is the note in the Cumberland *Civilian* of December 16, 1877, to the effect that 500 members of the Methodist Episcopal church at Lonaconing had signed the pledge. Such national organizations as the Good Templars, Rechabites, and National Total Abstinence Society attracted many. Local societies included St. Patrick's Temperance Society at Lonaconing and St. Michael's Total Abstinence Society at Frostburg.[88]

Fraternal organizations flourished. Scharf listed Masons, Odd Fellows, Knights of Pythias, Foresters, Free Gardeners, Knights of Honor, Red Men, American Legion of Honor, in addition to the temperance societies. In slightly different categories were the G.A.R., the Y.M.C.A. and the Irish Land League.[89] The friendly societies had their social side, of course, but mining families who joined them were motivated in part by other considerations. "The dangers attending these operations [mining] are very considerable; to such an extent is this the case, that one can rarely attain middle age without having realized in his own person some accident or remarkable escape from destruction. The benevolent and provident societies, therefore, attain large memberships."[90]

Although there was recreation for every taste, many of the social opportunities of the region were of an "improving" nature. Mount Savage had a reading room, established in 1856.[91] The Odd Fellows of Lonaconing opened a library in 1868, intending it also to be used as a reading and lecture room.[92] Frostburg citizens met in 1873 to discuss the organization of a literary society.[93] "The Barton Circulating Library, which is doing so much for the literary advancement and culture of the town, was established in 1880. It has copies of the works of the standard authors, and keeps files of many of the principal newspapers of the country. . . . Its admission fee is one dollar and ten cents, and monthly dues fifteen cents." [94] Frostburg had, at the time of Scharf's writing, the Orion Literary Sociable and the Welsh Literary Class,[95] and Lonaconing also could boast of a Literary Society during the same period.[96] As a matter of fact, Lonaconing had a literary tradition reaching back to 1839, when the George's Creek Coal and Iron Company maintained the "Lonaconing Library Institution," for which it ordered on one occasion the following "Reviews": "Westminster, Edinburg [*sic*], Quarterly, British foreign, Washington National Intelligencer, Phil. Sat.y Evening Post." [97]

Some forms of entertainment were linked to national backgrounds. The Irish had great celebrations on St. Patrick's day; the Scots annually observed the anniversary of Burns' birth; and the Welsh had their Eisteddfods. Ties with the old country were strong, and trips to Scotland and England were mentioned as casually in the newspapers as a journey to Baltimore or Philadelphia. The Baltimore *Sun* of December 12, 1873, for example, published a report from its Cumberland correspondent: "Last Thursday, eighteen persons, miners with members of their families, started from here for Scotland via New York. The number of persons now citizens here, visiting their native land in Europe this winter is greater than at any former period." Frostburg had three foreign steamship agents.[98]

A typical Burns Club Festival included a supper with "haggis forming an important item of the bill of fare," a series of toasts, and entertainment consisting of speeches and songs, followed by general participation in Scottish reels, country dances, quadrilles, schottisches, polkas, and waltzes.[99]

The Welsh employed by the George's Creek Coal and Iron Company celebrated St. David's day on March 1, 1839, with a

"Dinner by the Cymraeggyddion Society. . . . Healths drank [to officers and supervisors], . . . [the company's] prosperity, and that of the Temperance Society, all in good feeling and perfect decorum." [100] But this was a minor affair, not to be mentioned in the same breath with an Eisteddfod, which combined poetry, singing, composition, and recitation, with prizes for the best participants. A few excerpts from the English portion of the 1871 Eisteddfod at Frostburg will show the high standard of competition: a prize of ten dollars for the best essay on "The Past and Present of Frostburg"; a prize of six dollars for the best six English verses on the new Methodist Episcopal Church; a prize of eight dollars for the best anthem on the Bible verses "And when the chief Shepherd shall appear, ye shall receive a crown of glory that fadeth not away. To Him be glory and dominion forever and ever. Amen."; forty dollars for the choir of not less than twenty-five persons best singing "The Heavens are Telling," from Haydn's *Creation;* three dollars for the best rendition of "The Trumpet Shall Sound," from the *Messiah.*[101] The Welsh program was similar in content.

On the occasion of a later Eisteddfod, the *Frostburg Mining Journal* solemnly announced the arrival in America of "Mr. Crwdgimpas Ap Thomas, a native of Moclgwynstrwnstrell, and the only son of Twm-Shom-Dafydd-Leban, the inventor of the Brownwrw-Cymrongldrwstcwmdathuglwestly, or to give its simpler name, the Welsh harp, which Mr. Ap Thomas is said to play upon with great skill. He will appear at the Frostburg Eistedfodd in July next, and appearing in ancient Druid costume will play the 'Rhosellanerchrugog' and the 'Llanrhauadrmochant.' " [102]

All groups found pleasure in music, whether singing around the piano or organ at home [103] or playing in one of the many fine musical organizations formed by the miners. "Some of the best musical organizations in the county are composed of miners, notably the Second Battalion Band of Frostburg, so favorably spoken of by Baltimore papers during the annual parade of the State militia in Baltimore on Thanksgiving day." [104] Lonaconing in 1876 had a "flute band . . . composed of eighteen flutes and three drums." [105] Barton had a cornet band,[106] and Lonaconing had three cornet bands in the early 1880's.[107] Frostburg had a nineteen-piece band in the same period.[108] In 1884 the Union Mining Company made a gift of $300 to the Mount Savage cornet band and donated to the workmen of that town a large piece of land to be laid out as a public park, where the band was expected to provide music at least once a week.[109]

Vocal music was a tradition in all the bituminous fields.[110] In the beginning each nationality brought its own songs, but after the Civil War American coal miners began to create their own ballads. At least one Maryland ballad has survived, coming to us through the pages of the Baltimore *Sun* at the time of the 1900 strike.[111]

In the years before the Civil War, when military companies were a great fad, the miners of Eckhart formed the Eckhart Light Artillery.[112] Some of these men had seen service abroad. The editor of the Cumberland *Civilian* hoped they would "be ready to give Old Zack [President Taylor] a salute as he passes the mines." This artillery company was in great demand on ceremonial occasions and holidays, particularly the Fourth of July.[113] On a more solemn occasion, following the death of President Taylor in 1850, it fired minute guns and a national salute.[114] When the Chesapeake and Ohio Canal was officially opened at Cumberland in 1850, the Eckhart Light Artillery was there.[115] Frostburg had two similar organizations, the "Guards" and the "German Yeagers." [116] After the Civil War the military form of recreation had lost its glamor for those who had seen actual service. However, in 1870, the German Patriotic Association of Frostburg, exultant over the victories of the Fatherland in Europe, produced in celebration a procession complete with brass band and a number of men on horseback caparisoned as Prussian Uhlans.[117]

Holidays were occasions for relaxation from work and participation in some sort of celebration. The Christmas season was stretched out to include half of December and part of January.

The *Coal Trade Journal* of January 1, 1902, reported, "Shipments of bituminous coal have been light the past week, as mining operations have been largely suspended during the Christmas holidays. The last half of December throughout the Georges Creek Cumberland region is usually devoted to having a good time. With the coming of Christmas eve the mining pick and torch are laid aside and the Christmas and New Year celebrations are kept up until the early part of January."

The Fourth of July was another favorite holiday, observed with parades, the raising of liberty poles, reading of the Declaration of Independence, "pic-nics," speeches, and patriotic music.[118]

Needless to say, not all of the mining population was satisfied with innocuous and uplifting forms of entertainment, and some did spend "their hard-earned money in soul-debasing pleasure."[119] In spite of the temperance movement, it was easy to get a drink. "The number of grog shops that infest our town [Frostburg], covering every available space, and established upon every corner, is . . . disgraceful."[120] In 1887 Barton had 7 saloons; Eckhart Mines, 2 combination groceries and saloons; Frostburg, 20 combination saloons and restaurants; Lonaconing, 12 saloons; Pekin, 6 saloons; Vale Summit, 3 saloons.[121]

Among the more brutal sports were cock-fighting, pigeon-shooting, and prize fights, and one gathers that these were well patronized. In fact, a United Mine Workers' delegate, disgusted with Maryland's aloofness toward unionism, said on one occasion that only two things would bring the George's Creek miners together—"prize-fighting and chicken-sparring."[122]

Prize fights in the early days of George's Creek were murderous and strictly illegal. An indignant editor wrote in 1869, "A regular prize fight came off at Barton. . . . It cannot escape inquiry as to where were the officers of the law while this disgraceful proceeding was going on? . . . If no stop is put to such affairs, there is no telling to what they may lead."[123] The "roughs" of Frostburg celebrated the Fourth of July with a socially unacceptable "scrimmage, yclept a 'prize fight' between two of their number. The affair attracted a large crowd of spectators. Three rounds were fought, in which the claret flowed freely, and one of the parties nearly lost an ear, while the other carried away a pair of black eyes as a souvenir of the encounter." [124] Frostburg authorities had earlier passed an ordinance forbidding sparring exhibitions, but this was largely disregarded.[125]

On one occasion a point of honor was settled by weapons more sophisticated than bare fists: "A duel came off in the town of Westernport, in this county, on Monday last. The seconds had so arranged it that the challenging party supposed it was all in good faith, while the defensive party was let into the secret of said arrangement. The pistols were charged with powder only, and the challenged party being prepared with a little red paint, at the word 'fire' he fell covering his face with the paint. The challenging party, supposing he had mortally wounded his enemy, fled." [126]

Shooting matches involved live birds, not clay pigeons. In a contest between two miners at Lonaconing for a purse of $50, each man shot at twenty-one birds. One hit seventeen; the other, eighteen. This was described as probably the "most successful shooting ever done in the United States by laboring men." [127]

Cock-fights, like prize fights, were illegal, but "as all well know, our community is plentifully supplied with [fighting cocks], and the owners of them don't seem to harbor the idea that it is immoral or wrong to bring them face to face occasionally." [128] The Cumberland *Alleganian* of January 27, 1876, reported, "Yesterday afternoon a fight between Mount Savage cocks and George's Creek cocks came off at Barrellville. The match was for $300, the four successful cocks out of seven winning the money. There was a large attendance from Mount Savage and the George's Creek region, and there was a considerable amount of money staked by the backers of the cocks from each place. All the cocks were of the game species, and had on two-inch steel spurs when fighting. Four of the George's Creek cocks were killed, and Mount Savage came off victorious in the fight. We suppose that none of the authorities knew that this barbarous sport would take place or they would have prevented it." The Frostburg editor found "the chicken-fight at Barrellville" a "sort of entertainment in striking contrast with the exercises commemorating the Scottish bard" described earlier in this chapter.[129] Cock-fighting had been popular among early nineteenth-century miners in England,[130] and was not likely to be stamped out easily among their descendants in the New World. Pompey Smash in 1894 was reputed to be a "village which can successfully manage a cock-fight, and when beaten capture the birds of its competitors." [131]

There were, of course, many respectable forms of sport to be enjoyed equally by participants and spectators. The region succumbed to the baseball craze shortly after the Civil War, and by 1867 at least two of the mining towns, Frostburg and Hampshire Mines, had their own teams.[132] Foot-racing was another favorite diversion of English miners. Here in Maryland, on one occasion, $250 was donated as a purse for a footrace of 100 yards.[133] About a thousand persons attended this event, displaying "good order and good nature" in spite of heavy betting.[134] When the canal was frozen, the curtailed mining schedule left plenty of time for winter sports. Scottish miners introduced curling to western Maryland, and a curling match followed by an oyster supper was quite a social affair.[135]

Almost from the opening of the mines, professional entertainment was available for highbrow and lowbrow tastes. The George's Creek Coal and Iron Company diarist referred to the arrival of "Jim Crow and other buffoons" at Lonaconing,[136] and a few days later wrote, "But little going on today in consequence of some equestrian performance exhibiting at Frostburg, whither most of the Lonac^{g} population has gone." [137] Itinerant troupes of all kinds came to the area. One might thrill to Van Amburg and Co.'s "Great Golden Menagerie" with a "living lion loose in the streets unchained" and "a full cargo of living animals . . . direct from Africa." [138] Or one could go to the opening of the Frostburg opera house to see a presentation of "Rip Van Winkle." [139]

When all other diversion failed, one could always fall back upon a "pic-nic," which might be anything from a small school affair [140] to a "Monster Picnic of Coal Miners" for which work was generally suspended and cheap excursions were scheduled on the railroads.[141]

PART II

Chronicle of a Mining Class

Chapter 9

The Antebellum Years (1835-1860)

Corporate coal mining in Maryland began in 1828, a year more often remembered as the beginning of the Jacksonian era. The country was gradually recovering from the depression which had followed the panic of 1819 and from the minor setback of 1825–26. With tariff protection, the New England textile mills and the Pennsylvania iron works were not merely surviving but were actually expanding and adopting more modern techniques. Irish immigrants fleeing their impoverished homeland found ready employment on the construction crews of the canals and railroads which were vital for marketing agricultural and industrial products. The transportation enterprises most important to the development of Maryland coal resources were already under way: on July 4, 1828, President John Quincy Adams threw out the first shovelful of dirt for the Chesapeake and Ohio Canal; and Charles Carroll, last surviving signer of the Declaration of Independence, performed a similar symbolic rite for the Baltimore and Ohio Railroad.

Throughout the nation, eager entrepreneurs were putting their money into the manufacture of cotton and woolen textiles, iron, and machinery; into steamboats and sailing vessels; into turnpikes, toll bridges, canals, and railroads; into lumbering and mining. British capitalists also found overseas investments increasingly irresistible.

At this early stage, Maryland coal was thought of not so much as a generator of steam power but as a possible fuel for smelting iron. Indeed, several of the earliest companies incorporated for exploiting Allegany County mineral resources were interested primarily in the production of iron; their coal operations were ancillary to the principal operation. Among these corporations was the George's Creek Coal and Iron Company, whose directors reported, "A great question of state indeed of national interest has been determined in producing as we have shewn it can be done Iron with raw coal." [1] The company planned eventually to develop its coal mines more extensively, but there was no point in doing so as long as the coal trade was "kept in abeyance . . . by the conditions and prospects of the Chesapeake and Ohio Canal." [2] Consequently, the George's Creek Coal and Iron Company employed few colliers during its formative period.

Certainly in its first years, and possibly until the Civil War, this company ruled its Lonaconing holdings with absolute authority. Its charter and by-laws conferred upon the superintendent almost feudal rights in the administration of justice in cases of misdemeanor,[3] and the management had organized a "Military Company," essentially a forerunner of coal and iron police, "for the defense and protection of Lonaconing." [4] It is true that the defense force dealt mainly with drunk and disorderly persons, but its members carried muskets. However, the officer of the guard held written orders not to use firearms unless the guard or guard house were attacked.[5]

Control over company employees' lives extended even to the ballot box. On election day in 1839, the hands were allowed time off for voting, but members of the supervisory staff accompanied them to the polls with instructions "to control the vote of the Lonaconing voters in obedience to the Co interests" in the realm of taxation. In all other regards "no intention [was] professed or entertained to restrict their rights as suffragans." [6]

All employees of the George's Creek Coal and Iron Company were subject to the same discipline. The superintendent of the works prescribed a set of "rules for the government of all persons in the service of the company," and these rules were prominently displayed.[7]

The "Rules of the Lonaconing Residency" (Appendix B) provided that employees should work every day of the year except Sundays and Christmas day. The "great Bell of the Company" signaled the beginning and end of the working day, which ran from sunrise to sunset with suitable intermissions for meals. Absentees had to account for their not being on the job, and could be punished by "abatement" of their wages or, in extreme cases, by dismissal. Once dismissed, erring workmen must leave the company's property forthwith; company tenants were expressly forbidden to shelter anyone who had been discharged.

Company rules applied also to the conduct of its employees outside of working hours. "Intoxication at all times and places" was strictly prohibited, as was the sale of "distilled spiritous Liquor." Brawling, quarreling, fighting, and gaming were frowned upon; the company expected all persons in its service to "observe an orderly and decorous conduct" toward one another and toward the company officers.

The company superintendent determined house rentals, boarding house rates, and the tenure of leases. He also laid down general rules for the neatness and cleanliness of a tenant's premises, and particularly for the condition of his pig-sty. A company physician was responsible for providing medical service, for compiling vital statistics, and for reporting nuisances which might affect the health of the community.

Some small pleasures were denied: a man might not fire a gun —"a dangerous and unnecessary amusement"—nor might he keep a dog unless he had the special permission of the superintendent.

In 1837 the company began to erect a furnace and to dig the ore, limestone and coal necessary for the manufacture of iron. It could scarcely have chosen a less auspicious time. The healthy investment activity of the late 1820's had given way, between 1830 and 1836, to an orgy of speculation in western lands and internal improvements. The unsound structure collapsed in 1837, following the banking crisis which was an almost inevitable result of Jackson's Specie Circular in combination with the distribution of the surplus revenue. Hard money became scarce. At times the superintendent of the iron works at Lonaconing found it difficult to scrape up enough cash to pay off recalcitrant employees who had to be dismissed. Nevertheless, except briefly in 1837, British capital continued to be available, and the company was able "to persevere toward its accomplishment thro' the long period of languor which succeeded the shock [of the panic]." [8]

But in the summer of 1839 English credit again became tight. Money flowed out of the United States, and the banks again suspended specie payments. This time there was no quick recovery; the ensuing depression lasted until 1843. At the end of July 1839 the company's officers received "bad news . . . from England in regard to the money market," and early in August decided "to 'blow out' the furnace and place the establishment upon a more economical scale by which the monthly expenses may not exceed $6000." [9] A skeleton force maintained limited production for a few months, but at the end of the year the directors suspended all operations. The works did not reopen until 1842.[10]

The relationship between management and labor reflected the uncertainty of the times, the scarcity of cash, and the desire of management to proceed with the utmost economy. So far as the colliers were concerned, the problem was almost entirely one of wages; the coal workers were seldom involved in the drunkenness, quarreling, and rioting frequently mentioned by the company diarist.

Before the George's Creek Company's decision to pay by the ton, late in 1839, colliers' wages in western Maryland were determined by various methods. The established mines at Frostburg and Eckhart customarily paid one cent a bushel for coal dug and wheeled out to the pit mouth. A good digger could deliver 100 to 150 bushels a day.[11]

In beginning operations, the George's Creek Coal and Iron Company had proposed to pay "workmen at the coal" not by the bushel but "at the rate of 25$^{c.}$ per cubic yard of coal got out exclusive of fine coal, with an addition of 25 cents for every yard forward." [12] The men would not accept this wage and stopped work, so that by the middle of February 1838 only two colliers remained in the coal drift.[13] These two had been offered a substantial pay raise, but it was not enough. "The two Welsh colliers at the coal drift will not accept $3 per yard forward, & the contract is given to a Frostburg miner at that price." [14]

The departure of the Welshmen left the "Coal mine going on with one hand," but within a week another had been hired.[15] These two managed so well that by April a heading was going off the drift,[16] and with the addition of eight new miners the main and cross headings in the coal mine advanced rapidly.[17] Next the superintendent contracted for timbering the mouth of the mine.[18] Workmen encountered some black-damp, and "two miners commence[d] a windway in the coal . . . to furnish air." [19] During 1838 the coal mine was clearly still in the process of development, but it is evident from frequent references in the diary that small quantities of coal were being taken to Cumberland and Baltimore by wagons which had brought supplies to Lonaconing.

Besides curtailing operations during the financial crisis of 1839, the company proposed also to begin paying by weight instead of by measure. Eliminating payment by the yard forward would deprive the men of compensation for dead work and make their earnings dependent upon the amount of usable coal they could bring out. If wages were to be cut, this was a good time to do it. Many miners and colliers from Pottsville, Pennsylvania, were applying for work at Lonaconing, "the works at Pottsville being very dull." [20] The coal mine boss at Lonaconing was directed "to find 12 or 15 colliers who will consent to work the coal at 50 cts per ton," in an attempt to force this rate on all the coal miners.[21] The company's employees did not react favorably to the suggestion of a rate per ton for mining coal.

Therefore, the superintendent determined to keep only seventeen colliers to supply the furnace, discharging eleven.[22] Those retained refused to observe the same hours as the furnace hands, thus provoking the harassed management to order the furnace blown out and most of the colliers dismissed.[23] The superintendent, at least, contemplated only a short interruption of work, during which some desirable changes would be made in the furnace, and raw materials would be stockpiled "before resuming blast with a view to be independent of a strike among the colliers or miners." [24]

The George's Creek Coal and Iron Company officially set the wage for "Large Vein coal" at "50 cts the ton of 28 bushels" beginning September 10, 1839. Those unprepared to accept this price were warned to be prepared to settle any debts they might owe the company for rent, store accounts, and so on.[25] When the day of reckoning came, the colliers were still undecided, "but after some conversation . . . and an endeavor to procure some promise of increased wages should they not be able to get more than 2 tons of coal pr diem," they went to work.[26] No subsequent mention of colliers appears in the day-to-day journal kept by the superintendent of the company from 1837 to 1840.

The unusually detailed sources on the early operations of the George's Creek Coal and Iron Company allow us to see a good deal of Maryland coal miners' wages and working conditions under one company in the late 1830's. Relatively little information exists concerning the operations of other companies during this decade.

With the 1840's came the first serious strikes in the region. In 1846 the Eckhart Mines employed 52 miners working on two shifts and producing about 200 tons in a 24-hour day.[27] Wages are not specified, but were unsatisfactory from the miners' point of view. On April 29, 1847, the miners employed at Eckhart by the Maryland Mining Company struck, "making a peremptory demand for an increase of the prices now given, which, we learn, yield the miners from $1.50 to $2 per day." [28] The newspaper account continued, "The Superintendent and Engineer of the Company immediately decided to stop the works, and accordingly paid off all the miners and notified the occupants of the Company's houses to vacate their premises by Tuesday last [May 4] at 12 o'clock, M., in accordance with the condition of their respective leases."

When the miners refused to vacate after a full day's extension of their notice to quit, the sheriff of Allegany County brought in a posse, forcibly evicted at least seven occupants of company houses, and left deputies on hand to maintain order. The company superintendent was determined not to resume operations until perfect order was restored and a completely new set of miners had been obtained. Like Lonaconing, Eckhart Mines at this time was a company town. The corporate management could dismiss its employees en masse, if it wished, with only five days' notice to vacate their houses. However, since the Maryland Mining Company had to call in the sheriff, we gather that, unlike the George's Creek Coal and Iron Company, it did not have its own militia.

The strike was of short duration. Within a week the men involved had either left or were about to leave "without any difficulty." "Many new hands have been employed, and it is expected, that in a day or two the list will be complete. Operations on the usual scale will be immediately commenced." [29] Because this was still within the period during which mining could be done by a laborer with a pick, shovel, and wheelbarrow, there should have been no difficulty in getting new hands from among the canal and railroad construction crews.

Another strike possibly took place at Eckhart Mines in the spring of 1848. An old miner reminiscing in 1882 recalled a time when the company employed as strikebreakers a force of canal boatmen, "a wild set of fellows [who] at night made the neighborhood . . . ring with the noises incident to their fiddling, dancing, card-playing and rude sports." [30] By sending out a quantity of refuse coal left by the regular miners, they made it appear that they were producing effectively, and the strike came to a sudden end.

Perhaps because the imminent arrival of the canal at Cumberland seemed sure to bring many more of the already incorporated companies into the coal trade, the coal operators of the western Maryland region gave some thought to "the evils of rivalry and competition," and held a convention in Cumberland on June 27, 1849, to adopt "an uniform tariff of prices and system of credits, both here and at the East." [31] Their agreement sounds very much like the pooling arrangements made by railroads in the 1870's, but in this particular instance no cry of "monopoly" or "restraint of trade" was raised. In announcing the forthcoming meeting, the editor of the Cumberland *Civilian* suggested that the operators also give some thought to preventing strikes. "Fair and just compensation, and prompt cash payments, on the part of the employers—a code of wholesome but reasonable rules, to be well understood and subscribed to by the miners, at the time of their entering the service of a company, and the avoidance of unnecessary austerity by Superintendents, will generally be found sufficient to repress any spirit of insubordination among the men, and prevent those disgraceful exhibitions sometimes witnessed in mining regions." [32]

At this point in Maryland labor history any attempted combination of coal miners probably would have been interpreted as an unlawful conspiracy. An organization of coal mine operators, on the other hand, was allowed to form "A Board of Trade, [subsequently named the Superintendents' Association] with authority to fix the prices of the qualities of coal . . . *and with power also to regulate the price to be paid per ton for digging coal*, and such other matters as the different companies . . . shall deem advisable." [33] (Italics supplied.)

In 1849 the small number of miners employed in Maryland could not have formed a union strong enough to resist the wage-setting decisions of the combined companies. All they could do was to cushion themselves against adversity by forming miners' relief societies.[34] The Eckhart miners, for example, had a benevolent association to provide for "taking care of the sick, for aiding any that may be disabled by accident, for the burial of the dead, for the support of widows of deceased brethren, and for the education of orphans."

The operators' organization set the price of mining at 28 cents per ton, at which the miners made wages of $1.50 to $2.00 a day.[35] In 1850 the 60 miners employed by the Maryland Mining Company and the 40 employed by the Frostburg Coal Company, averaged $25 and $30 a month, respectively.[36]

During the winter of 1850–51 the owners of the Washington mine proposed to cut the price of mining from 28 to 25 cents per ton. The miners called a mass meeting and voted unanimously to reject this proposal. The story of the ensuing strike is told by Andrew Roy, a participant, in his *History of the Coal Miners*. Since miners then had to work two months and twenty days before they received their first month's wages, the company always had in its hands a considerable amount of their back pay. At the time they announced the reduction, the mine owners tried to force their employees to sign a bond by which they would forfeit all money due them if they engaged in a strike. Roy says that his father advised the strikers to send for William Clachan of Virginia to come and manage the strike. Clachan had formerly been a leader of the miners of Lanarkshire in Scotland.[37]

In a letter published in the *United Mine Workers' Journal* of August 20, 1908, Roy recalled further details of this strike, which lasted six weeks and ended in the company's withdrawing its bond and allowing the men to go back to work at their old rate. Roy is mistaken in describing it as the first strike in Maryland; it is quite probable that since he did not arrive in the United States until 1851, and was then only sixteen, he did not know of the suspension in 1847.

It is usually difficult to determine exactly what was meant by any given rate per ton during the early period in this mining region. Roy makes a broad statement that before the Civil War, wages in the Frostburg region were 28 cents a ton for run-of-the-mine coal and 40 cents for riddled coal. In riddling, the coal was put through a sieve to separate the coarse from the fine, and Roy says that the fine coal was treated as refuse and "thrown in the gob."[38] Occasionally we know precisely what was meant. Scharf says that in 1851 the George's Creek Coal and Iron Company paid 30 cents a ton for lump coal and 20 cents for fine coal, which, he concludes, gave the miners a rate of 25 cents a ton for "average digging."[39] In the absence of specific information, we have interpreted rate per ton as being for run-of-the-mine coal.

The Maryland Mining Company, one of the largest and most highly developed in Allegany County, failed in 1851. As soon as the news reached the company's miners, they all stopped work, concerned about their back wages and also about their future employment.[40] Shortly thereafter, word came from New York that the assignee of the company would pay the wages, and that operations would continue as usual. The miners actually were paid by an outsider, who "took an assignment of their claims, and notified the Assignee that he would require payment of him out of the revenues of the property at his earliest convenience."[41] Presumably these same miners continued to work under the new management.

In any event, miners' wages would have been safeguarded by a lien law passed in the 1847 session of the Maryland legislature, designed "for the protection of Miners, Mechanics, Laborers and others employed in Mines and Manufactories, in the County of Allegany, and for other purposes."[42] The law provided that after an employer had failed to pay wages over a period of ninety days, a receiver was to be appointed to dispose of enough of the delinquent's movable property to pay the claims. However, the employees would have to petition the court or judge to appoint such a receiver, listing the amount due each individual involved, and they would have to wait until the aggregate indebtedness was at least $500. In the case of the Maryland Mining Company, the action of the third party saved the miners the tedium of the legal process and at the same time kept the company's assets intact until the time of their public sale in 1852.[43] The miners' lien law was amended in 1852.[44] Under the revision, when an employer had been in arrears for thirty days in the amount of $25 or more, the court was to appoint a receiver to handle his affairs and pay his employees.

Most operators and miners prospered during 1852 and 1853. The increased demand for Cumberland coal created a corresponding demand for labor, and nearly all of the companies were erecting houses for accommodating additional men.[45] Competent craftsmen from Pennsylvania and elsewhere were advised to move to the region, where they would obtain "steady employment at remunerating prices." [46] The first year after the Chesapeake and Ohio Canal reached Cumberland, Maryland coal shipments increased more than 36 percent. With the exception of the panic year 1857, the amount of coal marketed annually from this field rose steadily, reaching a peak of nearly 789,000 tons in 1860.

The unwillingness of the operators to share their prosperity with their employees was a cause of recurring labor trouble. In June 1853 the miners of the Cumberland Coal and Iron Company determined to strike for higher wages.[47] Within a few days they were joined by the employees of other companies in demanding a six-cent increase in pay, bringing the mining rate to 36 cents a ton. The strike began on June 15, with prospects of an indefinite cessation of business.[48] The *Cumberland Miners' Journal*, in first reporting the pending strike, expressed a hope that the wage increase would be granted.[49] However, after the strike had gone on for only two days, the editor had been persuaded that the companies could not afford to pay more, and urged the miners to go back to work.[50]

For reasons not apparent, within ten days after the beginning of the strike, both parties agreed to compromise on 33 cents a ton, a raise of three cents a ton.[51] In spite of this partial victory for the miners, the editor of the *Miners' Journal* reported gloomily, "There is still, we are sorry to say, a considerable amount of grumbling and dissatisfaction, and . . . [we fear] that the difficulty is not yet fully settled." [52] The men had agreed not to ask for another increase without giving thirty days' notice of their intention, but they nevertheless very soon struck for two cents more, which was given them.[53] In other words, the mining rate early in 1854 was 35 cents a ton.

Although the miners of this region had no union until the Civil War period, they were tending to act less as the employees of a particular company and more as a group of laboring men with a common interest. Their cohesion was demonstrated very clearly in the long strike that began in January and ended in April 1854.[54] Management also presented a united front through the agency of its Superintendents' Association, which had been in existence since 1849.

In the latter part of 1853, the Baltimore and Ohio Railroad raised its freight rates on coal 50 cents a ton. The coal companies "cheerfully acquiesced in the increase of freight, believing that their interests were identical with those of [the railroad] company." [55] Moreover, the augmented transportation cost furnished a convenient excuse for reducing miners' wages. Some of the principal companies determined "not to pay more than 28 cents a ton for mining coal, and the superintendent of the Cumberland Coal and Iron Company was directed to carry this out or close the mines." [56] Whether or not this company was to act first in order to sound out the miners' reaction is not clear. At any rate, the reduction was duly announced at the mines of the Cumberland Coal and Iron Company, where trouble had occurred in 1853. The agent of the company gave notice that beginning on Monday, January 9, he would pay only 28 cents, a decrease of seven cents a ton. The employees considered the wage cut doubly unjust: "first, because provisions were very high, and getting higher every day—and, secondly, because the company were not giving them work over half the time, nor had not been for a month or more." [57] In retaliation they resolved not only to refuse the agent's proposal, but also to strike for 40 cents a ton, an increase of five cents.

At first only this company was involved. However, upon the urging of the Cumberland Coal and Iron Company men, the other miners in the region held a mass meeting, at which over 500 were present, and resolved to join the demand for a five-cent raise.[58]

Letters from partisans of both sides began to appear in the Baltimore *Sun*. One accused the Cumberland Company's men of resorting to intimidation and forcing the men in the other mines to leave their work.[59] Another writer in the same vein alleged that the men in the other mines had been content with their wages, and were now out of employment because of fear of coercion.[60] "A Miner" from Pompey Smash challenged the statements made by these correspondents, denying any threats of violence.[61] According to his letter, joint action was necessary if the men were to get a living wage. He pointed out that all practical miners agreed that five tons a day was a fair day's work. At 28 cents a ton, the miner would earn a daily wage of only $1.40, out of which he had to keep his tools in repair and buy the oil and cotton for his lamp.

While these exchanges were going on in the columns of the newspaper, the Superintendents' Association determined on a course of action predicated on the unwarranted assumption that the offer of 28 cents a ton by the Cumberland Coal and Iron Company had established the mining rate for the region, even though that offer had been summarily rejected by the miners.[62] The Association announced that "for *run of mine coal*, for which the miners formerly received 28 cents per ton, they are now offered 33 cents—an advance of 5 cents." Most of the mining fell in this category. To top off their supposed generosity, the superintendents offered increases in two types of work—galleries or headings and screened or riddled coal—which involved few miners and regarding which no demands had been made.

A few miners went to work at 33 cents, but the majority resolved to hold out for 40 cents.[63] A number of the younger miners without families began to leave the region to look for work in other states. By the end of February, about 250 had left the county.[64]

The companies circulated successive rumors that they would import foreign miners;[65] that unskilled laborers in the mines were producing five and a half tons a day;[66] and finally that they would close down completely until they could "commence under better auspices." [67] Early in March striking miners received notice to vacate the company houses occupied by them.[68] Some of the strikers resorted to law to prevent their eviction, but the cases went against them, and they were ordered to vacate or go to work.[69]

These stratagems had their effect on those miners who were already wavering, and some gave in. At Lonaconing, where a few miners had accepted 33 cents a ton, the strikers staged what the Baltimore *Sun* described as a "Riotous Demonstration," marching down the George's Creek valley "variously armed with rifles, revolvers, swords, &c." for the purpose of intimidating those at work.[70] According to the newspaper report, they attempted to enter the mine by force and were prevented "by the armed resistance offered to them by the officers of the company." They next allegedly threatened to burn one of the houses and to manhandle the superintendent and employees of the George's Creek Coal and Iron Company, but were "deterred by preparations by Mr. Smith [the superintendent] to resist and repel them.'" Roy describes the company armed men as "the militia," and says that the demonstrating miners voted "not to provoke a conflict with the soldiers, . . . but all were to return peacefully to their homes as good law-abiding citizens." [71] When three days later the strikers returned to Lonaconing, they were met by the sheriff, who had warrants for the arrest of several who had been recognized as participating in the earlier "riot." The eight men apprehended were "held to bail on the charge of entering into a conspiracy to prevent other miners from going to work in the George's Creek Valley." [72] Nothing in circuit court records indicates that these cases ever came to trial.

By the first of April the rumors circulated by the companies had greatly weakened the will to hold out. A committee of miners met with the Superintendents' Association and offered to compromise at 38 cents a ton, but the offer was declined.[73] The companies were still resolved not to give more than 33 cents, which they said was more than was paid at any other mines in the country "under the circumstances"—a weasel wording whose meaning is not clear unless the operators were referring to the more favorable working conditions in the Big Vein. On April 3 the miners in a mass meeting decided to accept the companies' terms, and resumed work after a suspension of nearly three months' duration.[74]

Needless to say, the contingent of foreign miners never arrived, and the companies did not stop operations for the rest of the season. Although the employers had not succeeded in cutting wages to 28 cents a ton, they had very neatly taken two cents from the rate they had been paying before the strike, and had left the public with the impression that they had magnanimously given the miners a five-cent raise. As coal started to move again, the clerk of the Chesapeake and Ohio Canal Company was informed by a Cumberland correspondent, "I should have written to give you the good news that the miners have been overcome and have all turned on to work; and I should judge from the way in which the coal is coming down that they are trying to make up for lost time." [75]

Within a month the Superintendents' Association capriciously agreed to raise the mining rate two cents a ton, thus restoring the miners' wages to the amount paid before the strike, that is, 35 cents a ton.[76] In the meantime the whole local economy had suffered. One wonders to what purpose the thirteen other coal companies in the area had so stubbornly supported, and perhaps abetted, the irresponsible act of the Cumberland Coal and Iron Company in insisting on an apparently unnecessary pay reduction.

Corporate control of the Maryland coal resources had already aroused misgivings in at least one observer who visited Allegany County at about this time. It was his opinion that "Companies generally monopolise large tracts of land and are rather disposed to *realize* from the fluctuations of the stock, in the market, than from the actual profits of the mines. . . . The mines are, in fact, made subordinate to speculative purposes; and as a consequence the whole region of country involved in their scheme is . . . the greatest sufferer." [77]

This writer's suspicions were in at least one case well founded; in 1854 one of the large companies, the Parker Vein Coal Company, was involved in a stock-watering scandal. For our purposes we need to know only how the company's financial difficulties affected the miners, who quit work and planned to avail themselves of the miners' lien law.[78] They filed their claims, and, as in the earlier case of the Maryland Mining Company, the liens were transferred to third parties, who advanced the money on them.[79]

Employment was good in 1854, and experts predicted that it would remain so for some time to come.[80] The coal trade started off briskly enough in 1855. Between January 1 and April 1, the amount of coal shipped was more than two and a half times that sent out in the corresponding period for 1854.[81] Many of the companies had good contracts or were in the process of negotiating them.[82] The manufacture of iron was "going on actively" at Frostburg and Lonaconing.[83]

In the four years preceding 1855, coal shipments from Maryland had increased, on an average, 100,000 tons each year. There was reason to expect a similar increase in 1855, but on June 7, the Cumberland *Telegraph* reported that no coal company was working to capacity and that some had almost suspended operations. Coal was selling at Baltimore for $3.50 a ton (75 cents less than the 1854 price) and the costs of mining, transportation and other expenses were said to absorb nearly the whole of that sum, leaving no margin for profits.[84] The editor of the *Telegraph* hinted that a reduction in wages was necessary if the companies were not to close their works and cease operating altogether.

So far as can be ascertained, this is the first time that the Maryland coal companies made any attempt to justify a wage reduction to the miners by explaining what was involved in a current recession or depression. In this instance, the companies presented a convincing case. The miners themselves discussed the issue and proposed a reduction of three cents, which would have brought the mining rate to 32 cents a ton.[85] The operators, however, asked that it be reduced to 30 cents.[86] The reduction to 30 cents announced on July 12 apparently aroused no overt opposition.[87]

Trade was dull for the rest of the year; shipments increased only 2,500 tons. The curtailment of orders for Maryland coal in 1855 was attributed to depression in the iron industry, to suspension of operations by several manufacturing establishments, and to the loss of bunkering trade occasioned by the withdrawal of Atlantic steamers to supply troops and munitions for the war in the Crimea. Some blame was placed on the reciprocity treaty with Canada, which made possible the importation of Nova Scotia coal to compete with the Maryland product.[88]

On subsequent occasions, when economic factors were clearly laid before them, the Maryland miners would again show that they were willing, as in this case, to undergo a wage cut or to postpone a pay raise, until times should improve. They would listen to reason, but they would not quietly accept any arbitrary or capricious dictum. And when they believed hard times had passed, they would argue for their rights—sometimes successfully, sometimes not.

When orders for coal began to pour in at the beginning of 1856, and business seemed promising for the year,[89] the miners of the Cumberland Coal and Iron Company and of some of the companies on George's Creek struck for a return to the old rate of 35 cents a ton.[90] This short-lived strike, said the editor of the Cumberland *Telegraph*, arose out of a misunderstanding; the miners thought that increased orders meant that the price of coal had risen, and that they should share in the companies' prosperity. On the contrary, he argued, coal was selling at a lower price than ever before, but "with a material falling off in the price of provisions and other things, the miners are now better able to mine coal for thirty cents per ton than they were one year ago at thirty-five cents." As soon as the miners learned these "real facts," they went back to work at the old rates, which continued to apply throughout the remainder of the decade.[91] As to the "facts": coal prices at Baltimore, reported in the Cumberland *Telegraph* of May 22, 1856, were $4.25 per ton for lump coal and $3.75 for run of mine, as compared with the 1855 price of $3.50 per ton for an unspecified grade of coal. Furthermore, it is most unlikely that the cost of provisions had gone down in this period of generally rising prices.

No labor disturbances of note occurred between 1856 and 1860. Employment conditions fluctuated with the state of the coal trade, and, unfortunately, with the state of the canal, which was subject to frequent breaks. In the panic year of 1857 the closing of a number of important New England textile mills undoubtedly accounted for at least part of the decrease of almost 124,000 tons in the Cumberland region's coal shipments.[92]

The return of prosperity in 1858 moved the editor of the *Telegraph* to new heights of composition: "At the mines there is manifested considerable activity. Additional miners have been put to work, and from each dark subterranean gallery there comes forth train after train loaded with the matchless wealth of the mines. Within, you hear the ceaseless thug [*sic*] of the pick as it comes in contact with the mineral, or the metalic [*sic*] ring of the shovel as the coal slides from it into the cars, ready to be brought forth and sent on its way to market. You behold lights flitting to and fro in the Eribus-like [*sic*] darkness like fire-flies, and ever and anon the shrill neigh of the horses as they pass in and out. All is activity and bustle." [93]

Chapter 10

The War and Readjustment (1860-1868)

The miners took an active part in the 1860 Presidential campaign. At Jackson Mines, near Lonaconing, they organized a Bell and Everett Club at a meeting which featured speeches, patriotic music by the Lonaconing Band, and the raising of a 125-foot pole flying the Union party flag.[1] At Pompey Smash, the Lincoln Club "raised the first Republican campaign pole ever raised south of Mason and Dixon's line." [2] Late one night members of an opposing faction (unidentified) cut down the pole, but it was re-erected and Thomas Brown, president of the Lincoln Club, subsequently defended it. Brown, "a Scotchman and a man of nerve," used a shotgun to discourage the second set of vandals, purportedly "a number of Irishmen." [3]

William Cullen Bryant reported to readers of his New York *Evening Post* "proof that the approaching election of a President was as much a matter of interest to the people of this remote region as to you in New York. A flag was flying on a prodigiously tall pole near the mouth of the mine. It was a Lincoln and Hamlin flag, they said, and the workmen had chalked the name Lincoln (only a little ill-spelled) on the sides of the trucks." [4]

The election returns of 1860 showed how little sympathy the voters of Allegany County had with the South.[5] Although the eastern and southern counties carried the state for Breckinridge,

Allegany County gave only 23 percent of its vote to the southern wing of the Democratic party, dividing the remainder as follows: Bell, 36 percent; Douglas, 29 percent; Lincoln, 12 percent. In the mining districts (Westernport, Frostburg, Lonaconing and Mount Savage) Bell and Douglas ran neck and neck, with 30 percent and 29 percent, respectively, of the votes. Lincoln got 522 votes, 342 of which were from the mining areas. The Cumberland *Civilian and Telegraph*, which had supported the Constitutional Union party, attributed the miners' vote for Lincoln to the tariff protection promised by the Republican platform, saying that it was given "without reference to the Slavery question." [6] However, whether the votes had been cast for Bell, Douglas or Lincoln, they were all in a sense votes against secession, and at the outbreak of hostilities the sentiment in Allegany County was pro-Union.

While secessionist riots were rocking Baltimore,[7] "in the Cumberland Coal District . . . the Miners, &c., [were] volunteering in the Pennsylvania [military] Companies, on the side of the Stars and Stripes and against treason." [8] Furthermore, within two days after receipt of the news that Massachusetts troops had been attacked in Baltimore, residents of Lonaconing formed a military company and offered its services to the Secretary of War at Washington.[9] The "Detmold Riflemen" were part of an Allegany County regiment sworn into the Union service late in August 1861.[10] The Second Regiment, Potomac Home Brigade Infantry, Maryland Volunteers, was also organized in 1861.[11] Andrew Roy's old mining friend, Andrew Spier, was a lieutenant in this outfit, and Roy says: "The Company which Lieutenant Spiers [*sic*] belonged to was raised in the Frostburg mining region, and was composed of miners, with all of whom I was personally acquainted. They were all stalwart Republicans, when it took nerve to be a Republican in a slave state." [12] Members of the Second Potomac Home Guard took part in twenty-three skirmishes and battles, serving in Maryland, the Valley of Virginia, and the Gettysburg campaign.[13]

In 1862 Allegany County was requested to furnish 872 men

for the army. Over 1,400 volunteers, many of them from the mining area, offered their services.

Initially the depletion of the mining force was not as detrimental as it might have been, for the mine operators were not in a position to offer full employment. The first year of the war brought the Cumberland coal trade close to ruin. In 1860 the various companies had marketed almost 789,000 tons, the largest annual amount recorded up to that time.[14] Because of transportation problems, shipments fell to slightly less than 270,000 tons in 1861.[15] The slump dramatically illustrated the dependence of this region on the Chesapeake and Ohio Canal; the amount shipped in 1861 was almost exactly that shipped in 1851, when the trade was just opening on the newly-completed waterway.[16] In addition, the coal companies relied on the Baltimore and Ohio Railroad not only as a carrier but also as a consumer of their product, and the railroad line, like the canal, was perilously close to the Virginia border. No matter how much coal was dug, no matter how free from labor disputes the region remained, the fact is that for long periods no coal could be moved east after the transportation lines had been cut by Confederate raiders.

Hostilities had scarcely begun when the Pottsville *Miners' Journal* predicted, on April 27, 1861, "In the Cumberland Coal District, the business will of course be greatly checked by the suicidal course of the people of Maryland," referring probably to the panicky destruction of the railroad bridges around Baltimore. And on June 1 this paper announced, "The Cumberland Coal Trade appears to be entirely suspended for the present by Railroad as well as Canal. The dislodgement of the Rebels from Harper's Ferry, which will soon take place, will re-open the trade by Railroad from Cumberland again, but not by Canal. . . . Since the stoppage of . . . Cumberland Coal, Broad Top [Pennsylvania] . . . is coming into demand to supply its place."

The correspondence files of the Chesapeake and Ohio Canal Company bear witness to its wartime difficulties and their effect on coal transportation. On May 6, 1861, the canal agent at Cumberland informed the Company's chief clerk, "Coal can be shiped in a few days from Cumberland if we have no trouble in Virginia." [17] A week later he wrote, "The Coal Companys are not willing to Boat on account of the troubles along the line by the Virginia people and at Alexandria. I cannot tell what we are to do. The Canal is now in order, yet we will have no Boating as things now stand." [18] At the end of the summer, according to the Pottsville newspaper, "The Chesapeake and Ohio Canal is now navigable to Alexandria, but the Virginia rebels amuse themselves by shooting at the boatmen." [19]

In its customary year-end summary of the coal trade, this same source reported for the Cumberland area: "Since the war broke out the trade from this Region has been almost entirely suspended by the destruction of the Baltimore and Ohio Railroad, and also by the obstructions placed in the Canal. Since July last, no Coal has been transported over the Baltimore and Ohio Railroad, and no returns, of course, have been made." [20]

In the face of these difficulties, the Maryland coal industry might well have been expected to cease operations altogether, leaving the market to the Pennsylvania anthracite and semi-bituminous fields which were not plagued by the same transportation disruptions. One can merely speculate on why the operators did not give up. Possibly the greatest incentive to continued operation was the realization that the demand for coal would increase with the war effort and that profit would be great once the avenues of trade were reopened. Moreover, since coal mines deteriorate rapidly when not being worked, it was obviously advantageous to keep them open as long as possible within the bounds of economy. So far as ascertainable, no labor trouble occurred in the Frostburg or George's Creek mines during 1861. Those miners who had not volunteered for military service apparently were content with their wage of 30 cents a ton.

The Cumberland newspapers published during the war concerned themselves very little with affairs in the mining region, limiting themselves mainly to war news and to belaboring Democrats, Copperheads and the like. Fortunately, however, the records of the Chesapeake and Ohio Canal Company contain a small body of correspondence from A. C. Greene, one of its directors, who was also superintendent of the Borden Mining Company. These letters give us some interesting information on the progress of the coal trade, labor conditions at the mines and on the canal, and the frequent interruptions to navigation.

Greene wrote on April 7, 1862: "There are symptoms of a Strike among the Boatmen here—and it will not be possible for me to attend the meeting on the 10th inst. All the Coal Companies are willing to pay an advance on previous rates, but the demands of the Boatmen—so far as they make any—viz. 150 @ 175 [$1.50 to $1.75] per ton are altogether inadmissible—we could not compete with the Penna Coals. The Penn. RR Co. has reduced their charges upon Broad Top *30¢* per ton expressly to head off Maryland Coal."[21] It apparently did not occur to Greene that the boatmen who were serving as targets for Confederate snipers deserved extra pay.

Early in the war the federal authorities had seized a number of canal boats on the supposition that they would be useful to the military. Greene complained bitterly:

> The want of them [boats] is ruin, not only to the Canal Co. but to the poor laborers of this County who have been nearly a year out of work. In the meantime the Penna interests are moving heaven and earth to maintain themselves on the ground they were enabled to occupy by our disasters last summer. The Central RR has already abated 30¢ per ton in its tolls on Broad Top Coal—(the special competition of Cumb.) and we understand are prepared to go even further in order, if possible, to exclude us from market another year—when an attempt to regain it on our part will be almost impossible. The very existence of the Canal is trembling in the balance. The Boats in Govt hands number nearly, if not quite, *one hundred*—and they *cannot possibly* be replaced this season. Neither the Capital nor the courage to undertake it can be found.
>
> I cannot comprehend why the authorities at Washington continue to hold these Boats—so persistently—when they appear to have absolutely no use for them. If the Boats were actually and in-

dispensably necessary to the Govt we should submit but it does not appear that anything stronger than red tape, or other indifference, keeps them where they are.

It does seem to me, therefore, on the whole, that if the Board of Pub Works and the Governor [of Maryland] would unite with us in a representation of the facts to Secretary Stanton—that they would be released. To retain them would look as if the Govt itself had joined the Pennsylvania conspirators to crush out and destroy the Maryland Coal trade—at one blow.[22]

However, the canal was open, and that in itself was a cause for thanksgiving. Greene resigned himself to the shortage of boats and wrote, in a letter indicating an upswing of production at the Maryland mines: "Preparations for the resumption of trade on the Canal are now complete on the part of the Coal Co's and I hope and trust, we may yet have a season of large prosperity. The number of Boats has sadly diminished, but with *uninterrupted navigation* there are enough to earn as large a revenue, perhaps, as the Canal has ever derived from the Coal trade." [23]

The resumption of trade did not proceed as planned. Spring floods damaged the canal, and we can imagine Greene's state of mind as he told the company's chief clerk, "The concurrence of circumstances against the resumption of Canal Trade this Spring are positively *infernal.* Not only is the Canal deprived of Revenue—but the trade itself is imperilled by these untoward events. I have great fears that the largest and most important buyers of coal will despair of getting any thing from us this year—and contract with the Broad Top people, who are straining everything to retain their footing in market, acquired last year. I have about *1800* tons afloat on the way to Georgetown—and, I suppose, there is fully 7000 tons in the same situation. I trust every exertion will be made to restore navigation as soon as *any way possible.* Rely upon it—very much depends upon our being able to deliver coal early in May. The Rail Road is also out of the ring owing to the Bridge breaking at Harpers Ferry." [24]

In June 1862 Maryland coal miners' wages were advanced to

40 cents a ton.[25] The Cumberland and Pottsville newspapers do not report any instance of a strike or even of a request for the raise, nor does Greene mention it in his letters. At this particular time, fate seemed about to favor the Maryland operators. Great floods "stopped the transportation of every Anthracite mine in the United States for a month, and of some—as of the Lehigh region—for three or four months."[26] Greene predicted that "damages to the Penna works may throw upon our coal field an unusual demand," and stressed the fact that it was "of great importance that we should be in condition to meet it." In addition, he said, "The break in the [B&O] Rail Road makes it of the [utmost] importance that the supply of Coal should go down by Canal so that the demand for the article can be supplied—*moderately*—at least."[27]

The canal was still out of commission on June 21, an exasperating state of affairs since, according to Greene, "The disasters to the Penna coal lines have caused attention to be turned this way, and it is a thousand pities that we are not in condition to meet the demand." However, the railroad was open again, and Greene promised "to be in Washington next week at the meeting of the Board provided Gen Jackson permits the cars to run upon the B&O RR Co."[28]

Lack of communication between canal officers and the coal companies made shipments uncertain. It was impossible, said Greene, to know when boats were going to be available, and in at least one instance "the Coal Companies have maintained their expensive force in idleness expecting the arrival of Boats hourly. The Rail Roads too have had engines and trains on the Road, at large expense doing nothing—but, in the absence of advices, expecting the Boats."[29]

The miners received another increase—this time of five cents a ton—in September 1862.[30] Again available contemporary sources do not mention this pay raise. In fact, it is difficult to understand why an increase should have been offered or granted. The Cumberland *Civilian and Telegraph* of September 18 mentions suspension of navigation on the canal and damage done to the railroad, and says that large numbers of miners were thrown out of work. The Pottsville *Miners' Journal* of September 27 confirmed and expanded the report: "The raid made by the Rebels into Maryland enabled them again to destroy Coal transportation on the Baltimore and Ohio Railroad by the destruction of the bridges at Harper's Ferry and on the Monocacy. They have also destroyed navigation on the Chesapeake and Ohio Canal by damaging the works for a distance of 25 miles. This cuts off the whole Coal Trade from the Cumberland Region again."

In November, Greene complained once more, "There is no regularity whatever in our mails beyond this, that they are always three or four days behindhand." [31] The difficulty, under the circumstances, of separating fact from rumor led sometimes to unfortunate results. In Greene's words: "The neglect of the Supts. [of the various divisions of the canal] to advise this terminus of the condition of the Canal has been of very great disadvantage—We have been left to rumor wholly, and in consequence have been unable to form any judgment of the actual condition of things. Last Friday a report reached Cumb that the army had built four bridges across the Canal about Berlin so low as to stop Boats from moving in either direction. The consequence was that Boats which were about to load hauled off again and the news reaching the mines—there was on Monday a general stampede of our miners for Penna. They very naturally concluded that if the army had got hold of the Canal there was an end of it for this season at least. Can we not enforce the Resolution of the Board requiring information to be communicated? Some few Boats are loading today—but none have arrived from below. Proper energy in the maintenance of navigation may yet afford considerable business this season."

However, in December Greene was so busy sending down by canal "all the coal I possibly can," that he was again unable to attend the board meeting.[32] In their competition for the limited number of boats, the coal companies were paying the boatmen

$2.50 per ton,[33] although in April Greene had said that $1.75 a ton was more than the operators could afford.

The Pottsville *Miners' Journal* summarized the Cumberland coal trade for 1862 as follows: "Since the war broke out the trade from this Region has been almost entirely suspended. The Railroad only carried Coal in the months of May, June, July, August and September—and the trade on the Canal was interrupted entirely in the months of May and October, and only partially resumed in the other months. They are however looking forward to brighter prospects for this year." [34] In spite of its handicaps, the Cumberland region shipped about 318,000 tons in 1862, an increase of roughly 48,000 tons over the previous year.[35]

Prospects for 1863 were bright enough to afford the miners another five-cent raise in January.[36] Greene was optimistically expecting that "every Boat on the Canal will be actively employed this season," and was "sure that the high rate of freight lately adopted by the Railroad Co will stimulate the building of new [boats] to a large extent, provided the condition of the Canal is such as to warrant the expectation of steady navigation." [37] The editor of the Cumberland *Union* reported on June 13, "In our own county, the demand for laboring men in the various branches of mining and railroad operations, far exceeds the supply, and the workmen are receiving higher rates of pay, we believe, than ever previously realized for such services."

One of the reasons for the increased orders in the Maryland field was that the Broad Top mines were worked only nine months in 1863. According to the Pottsville *Miners' Journal*, "The [Confederate] raid into Pennsylvania checked transportation for one month, and a miners' strike [in that same state] lasting about two months, . . . also suspended shipments for that period." [38] The Cumberland trade increased to more than 748,000 tons,[39] only a little less than the amount shipped in 1860, and 430,000 tons more than was shipped in 1862. It is amazing that so much coal went out while transportation difficulties continued. Enemy activity in early July halted both canal and railroad for more than a month,[40] and in September Greene wrote hotly to the clerk of the canal company:

> You cannot have failed to learn that the Canal is again practically closed by the neglect of the Government to afford the Boatmen protection against robbery of their teams by the Virginia guerrillas. The few Boats which arrive at Cumberland decline to load and tie up their Boats being unwilling and, in fact, unable to risk the loss of, in many cases, everything they have. We all understood that the military authorities promised to secure the navigation, but a large number of Boats have since been disabled by the loss of their teams —and have stopped.
>
> I think it would meet the unanimous approbation of this whole community if Moseby [*sic*] or White or whoever leads these incursions should ride into Washington some fine night and carry off with them to parts unknown Gen Halleck, Sec Stanton, and every body else whose duty it was to prevent these shameful raids—even for their own sakes. If we could only have some heavy rains to raise the river enough to render it unfordable, we might hope to do some business, but our old fashioned drought is upon us—with no prospect of a let up until winter.[41]

In February 1864 the canal board notified its superintendent at Cumberland that navigation was to be suspended from February 1 to March 1 so that repairs could be made.[42] However, when the canal reopened, the officer at Cumberland reported: "The canal has been navigable ten days. No Boats been loaded. two or three that was loaded last December has started down. The Boatmen are holding back for an increase on freight. They are haveing trouble to get hands. I found it necessary last month to advance the pay of Laborers, and will have to continue it to obtain old hands." [43]

Shortly thereafter, the Washington [D.C.] *National Republican* published the following under the heading "Dead Lock at the Cumberland Coal Mines": "A gentleman who arrived here to-day from the Cumberland coal mines in Maryland states that nearly all the miners and boatmen have struck for higher wages. The company refuse to accede to these demands from the fact that to do so would necessarily carry up the price of coal to such a high rate that there would be comparatively little demand for it, especially when brought in competition with a superior coal selling at about the same price. There is very little coal at the mines now, hence this strike is the more unfortunate. If the miners insist upon their demand for increased wages, it is said that the company will discharge the hands and employ others. In such an event, resistance is threatened, but as some of Gen. Sigel's forces are near at hand, it will not amount to much."[44] The Baltimore *Daily Gazette* and the Baltimore *Sun* reprinted this news item on March 19, 1864, giving the *National Republican* as their source. In all probability the item refers to the employees of the Cumberland Coal and Iron Company, which was the principal operator at this time.

On March 18 Greene wrote from the Borden Mines, "We shall have a season of great trouble, I fear, in our business. The Boatmen have gone to work at *two dolls freight* 'for the present.' I do not doubt they will make a demand for more, later in the season. The miners are concocting a strike for *25¢* a ton additional! So we go. These demands cannot be granted and I have hardly a hope of avoiding serious suspension of business."[45] Borden miners had not struck by March 23, when Greene described them as "uneasy and restless," and said, "I have no doubt they are fixing for a strike. How it will end I cannot now tell. The prospect is not flattering for the season's business."[46]

Fincher's Trades' Review reported briefly on March 26: "A strike in the Cumberland Coal mines has been suppressed by military interference. Twenty-five of the disaffected men were sent to Fort Delaware." Wieck cited this statement in his history of the American Miners' Association as evidence that a strike did occur in Maryland in 1864 and was so disposed of.[47] However, the Official Records of the War of the Rebellion do not refer to any such military action, and a search of Civil War records at the National Archives has not revealed orders for the arrests or any list of civilian prisoners taken in this area at this time. Furthermore, as far as can be determined from a search of the Fort Delaware records in the same depository, no such prisoners were incarcerated there. It is highly doubtful that military action against strikers would have escaped notice by the two Cumberland newspapers, which lost no opportunity to castigate Copperheads or condemn any activity which might have aided the enemy. Andrew Roy, who spent a period of convalescence from war injuries at the military hospital in Clarysville, and who knew most of the Frostburg miners personally, makes no mention of either strike or arrest in his *History of the Coal Miners.*

It is reasonable to assume that Fincher's information was faulty and perhaps described some incident in the Pennsylvania anthracite region, where violence and Molly Maguirism were rampant during the war.[48] The Pottsville *Miners' Journal* recorded strike after strike in Pennsylvania from 1861 through 1864. The military were stationed in anthracite coal towns for several months after August 1863, and "men who have stopped Collieries by threats and violence [were to be] arrested and handed over to the military authorities for trial and punishment." [49] The assumption that Fincher was in error, possibly as regards the location of the arrests, is bolstered by the fact that Major Generals Couch and Sigel visited Pottsville a little before the alleged Cumberland incident to investigate what action they should take "in putting a stop to the lawlessness that prevails in this [anthracite] Region." [50]

Whatever the actual situation at the Maryland mines, by March 24, 1864, the boats were loading and leaving Cumberland at the rate of 15 or 20 a day.[51] The war effort drew heavily upon the eastern coal fields. "More coal was demanded by private manufacturers and carriers engaged in making and transporting arms, munitions, and supplies for Government," and the Government itself increased its consumption of coal from 200,000 tons in the first year of the war to a million tons in the fourth year.[52] In April the Maryland miners were given a raise to 60 cents a ton, and in June the rates increased still further to

75 cents, thus making up the 25-cent increase which Greene had said the companies could not afford.

On the subject of what the coal companies could afford we have one or two pertinent bits of information. The Central Coal Company was organized in 1863 with 20,000 shares of capital stock at $100 a share. In 1864, the first year of operation, this company paid dividends of 3.5 percent on the capital stock; in 1865 it paid 5 percent.[53] The Hampshire and Baltimore Coal Company, although able to pay only small dividends during the first three years of the war, reported to its stockholders in 1864: [54] "The prospect of business for the current year is good. The price of anthracite coal being so high will induce an increased consumption of our coal, which now costs considerably less than anthracite in New York and Boston. And the great activity of the engine-shops, ship-yards and rolling mills, supplying the government with iron gunboats, &c., and for whose purposes our coal is the most suitable, also causes an increased demand. In addition, the high rate of exchange excludes, in a great measure, foreign coal from the market. On the other hand, the limited supply, arising from scarcity of labor and transportation facilities, combines with the former in enabling us to sell our produce readily at good prices." *

* The price of anthracite coal to the consumer advanced nearly 200 percent between 1862 and 1865. The Pennsylvania anthracite operators blamed the high price of their coal (large amounts of which were supplied to the government by contract) on labor troubles at the mines. However, William L. Sylvis, president of the Iron Moulders' Labor Union, in *Fincher's Trades' Review*, October 29, 1864, said that the coal companies circulated rumors of strikes and violence to get military interference and to excuse their high prices. In many cases the operators themselves closed their mines to keep prices high and then said that their miners were on strike. Sylvis, who had just returned from a tour of the mining region, concluded: "And they [the operators] are now engaged in the very Christian work of making the people believe that the miners are on a strike; and just as soon as the supply becomes short, and all danger of a further decline has been removed, they will say to the authorities, the supply of coal is short, our miners are on strike, and are banded together for the purpose of keeping the mines closed, and thus crippling

The coal company agents were not inclined to quibble when they met in September 1864 to agree upon joint action "in reference to the demands of the miners and boatmen for an advance upon the old rates. The result of the conference was that the demands of both miners and boatmen were acceded to—the price for mining being fixed at $1 [per ton], and transportation at $3 per ton." [55]

Greene reported at the end of September: "The Rebs are stealing the Horses from the Boats close to Cumb[d]—Two Boats have been robbed within ten miles of Cumb[d] and *last night* a gang of McNeal's [*sic*] men crossed at Black Oak bottom, passed over Will's Mountain into the valley of George's Creek and swept the Coal mines of their horses. The American Co. lost *sixteen.* In the meantime Gen. Kelly sits in Cumb surrounded by thousands of men, but takes no steps to prevent or punish these maraudings. I wish you could impress upon Stanton the facts and the Consequences resulting from them—the Canal trade is paralyzed by reason of them." [56] This "paralysis," no doubt, accounts in part for the fact that shipments in 1864 were about 90,000 tons less than in 1863.[57]

In spite of their increased labor costs at the mines and on the canal, and in spite of the activities of raiders, "The Coal Companys talk in good Sperits [;] thing[s] look well here." [58] The miners in turn passed on some of their benefits to those less fortunate and donated "one day's digging of coal, free of charge" to help the poor in Baltimore.[59]

An interesting sidelight on the desperate need for coal during the war is a note from the President of the United States to J. W. Garrett, president of the Baltimore and Ohio Railroad. Lincoln wrote on January 10, 1865: "It is said we shall soon all be in the dark here, unless *you can* bring coal to make gas. I suppose you would do this, without my interference, if you could; and I only write you now to say, *it is very important to us;* and

---

the government. And straightway they will be forced into the mines at the point of the bayonet, to work at whatever price the operators may please to give."

not to say that you must *stop* supplying the army to make room to carry coal. Do all you can for us in *both matters*." [60]

The end of the war shocked the Maryland coal operators almost as much as its beginning. The canal agent at Cumberland informed his superiors late in the spring of 1865 that "about all the Coal Cos. have stoped Shipping Coale—on account of the high rates of freight mining tools &c &c. The Canal is doing nothing now No Boats loading all stoped or will in a day or So —Everything must come down to enable the Coal Cos to make further shipments of Coal from here and Compete with other Coales now in the northern market. So Says the Coal men here." [61]

At least a temporary curtailment of operations was to be expected, since the reduction of choice government contracts for manufactured goods would have inevitable repercussions in the coal trade. With the restoration of normal competition, the various regions would look for ways in which to decrease their costs and attract customers. In Cumberland, the representatives of the several coal companies agreed to suspend operations while they were determining a further course of action.[62] The mines closed early in May, ostensibly because of "the rapid and heavy decline of prices for coal in the Eastern cities," and the company agents agreed that "a corresponding reduction in the cost of producing and transporting coal to tidewater is therefore absolutely necessary to the revival of the trade." The essential economizing would involve reductions in freight (including boatmen's charges) and mine labor.

Since September 1864 the Maryland miners had been receiving $1.00 a ton. The Cumberland *Alleganian* of May 17, 1865, announced that the operators in conference had decided to cut this wage to 60 cents a ton. At the same time, they proposed to reduce boatmen's fees from $3 to $2 a ton and to ask for lower tolls on the canal. The Baltimore and Ohio Railroad had already taken $1.50 a ton from its charges on coal.

A week later the mines of the larger companies were still closed. A few of the smaller companies with contracts to fill were shipping coal. "The miners generally . . . [were] indisposed to accept the rates adopted at the Superintendents' meeting, while the Companies continue[d] to urge their inability to pay higher figures." [63] Even the small amount of mining still being done stopped when the canal was drained to repair flood damage.[64] The wage issue remained unresolved, the miners holding out for their old rate, while the companies affirmed that they would not operate at all unless the rates were reduced to 60 cents.[65] "The same state of affairs exists in all the mining regions," said the Cumberland *Civilian and Telegraph*, "but we are glad that, so far, we have heard of no acts of lawlessness, as in Pennsylvania and Ohio." [66]

After what amounted to about a month's lock-out, both sides yielded a little and compromised on a mining rate of 75 cents a ton.[67] Unfortunately, transportation difficulties prevented immediate resumption of full-scale production. The B&O's troop-carrying activities temporarily precluded its transportation of coal, and consequently the miners for a brief period were on half-time.[68]

Although the six largest coal companies complained jointly that "the coal trade is almost prostrate by the suspension of all industries, consequent upon the sudden transition from War to Peace," [69] the fact is that never before in the history of the region had so much coal been shipped as in 1865—over 900,000 tons.[70] In these flush times the demand for labor in coal and iron absorbed all available workers and even infringed on the supply of agricultural workers.[71]

When the shipping season approached in 1866, after the usual winter closing of the canal, the coal companies and their employees were again at odds over the price for mining.[72] The companies had agreed among themselves to reduce wages to 50 cents a ton, giving as their reason the decrease in the price of coal on the market. The miners, after consultation, refused to work at the lower rate. Operations were not completely at a standstill, for several of the companies carrying out existing contracts kept on paying 75 cents a ton. However, it was understood that as soon as they had filled their orders, these companies too would stop operations unless the miners would accept the reduction.[73] Again the parties compromised, this time at 65 cents a ton, a rate which was maintained, in spite of company efforts to whittle it down, until early in 1877.[74]

Business resumed briskly, and the editor of the *Civilian and Telegraph* prophesied: "If Congress does not ruin us by a reduction of the tariff on bituminous coal, there will be such an extensive trade during the remainder of the season as has never been known in the history of the canal. Some of our companies will ship over a thousand tons per day." [75] The tariff of 1861 had placed an import tax of one dollar a ton on bituminous coal, with the expectation of limiting competition in the market to domestic producers.[76] What Congress proposed to do in 1866 was to place a duty of $1.50 a ton on anthracite coal and only 50 cents a ton on bituminous coal.[77] In the eyes of the Maryland coal interests, this proposal would "legislate to the ruin of our immense coal regions, and for the benefit of northern manufacturers," since bituminous coal from Nova Scotia could be "imported to the northern cities at far less cost than it can be produced from our coal fields." [78] Republican Congressman Frank Thomas, representing the Allegany County region, averted catastrophe by securing an amendment to the tariff bill, raising the import tax on soft coal to $1.50 a ton.[79]

Lower wages, freights, tolls, and the preservation of tariff protection combined in 1866 to save the operators from the ruin they had predicted, and made possible the establishment of a new record of over a million tons in shipments.[80] Company incomes were handsome. The Central Coal Company's profits, to quote one example, "were 50 per cent over the year 1865, but the dividends were passed, and the profits reserved to cancel large items for improvements, and in payment of additional and valuable real estate purchased by the Company." [81] This corporation paid dividends of 7 percent in 1867 and 5 percent in 1868 and 1869.[82]

The matter most talked of in the mining region in late

1868 and early 1869 was the murder of Douglas Love, boss miner for the Central Coal Company, at his home in Lonaconing on December 6, 1868.[83] This killing had many of the earmarks of the Molly Maguire terrorism then rampant in Pennsylvania, and although no connection was suggested in the press at that time, a legend still persisted in Lonaconing as recently as 1950 that the Mollies had been responsible.[84] The legend may have arisen because one of the five men arrested and indicted for the murder was named Maguire. The state subsequently found itself without sufficient evidence to prosecute, and the prisoners were discharged after insisting, through counsel, that their objection to the *nolle prosequi* be entered on the record as part of the proceedings. The case was never solved.[85]

The region had added tremendously to its mining population during the war and immediate postwar years; in the decade 1860–1870 the number of coal miners more than doubled. Production continued to rise in 1867 and 1868, and in 1869 was materially affected by the "protracted strike in Penna., [when] there was such a scarcity of coal in the general market that large quantities of Cumb. Coal were taken."[86] The news sources available for the last three years of the decade do not mention strikes or even any controversy between coal-mine management and coal-mine labor in Maryland. Thus with a minimum of postwar readjustment, the Maryland coal industry by 1870 was ready to begin the remarkable expansion which continued well into the twentieth century.

Chapter II

Big Business: Small Union (1869-1875)

In terms of what it would eventually mean to the miners in western Maryland, the most significant event of the 1860's was the formation of the Consolidation Coal Company.[1] Chartered in 1860 by an extraordinarily competent group of northern business men the company in 1864 marketed $5,000,000 in capital stock and acquired all the properties of the Ocean Steam Coal Company, the Frostburg Coal Company and the Mount Savage Iron Company. The newly-formed corporation began its career with thousands of acres of coal lands, the railroad and wharf facilities of the well-established Frostburg Coal Company, and control of the Cumberland and Pennsylvania Railroad, a crucial link in the delivery of coal to the Chesapeake and Ohio Canal and the Baltimore and Ohio Railroad. By 1869 Consolidation was the largest shipper in the region.[2]

The advent of this new and formidable competition changed the whole character of the coal trade of the George's Creek-Frostburg area. Up to this time, the older companies had co-operated through their informal association organized in 1849 and had tended to act as a unit in trying to keep their production costs down. In this way they secured lower rates from their carriers and banded together to determine what wages they would pay. After 1869 this association weakened, and rivalry became "ruinous and . . . uncontrollable." [3] Seeing their profits vanish in an era of price-cutting, the five largest concerns in 1869 proposed a grand alliance of all twenty-one existing companies under a single management in order to reduce expenses and eliminate cut-throat competition. The majority of the operators rejected this alliance.[4]

Instead of a single organization acting in complete unanimity, two combinations came into being in 1870. The Consolidation Coal Company, through its merger with the Cumberland Coal and Iron Company, acquired control of more than half the coal lands and all of the local railroad facilities in the area,[5] and its capital, according to a trade publication of the period, rose to nearly $12,000,000.[6] The second combination was that of the Central Coal Company and the Maryland Coal Company, which together had mined more coal in 1868 than any single company in the region.[7] For the most part the miners felt that the amalgamation of corporate strength was "inimical to their freedom and interests." [8]

Throughout the decade 1870–1880, the efforts of the companies to outdo each other and to secure the lion's share of the business resulted in practices which affected the miners adversely in some ways, favorably in others. Perhaps the most significant effect was, in the words of the mine inspector, that "the irregular and illegitimate mode of business pursued by some of the companies, rendered all community of interest between the several companies of the region impossible, and made it impracticable for them to combine for mutual protection against the arbitrary exactions of labor organizations." [9] It was this decade which saw the beginning of the first effective labor organization among the miners of the Cumberland region, an accomplishment made possible by the preoccupation of employers with problems of competitive marketing.

Unionization had been attempted earlier in this coal field. The short-lived American Miners' Association, organized by bituminous coal workers in St. Louis in 1861, had claimed Maryland as one of its strongest states,[10] but the association did not appear as a factor in the Maryland wage disputes of 1865 and 1866. In any event, it gradually deteriorated throughout 1867 and ceased to exist in 1868.[11]

Even while the American Miners' Association was fading away, the immigrant miners who came to the United States from the British Isles during the Civil War period [12] were keeping the idea of unionism alive. Many of these men had been leaders in English miners' unions, which by the late 1860's had succeeded in getting Parliament to pass measures on safety, education, and social improvement.[13] Maryland miners, particularly those of Scottish origin, were well informed of the lobbying of Alexander McDonald and the Miners' National Association of Great Britain for a better mines act, true weighing, a minimum age of twelve, an eight-hour day, weekly pay in current coin, and abolition of the truck system.[14] When McDonald visited Lonaconing in 1867, the George's Creek miners presented him with a substantial sum of money for assistance to miners in England, Scotland and Wales [15] and on his return in 1869, the Lonaconing miners donated another purse as "an offering of gratitude from those who had been guided by Mr. McDonald's counsel in the old country." [16] On both occasions McDonald urged the Maryland men to join the union movement in the United States.

Nevertheless, so far as we have been able to determine, after the demise of the American Miners' Association the Cumberland region had no organization until 1872. In that year, the miners' and laborers' association which originated in the anthracite region of Pennsylvania was said to have extended its membership to the bituminous miners of a number of states, including Maryland.[17] However, it is not until 1873 that we have any detailed information on the union movement among western Maryland coal miners, and the manner of its development seems to rule against an earlier affiliation with the Pennsylvania workers.

The spring of 1873 was an opportune time for the men to organize. In the years between 1869 and 1873 the operators had prospered and had gradually built up their production. In 1873, the high point before the effects of the panic were felt, the region shipped a record total of well over 2½ million tons.[18] Production had hesitated slightly in 1870,[19] and the superintendent of the Cumberland and Pennsylvania Railroad (part of the Consolidation Coal Company's holdings) had suggested that miners' wages be reduced to 45 cents a ton.[20] But the rate remained unchanged at 65 cents a ton, one of the reasons being a recognition of the fact that the closing of the canal in winter drastically shortened the working year.[21]

In the spring of 1873 some of the miners became convinced that the companies were trying to economize by paying them for less coal than they actually delivered to the mouth of the mine. At the Hoffman mine of the Consolidation Coal Company the men complained not only that the scales were incorrect but also that the inadequate size of the cars caused a loss of coal in transit to the scales.[22] They stopped work until the scales had been checked and found accurate, and the superintendent had promised to enlarge the cars to carry a full two tons. Employees of the Midlothian mine of the New Central Company struck at about the same time, complaining of short weight and demanding that the scales be adjusted, that the miners be granted the privilege of placing a check weighman at the dump, that the weight of each car be marked upon its side, and that they be permitted to load each car to its full capacity.[23] The company president granted all these demands except the last, refusing to allow more than two tons to a car.*

To bring their grievances to the attention of the public, the Midlothian miners placed an advertisement in the *Frostburg Mining Journal* of April 26, 1873. After citing examples of dis-

* According to the report in the Cumberland *Civilian and Telegraph* of May 1, 1873, "The rule of fixing the capacity of the cars . . . is one in vogue upon all railroads and as the Company is compelled to stand the loss in overloading in their dealings with the railroads it is no more than just that they should have the benefit of the excess sent out from the mine." B&O coal cars held 10 tons. If a coal company loaded less, it still paid for 10 tons; if more, it paid a premium rate on the overload. Obviously it was a convenience for the mine management to have coal come out in two-ton loads, but it was also obviously an injustice not to pay the miners for their actual tonnage.

crepancy between the amount of coal sent out and the amount the company paid for, the advertisement continued: "Next, we call your attention to the UNION principle. We are endeavoring to form a MINERS' UNION along George's Creek and vicinity for the better protection of our interests. In furtherance thereof we invite a delegation from each and every mine to meet us AT OCEAN MINES PLATFORM, ON Monday, April 28, 1873, for the purpose above indicated."

This meeting took place at the Ocean Mine with apparently no interference from its operators, the Consolidation Coal Company.[24] A small group representing only seven mines elected Thomas Brown of Pompey Smash as chairman, listened to speeches favoring the establishment of a union, and called another meeting for May 3.

Local editors were not enthusiastic about the proposed union, apparently believing that those who initiated the movement intended "to advocate a fusion with the Pennsylvania miners." [25] The Cumberland *Civilian and Telegraph* reported that the meeting was managed by "a couple of young men from the Pennsylvania region" and commented that the miners would gain nothing by joining an association which would actually "put them at a disadvantage by making them parties to strikes, &c., in which they have no interest [i.e., in Pennsylvania]." [26] The Cumberland *Times* was also opposed to any outside labor influence.[27] The editor of the *Frostburg Mining Journal* felt that he could approve of "a mutual, independent, *home* organization" not created purely to gain higher wages but having benevolent ends as well, but he, too, spoke strongly against a union of Maryland and Pennsylvania miners.[28]

The meeting announced for May 3 was held "in the pine woods, near the Allegany cemetery," with a good attendance in spite of rain.[29] The 300 miners present adopted a resolution calling for the election of two delegates from each mine to meet in Frostburg on May 10 and draw up a "code of laws for the government of this Union." [30] The Cumberland *Alleganian* reported, "The sentiment was universal that the organization contemplated should have no connection with other Unions outside of this county." [31]

On Saturday, May 10, the miners' committee met at Frostburg to draft a constitution.[32] Only six mines sent delegates: New Hope—William L. Evans and James B. Thomas; Ocean No. 1—John P. Cavanaugh and William R. Oats; Ocean No. 2—Z. E. Jones and Samuel D. Phillips; Johnson—E. M. Davis and C. Crooks; Miller—John Lindsey and William Bell; Pompey Smash—Peter Cain and Thomas Brown. E. M. Davis was elected chairman, and Thomas Brown, secretary. The committee completed its work with resolutions to the effect that the constitution and resolutions be printed and copies be furnished to each mine; an address to the miners of Allegany County be furnished the *Frostburg Mining Journal* and the Cumberland *Daily Times;* and a mass meeting of Allegany County miners be held at Knapp's Meadow, near Midland Mines, on Saturday, May 24, 1873.

The constitution which the committee drew up (Appendix C) shows that what the union proponents had in mind at this time was a purely local organization, following the general pattern of miners' and laborers' benevolent associations then being formed in the bituminous fields of Pennsylvania, Ohio, and Illinois.[33] Those associations had no state or national head, and were created primarily for assistance to injured members and to widows and orphans of those killed in the mines.

The Maryland organization was to be known as "The Miners and Laborers' Protective and Benevolent Association of Allegany County, Maryland." Its purpose was "to promote the best interests of the Miners and Laborers socially, politically and morally, for the protection of their health and lives; to spread intelligence among them in general on all questions affecting their labor; to abolish as far as possible, all strikes; and to adopt, in their place, ARBITRATION." [34] Any mine worker in Allegany County at least sixteen years old might become a member by paying an "entrance fee" of fifty cents and signing the constitution. The dues of fifty cents a month were to be used only for defraying "the expenses incurred in carrying on the Association"; in other words, a strike fund was not provided for. The constitution also set out the duties and manner of election of officers on both local and county levels. Each local was to manage its own affairs and pass its own by-laws providing they did not conflict with the overall constitution. Disability benefits were established as five dollars a week for six months and $2.50 a week thereafter until the injured man was able to go back to work. In case of the death of a member, the association would pay $30 toward funeral expenses.

The aims of the union were very general. The constitution did not mention wages or suggest a shorter working day, and the emphasis on arbitration indicated a willingness to be reasonable in case of labor disputes. As a miner signing himself "Felix" wrote to the Cumberland *Daily Times* on May 6, 1873, "Nothing will be left undone to reconcile all differences without resort to the *dernier resort* of a general strike."

Since the organization was also in no way secret, apparently nothing in this embryo labor movement gave concern to the operators. On the contrary, the companies allowed the men to hold their mass meeting on a Saturday (ordinarily a working day), and the Cumberland and Pennsylvania Railroad, controlled by the Consolidation Coal Company, ran a special train from Frostburg to Knapp's Meadow on the day of the meeting called to consider the constitution of the proposed union.[35]

Leaving Frostburg with 300 miners aboard, the train picked up additional groups at Borden Shaft, Ocean, Miller, and Midland, and arrived at its destination with more than 400 passengers. Knapp's Meadow, about two miles north of Lonaconing, had been chosen as a central meeting place to attract the lower creek men, who had not shown much interest in the union movement. Even on the day of this well-advertised meeting only a small contingent came up from Lonaconing and the settlements to the southwest.[36]

A roll call established that delegates were present from Big Vein, Pekin, Caledonia, Consolidation No. 2, Midlothian, Johnson, Furnace, Iron Ore, Pompey Smash, Ocean No. 1, and Midland, and non-delegate attendants from Pickell, Hoffman, Jackson, and Borden. Lonaconing, Detmold, Swanton, Potomac, Phoenix, Franklin, and Hampshire and Baltimore mines sent neither delegates nor observers. At the time of the meeting at least two mines were on strike because of weight disputes—Pickell and Big Vein. The region had about 2,700 coal-mine workers in 1870. Therefore, in spite of the gaps in representation, the turn-out evidenced a fair amount of interest.

However, it is easy to understand why many Maryland miners were cautious about associating themselves with a union in 1873. They could not have been unaware of the attempts of the Pennsylvania anthracite operators to destroy the Miners' and Laborers' Benevolent Association in that state.[37] Neither could they have failed to know of the activities of the Molly Maguires, which the anthracite operators took pains to link with the union movement. In this connection, Irish miners in Maryland may have held back until they were sure that the proposed organization would not fall into the category of "secret societies" frowned upon by their Church. Whatever the reasons, some miners outspokenly opposed efforts to unite them. At Borden Mines, for example, "much prejudice existed . . . regarding the formation of a Union. Many men there said that a Union in this region would be like a pit of rotten potatoes." One of the Borden men present said of his fellow-workmen, "They will not listen to reason, and will stand out against the Union. We can't get them out, every effort has failed." [38]

Then, too, although the initiators of the union had disavowed any intention of an alliance with Pennsylvania miners, the possibility was still present and unwelcome. The problem which continually plagued union organizers in bituminous coal had already arisen; miners in competitive areas had learned that they benefited from the extra orders which came in when a rival area went on strike or was otherwise suspended. This had been true in Maryland and Pennsylvania during the Civil War, and Maryland had prospered during postwar strikes in Pennsylvania. Even at this early meeting of a labor organization "independent of all others"[39] there was some inkling of a desire to have at least an exchange of courtesies with unions elsewhere. One of the leaders asked specifically "whether a man can get a card at this Union which will admit him to Unions in other parts of the country."[40] The chairman assured him that this could be arranged later.

Because of the great duplication of names in the mining areas, only eighteen of the union leaders and delegates can be definitely identified in the 1870 manuscript census. Most of them were married men with children. Six owned real estate ranging in value from a low of $250 to a high of $800. Nine reported personal property worth at least $100. Five were born in England, five in Ireland, three in Scotland, three in Wales, and two in the United States. Their ages ranged from twenty-five to fifty-eight.

From the point of view of later developments, the most interesting person connected with the 1873 movement was Thomas Brown of Pompey Smash. A native of Scotland, he came to the United States as a boy. After settling in Allegany County, he remained there until his death in 1905. In 1873 he was forty-seven years old, married, with four children at home. Two of his sons, aged fourteen and sixteen, worked with him in the mines. He owned real estate valued at $800 and reported $100 in personal property to the census enumerator in 1870. This was not his first venture at getting better conditions for the miner. According to Andrew Roy, Brown had led the strike in 1853,[41] and Brown himself spoke of the earlier trouble with the Cumberland Coal and Iron Company.[42] The letters signed "A Miner," sent to the Baltimore *Sun* from Pompey Smash during the strike of 1854,[43] probably were written by him; certainly they reflect his belief in moderation and reasonableness. He was a self-educated man, "a prolific and careful reader" with a knowledge of a great variety of subjects, and his home library was extensive. Considered an authority on coal mining in the region, he had been put in charge of opening Consolidation's Hoffman mine. Throughout his life he enjoyed writing letters to the various area newspapers. He was also a persuasive orator.[44]

In his address to the miners of Allegany County published in the *Frostburg Mining Journal* of May 17, Brown wrote movingly of the need for unity and the discarding of all differences of nationality or politics. As the principal speaker at the meeting of May 24, he discussed the miners' grievances, but warned that they could not be remedied by strikes. He pointed to the benefits of union organization in England, Scotland, and Wales and to the fact that a miner (Alexander McDonald) had been elected to Parliament. Maryland miners too, he said, ought to unite, seek their remedies in the ballot-box, and send a man of their own to the General Assembly to secure legislation for the proper ventilation of the mines.[45] In addition, Brown attacked the lien law, asserting that under its provisions the amount of movable property available for sale to meet miners' wages in arrears was utterly inadequate.

The chief accomplishments of this meeting were the adoption of the constitution and the decision that each mine be authorized to establish a "lodge" and appoint delegates to a county convention to be held in Frostburg on May 31. Each delegate to the county convention would cast one vote for every ten men employed in the mine which he represented.[46]

When the county convention met on May 31 delegates were present from additional mines which had not been represented at the earlier meeting.[47] The following officers were elected: Joseph G. Kelley, president; George Hoskin, vice-president; Thomas Brown, secretary; A. M. Sample, corresponding secretary; Peter Cain, treasurer. (Both Cain and Brown later served as mine inspectors.) In order to meet initial expenses (mainly for printing and distributing copies of the constitution and bylaws), the delegates voted a levy on the membership.

After a long discussion of strikes, the delegates adopted a resolution "that any member of this Association striking without the cognizance and approval of the General Board shall forfeit all claim to support in any way from this Association during such strike."

By the time the executive committee met again on June 14, lodges had been organized at Pompey Smash, Midlothian,

Ocean, Miller, Midland, Furnace, Pekin, Caledonia, Hampshire, and Virginia mines.[48] Each of these had an identifying number.* The committee adopted by-laws, ordered them to be printed and distributed to the union membership, and then adjourned until August 2.

On August 2 it made plans to secure a charter from the legislature and to hold a mass meeting at Barton later in the year.[49] It also instructed the secretary "to ascertain from unions in Ohio and Pennsylvania whether pass cards from this Union would be recognized."[50]

On August 30 the executive committee raised the entrance fee to one dollar, chose Thomas Brown as a candidate for the state legislature, and completed arrangements for a mass meeting to be held at Barton on September first.[51] With special trains providing transportation from Frostburg and Westernport, about 500 persons turned up at this meeting. The Frostburg, Barton, and Westernport contingents brought along bands, the Eckhart Mines furnished a drum corps, and in the flurry of speeches and excitement the miners from the lower end of the region were finally persuaded to organize.

The new union faced its first real problem in September 1873.[52] The miners at Hampshire mines of the Hampshire and Baltimore Coal Company had asked for scales at what they considered the proper place for weighing the coal. The company had refused. The union committee investigating condemned the company action as "the hand of the tyrant, oppressor, and robber" and recommended that the Hampshire miners be helped "morally and financially." Accordingly, the organization instructed its members at the Miller and Midland mines of the same company to notify their employer that they would strike on September 23 unless the grievance of the Hampshire miners was redressed. Union men employed by other companies offered

* These were as follows: Johnson and Midlothian, No. 1; Pompey Smash, No. 2; Ocean, No. 3; Furnace (coal and iron), No. 4; Miller and Midland, No. 5; Pekin, No. 6; Caledonia, No. 7; Hampshire, No. 8; and Virginia, No. 9.

to share their work with the strikers, and in addition each member was assessed one dollar to build up a strike fund.

In a matter of days, arbitrators settled the question of the scales.[53] Thomas Brown acted for the miners; Mr. H. Brown, for the company. These two chose a third party, a Mr. Simms of Piedmont, who decided the case in favor of the company. Said the editor of the Cumberland *Alleganian*, "The good sense displayed in this adjustment of a difficulty cannot be too highly commended."

However, during the strike on the weight issue, some five or six non-union men went to work at Hampshire "without the consent of the majority." The union men, without consulting their executive committee, refused to work "with or by any such men" and asked the superintendent to discharge them. In this instance the executive committee of the union, meeting at Frostburg on October 4, "disapproved of the proceedings of the Hampshire men in their last strike as they did not proceed according to . . . the Constitution." Furthermore, it instructed the Hampshire men to resume their work immediately.[54]

It was evident from its first year-end report that the Miners and Laborers' Protective and Benevolent Association had accomplished a good deal in the less than eight months of its existence. Reports from nine of the fourteen lodges showed that at the beginning of 1874 the union had at least 532 members in good standing. (There is no way of determining the membership at the remaining five mines.) It had paid out some $1,480 for injuries and funeral expenses and still had $830 left in the treasury.[55] It had taken action in disputes between its members and various companies involving weight allowances, but had managed to avoid a strike in 1873 by practicing arbitration as stressed in its constitution. The arbitration committee also acted concerning scales at the Potomac mine in December 1873.[56] When a question of the proper loading of cars arose at Phoenix and Franklin mines in January 1874, the union urged limiting the load on the cars to two tons each, and if necessary putting a checkweighman at the scales.[57] All of this the organization achieved in an atmosphere of general prosperity. Coal production for the whole region rose 12 percent during 1873, and the Maryland Coal Company told its stockholders, "The demand for Cumberland Coal for steam and iron purposes, is increasing more rapidly than is reflected in the actual increase of production for the year." [58]

During this period relationships between workers and employers seem to have beeen reasonably cordial. The miners at Midlothian (New Central Company), for instance, during the normal winter slack season, publicly thanked the superintendent for dividing the work among all of his men instead of limiting it to those who occupied the company houses.[59] However, there may have been some company recrimination against union officers. On April 5, 1874, the members of the arbitration committee were instructed to visit the agent of the Caledonia Mine "in the interest of Mr. John Mansfield [the union president], who had been discharged, and respectfully ask that he be taken back to work." [60] We do not know the outcome of this visit; Mansfield was still holding his union office in June.[61]

The "frenzied prosperity" of 1870–73 encouraged the formation of a number of new national trades unions, among them the Miners' National Association headed by John Siney.[62] Although Maryland miners did not participate in the organization of the national association in October 1873,[63] within four months the executive committee of the Allegany County local recommended affiliation with the larger group.[64] Richard F. Trevellick, one of the prominent figures in the national labor movement, addressed the miners of this region on several occasions in the fall and winter of 1873. In his capacity as president of the National Labor Union he spoke twice at Frostburg and made single visits to Lonaconing, Barton, and Westernport in October.[65] The miners' union arranged these meetings, and Trevellick specifically thanked Thomas Brown, E. M. Davis, and Thomas Rooney for their assistance.[66] Trevellick was later scheduled to address miners and workingmen at Frostburg, Pompey Smash, Lonaconing, Barton, Westernport, and Piedmont between December 15 and December 20, 1873.[67]

Although the exact date is not known, the affiliation of the Miners and Laborers' Protective and Benevolent Association of Allegany County, Maryland, with the Miners' National Association took place some time during 1874. On May 23, lodges of what was vaguely referred to as the "Miners' Union" from Allegany County, Maryland, and from West Virginia met in Frostburg to hear addresses by John Siney, president of the Miners' National Association, and David Harris, president of the United Sons of Vulcan (iron puddlers).[68] The occasion was the first anniversary of the organization of the Allegany County union. The *Frostburg Mining Journal* of May 30, 1874, described the meeting as "not very strong as regards members" (although 500 seems a good attendance), but quiet and orderly and "devoid of that boisterous offensiveness which sometimes characterizes such demonstrations in great cities."

The New Central Coal Company tried to keep its men from participating, on the pretext that it needed them to dig coal for filling orders.[69] Siney was greeted with laughter when he said he was glad to learn that business "was so brisk along the creek that the men could not be spared a half day on Saturday to attend a meeting." [70] Among the speakers, Edward Davis, one of the founders of the county union, referred to opposition from many of the companies, but encouraged the miners to support their association, which "had accomplished all the local organization can accomplish." "At our back," he said, "stands the National Association ready to receive us under its protecting care, and contribute to our efforts to preserve our organization in the defense of our rights." [71]

Siney in turn urged the miners to join the union, pointing especially to the financial benefits in case of accident. He also took particular care to correct any impression his listeners might have that the national association wanted them to join so that "the miners of Pennsylvania may drag you into a strike some day by which they may recover the trade you have taken from them." In any event, he added, "I have yet to learn that you have taken any trade from us in Pennsylvania." [72]

Curiously enough, Hoffman Lodge No. 10 disbanded in June, and its funds were distributed pro rata to its members.[73] The most probable reason for this action was that about 100 Hoffman miners had joined the Miners' National Association soon after Siney's visit in May and in advance of the affiliation of the entire membership of the Allegany County union with the national organization.[74] Hoffman later reorganized under the new auspices.[75]

By October of 1874 the Allegany County miners were certainly allied with the Miners' National Association, for the *Frostburg Mining Journal* of October 10 reported a meeting of the First District of Maryland, "National Miners' Union," held to elect delegates to the convention to be held in Cleveland, October 27. Although before joining the national organization, the Maryland local association had lodges at fourteen mines,[76] at the time of the district meeting there were only five lodges, four of which were represented as follows: Johnson, 10 delegates; Hoffman, 11; Pompey, 8; Borden Shaft, 8. The remaining lodge, Bowery, had no delegation at the Frostburg meeting. John Betz was selected to attend the national meeting, with an allowance of $2.50 a day plus traveling expenses. At the convention he served on the committee on credentials and later was elected one of the organization's vice-presidents.[77] In his history of the United Mine Workers, Chris Evans identifies Betz as a member of the committee to amend the constitution of the Miners' National.[78]

The depression that began in September 1873 curbed the activities of labor unions in general, but the Miners' National Association managed to survive until late in 1876.[79] In 1875, when it reached its highest development, the Miners' National had over 35,000 members, with 431 in Maryland.[80] From coal production figures we can see how the sustained membership in Maryland was possible. Although the demand for coal throughout the nation fell 50 percent following the panic in 1873, shipments from the George's Creek region in 1874 and 1875 were almost up to the record 1873 level.[81] Roy speaks of the "unparalleled depression in the coal trade" in 1874,[82] but miners were still in "heavy" demand in Maryland in the spring of that year,[83] and the Cumberland *Alleganian* of May 13 did not appear unduly concerned that "the coal trade of this vicinity so far during the present season has not been so brisk as could be wished." Only one company in the region tried to reduce wages. The superintendent of the Bowery Furnace notified his coal miners that he intended to pay them ten cents less a ton—55 instead of 65 cents—but when they refused to work for this amount, he restored the wage to the old rate.[84]

At the beginning of October 1874, however, the *Frostburg Mining Journal* reported that the coal trade was slack, with all companies except the Maryland Coal Company on half time. The miners, who had been making from $75 to $100 a month, were now earning only $50 and were unable out of this amount to add to their savings for the winter. In spite of the depression, said the *Journal* editor, the cost of living had not come down, and provisions and rents were still high.[85] An oversupply of labor made matters even worse.

The secretary of the Miners' National described relations between Maryland operators and men as "harmonious." The fact that the operators had not reduced wages had attracted men from other regions and contributed to the overcrowding in the Maryland mines. Disturbances in areas where employers tried to force their miners to accept lower wages had resulted in lost contracts, keeping the George's Creek mines running on the strength of the extra business they picked up. But the end result was that the Maryland men, still receiving 65 cents a ton, averaged less than half time as the work was shared, and by the end of the year were making a "miserable" living.[86]

Reports of conditions at the end of 1874 may have reflected nothing more than the usual let-down at the approaching end of the season. Production in 1875 was almost normal. Cumberland coal helped to supply the deficiency in the anthracite market during the long strike in the first several months of 1875.[87] Furthermore, during the strike which began in the Clearfield region in April 1875, the Clearfield operators bought Maryland coal in order to fill their contracts,[88] paying from 15 to 30 cents a ton extra in preference to granting the 10-cent increase asked by their own miners.[89]

The *Miners' National Record* described the Clearfield strike, which was called against the advice and contrary to the wishes of the national organization, as "ill advised, ill considered, and ill managed." [90] Its immediate result was the arrest and imprisonment of a number of men. Siney, who was arrested in May on charges of conspiracy, was not tried with the others in June, but was allowed out on bail until October, when he was tried with organizer Xingo Parks. The court found Parks guilty of unlawful assembly and conspiracy, but acquitted Siney.[91] The long-range result of the Clearfield strike was the gradual crumbling of the influence of the Miners' National until the association finally collapsed at the end of 1876 following a series of unsuccessful strikes throughout the country.

The decisive action of the operators, aided by the full cooperation of the legal powers of the state, killed the union in Clearfield. In Maryland the organization died more slowly at the hands of the miners themselves. The prosperity brought by the strike entirely outweighed any considerations of solidarity. A Frostburg correspondent wrote to the *Miners' National Record* in June, "It is no use telling these men that they are filling the Pennsylvania contracts and ask them to stop. They would only laugh at you. Some of them are glad to hear of strikes in Pennsylvania." [92] Siney visited Maryland in July while he was awaiting his trial, and held meetings at Frostburg, Barton, Lonaconing, and Bloomington (Garrett County) in an unsuccessful effort to keep the men interested in the union. Later, explaining his failure, he pointed out that while strikes were going on throughout the country, "the men and employers of George's Creek were getting fat." The George's Creek men refused to have anything to do with men of Pennsylvania, Ohio, Indiana, and Illinois, who, they said, were "never satisfied" except when on strike. He placed the blame for disintegration of the organization in Maryland squarely on the various factions which fomented distrust and spread anti-union propaganda. Rather surprisingly, he commented that he had met no superintendent or mine boss along George's Creek who was opposed to the union, and said further, "It cannot be alleged with any show of truthfulness that the employers are opposed to the organization." [93]

Siney's charge that the Maryland men were not even willing to contribute toward the support of their "less fortunate brethren" seems unjust in the face of their raising a fund of over $300 for the relief of destitute miners in Neshannock, Mercer County, Pennsylvania, who were subsisting on weekly rations of "four lbs. flower and two lbs. corn meal per adult, tea sugar, etc. being out of the question." [94]

At least one lodge, No. 204 at Eckhart Mines, was still affiliated with the M. N. A. in September 1875, when it protested an increase in monthly dues "for striking purposes." [95] Saying that strikes seemed to be the main object of the majority of the rank and file in the national organization, the local suggested instead "productive co-operation" as a remedy for the miners' troubles. John Betz, president of the lodge, and Francis Thompson, secretary, signed the letter of protest. By November Betz' name had disappeared from the list of state vice-presidents of the Miners' National. Since no substitution had been made by December, one assumes that Maryland had withdrawn entirely from the association.[96]

Like many other laboring men in this period, Maryland miners explored the possibilities of consumer cooperatives as a means of cutting living expenses. The movement was particularly strong in Lonaconing, where the union, on the other hand, had received only tepid acceptance. Advocates of a cooperative society held their first meeting in Lonaconing on March 9, 1874.[97] Apparently none of the twelve storekeepers in town objected to the proposed organization.[98] Plans were completed on March 17.[99] A committee composed of James Frew, Adam

Thompson, Adam McMillan, John McFarland, George McLuckie, James Murray, and Walter Ireland drew up rules for the association. The Lonaconing Co-Operative Society would have a president, a secretary, a treasurer, and twelve directors. The treasurer would also be the chief manager of the store and would hire the clerks. The directors would meet every week to examine the merchandise, fix prices, and audit the accounts. Nontransferable shares in the enterprise were to sell for $10 each, with a limit of twenty shares to a person. Since the society would need a charter, a committee was appointed to draw up a petition and present it to Judge Pearre of the circuit court at Cumberland. James Murray, William Thomas, David Laird, Joseph Laird, Jr., and Adam Thompson were named for this purpose. Stock subscriptions taken at this meeting amounted to $3,660.

By the end of March the cooperative had received its charter and was officially in existence, with 130 stockholders and a subscription of $8,000. The members had elected a president, James B. Nichols, and a secretary, Thornton I. Crow, but had not chosen a treasurer. The board of directors included James Murray, Joseph Laird, Jr., Adam Thompson, George McLuckie, James Kirkwood, James Frew, David Laird, John Miller, George Footer, David Dunn, William McFarland, and Robert Anderson.

According to evidence in the 1870 manuscript census, all the moving spirits in the cooperative were miners and of Scottish birth.* They were described as "representative men of our town" and long-time residents, most of them for fifteen years or more. John Douglas, the Scottish superintendent of the George's Creek Coal and Iron Company, expressed himself as being in "full accord" with the cooperative movement and offered the use of the company schoolhouse free of charge whenever the members wanted it.[100]

Cooperative stores in British mining communities dated back

* The store was known unofficially as the "Scotch Co-operative Store" to distinguish it from the "Irish Co-operative Store" which was also thriving at Lonaconing. *National Labor Tribune,* September 25, 1875.

to 1861, and very early showed that they could be operated profitably. More important, they made it possible for the miners to escape from the credit system offered by village grocers.[101] In later years the cooperative store was often the "strike larder" for British miners, but whether or not this was true in Lonaconing during the strikes of the 1880's cannot be determined. The Lonaconing store required cash payments of nonmembers but apparently allowed credit to members.[102]

The George's Creek region, where the population was composed almost entirely of laboring men, seemed an ideal location for this type of enterprise.[103] Possibly because it purportedly began to sell dry goods and breadstuffs at prices 30 percent less than those of other merchants in town,[104] the miners' cooperative store at Lonaconing completed its first six months of operation with a net profit of $2,102.66 on total sales of $35,749.78, and paid a 6.75 percent dividend.[105]

The Lonaconing cooperative store survived the difficult times which resulted from the panic of 1873. At the end of 1880 it was still in business, with an inventory of $6,600 and fixtures valued at $2,000. During the six years of its existence it had paid out dividends and interest of well over $18,000,[106] an enviable record in view of the general depression which settled on the Cumberland coal region in 1876 and 1877.*

* Although the cooperative was in serious financial difficulty in 1893, it recovered, and between 1893 and 1918 paid back to its stockholders in dividends and rebates over $100,000. *United Mine Workers' Journal*, February 28, 1918. The store remained in business until 1921.

Chapter 12

Depression: Emergence of the Knights of Labor (1876-1880)

Following the Civil War, the United States entered a period of unprecedented prosperity. Profiteers invested their gains in every conceivable enterprise; apparently limitless natural resources awaited exploitation; southern and western markets called for increasing supplies of manufactured goods, and the tariff policy favored domestic industry. Capital was abundant. In four years after the war, government payments alone added roughly three-quarters of a billion dollars to the spending power of soldiers and contractors, and European investors poured over one and a half billion dollars into America. Speculators showed an overwhelming interest in railroads, sinking more than a billion dollars into this anticipated source of income.

This uncontrolled activity could not continue indefinitely. The era of too-rapid expansion came to an end with the failure of Jay Cooke and Company, a prominent investment house, in September 1873. The depression which ensued, reaching its greatest depths in 1877, was the most severe which the country had so far experienced. Before the upswing in 1878, two-fifths of the nation's railroads had gone into receivership and more than 22,000 businesses had failed.

The development of the Maryland coal industry was symptomatic of the age. Although the operators had complained mightily from time to time that they were being impoverished by high wages, the unreasonable cost of transportation, and competition from other areas, their difficulties before 1876 had been more imaginary than real. Their only serious set-back since 1857 had occurred during the first two years of the war, when shipments by rail and canal had been greatly curtailed, but they had quickly recovered as soon as they were able to move their coal. Throughout the late 1860's and early 1870's they had expanded production and had paid dividends. Even in 1874 and 1875 coal production had dropped only slightly in the Cumberland field. Furthermore, the companies had enjoyed more than twenty years of labor peace. Since the general strike of 1854 only minor disputes had arisen from wage or weight disagreements.

The rude disruption of this peaceful productive environment in 1876 stemmed from a complicated set of circumstances resulting in "a loss of trade amounting to twenty-one per cent. of the output of the previous year, and over thirty-one per cent. of that of 1873." [1] The employers quarreled among themselves; the miners struck in 1876; and in 1877 strikes of railroad workers, canal boatmen, and miners in Allegany County added to the labor unrest which in that year convulsed the nation.

In western Maryland the economic reverses were for once not attributed mainly to the miners and their wages. Nor was the blame placed on the general depression of business. Instead, various interests in the region—newspapers, laboring men and competing operators—turned upon the Consolidation Coal Company and censured it for conditions existing in the Maryland coal trade.[2] Because of attacks upon it, the company contemplated moving its offices from New York to Cumberland or Mount Savage, "believing that such a move would allay local prejudice against 'outside capitalists.'" However, this proposal merely touched off a stockholders' fight so bitter that directors who voted for the removal were turned out of office. The company maintained its headquarters in New York until 1877, when it transferred them to Baltimore.[3]

Controlling strategic railroads over which Cumberland coal

must move to market, Consolidation faced repeated charges of unfair competition arising from its freight rates. In its annual report for 1876, the Maryland Coal Company criticized Consolidation also for its "failure to discern a mutuality of interest" with the other companies and refusing to cooperate with them in reducing miners' wages, which it claimed were "nearly double" those being paid in competing fields.[4]

The Maryland Coal Company report also made it plain that competition from the Clearfield, Pennsylvania, coal region was emerging as a serious threat to the Cumberland operators. Clearfield, about a hundred miles north of Cumberland, marketed a type of coal similar to the highly-regarded George's Creek product and had the advantage of being closer to the New Jersey ports and the New England manufacturing centers. Although the Clearfield mines had not opened until 1871,[5] by 1876 it could be said that this "rival bituminous . . . region, relieved from the burden [of higher production cost] which afflicted the Cumberland field, and aided by the liberal policy of the Pennsylvania Railroad Company, has increased its production, and extended its trade in our markets." [6] The growth of this competition is worth noting, for from this time on it played an important part in the wage negotiations of the western Maryland miners. Clearfield miners in 1876 were receiving only 40 cents a ton as against the Cumberland miners' 65 cents.

Since the Consolidation Coal Company refused to join its competitors in a move to reduce miners' wages in January 1876, the other companies united in petitioning the Maryland legislature to order a reduction of the freight rates charged by the Consolidation-controlled Cumberland and Pennsylvania Railroad. Two other matters vitally important to coal operators, miners, and the economic interest of the entire region were also before the January session of the legislature. These were a proposal to levy a tax of two cents a ton on coal produced in the Maryland mines, and a request from the miners for a law providing for ventilation and mine inspection.

Once the legislature had adjourned, mining matters in Allegany County came rapidly to a head. The inspection law will be discussed in the next chapter, but its passage should be noted here. The General Assembly rejected the proposed tax on coal and ordered reduced rates on the Cumberland and Pennsylvania Railroad,[7] thus freeing the companies from some of the burden they had complained of. Several operators nevertheless announced that they would have to cut wages. The New Central, American, and Maryland coal companies posted notices at their mines that on and after April 10 they would pay miners 55 instead of 65 cents a ton and would reduce the wages of other mine labor proportionately.[8] The Cumberland *Alleganian* supported the reduction as a step necessary to meet competition, apparently seeing no inconsistency between this point of view and a news item which it had published on February 15, reporting that the Maryland Coal Company had recently paid a dividend of $1.50 a share, "equal to seven and a half per cent. on the par value of the stock."

Company employees discussed the proposal "in a very temperate manner."[9] Handbills posted at the various mines announced a mass meeting to be held at New Detmold (near Lonaconing) on April 10. All men who had not been notified of the wage cut were invited to send a delegation of three from each mine still working. Those who had already been told that their wages were reduced were urged to come to the meeting "enmass." This notice was signed "By order of the Miners."[10] A supplementary notice read: "Attention Miners.—We have been notified of a reduction of 10 cents per ton after Monday next. This, along with light weight, will reduce us [to] at least 45 cents per ton. Now, will we submit? We all say no, but will stand for full weight and present wages." This was signed simply "Miners."[11]

The second poster confirms an editorial statement which appeared in the *Frostburg Mining Journal* of March 25: "The miner remembers that he is digging coal now at a rate somewhere between 50 and 55 cents per ton through the indirect arrangement which allows him nothing for overloading and 'docks' him for underloading. In this way the game plays out of his hands both ways and naturally enough he protests against a further *nominal* reduction of fifteen cents which *really may* amount to twenty cents." In their scramble for the diminishing amount of business, the Maryland companies were not alone in resorting to unscrupulous devices for cutting costs at the expense of their employees. The bituminous coal industry was a classic example of unrestrained competition.

The meeting of 1,200 miners at Lonaconing on April 10 proceeded without disorder.[12] Reports from individual mines immediately revealed a marked lack of unanimity on the part of the companies and an equally marked solidarity on the part of their employees. Miners at Blaen Avon, Pompey Smash, George's Creek, Atlantic, Swanton, and Potomac mines volunteered financial support for a strike—anywhere from 10 cents a ton to half their wages. These mines had received no notice of a reduction. Miller, Midland, New Central, Caledonia, Phoenix, Franklin, and Hampshire mines—all notified of the cut—awaited the action of the meeting. Consolidation's Hoffman and Ocean miners had not been informed of a cut, nor had Midlothian and Pickell miners. Men from the Potomac mine said that their superintendent had announced payment of the old rate until the matter was settled.

A motion to submit the question to arbitration was defeated, and the final decision was that all miners who had received notice of the 10-cent reduction would strike for 65 cents and would receive assistance from those still employed at the old rate.

On April 10 Consolidation too announced that it would cut wages to 55 cents, whereupon its employees stopped work.[13] On April 14 Consolidation restored the pay of its miners to 65 cents a ton, and they went back to work.[14] The retraction infuriated the other companies, which were already convinced that Consolidation was doing its best to cripple them, intending eventually to "have a monopoly of the market and charge what it pleases for its wares." [15] Their feeling was reinforced by Consolidation's refusal to obey the mandate of the state legislature to modify freight rates on the Cumberland and Pennsylvania Railroad.[16] As an additional slap, the coal company insisted that carriage charges be paid directly to it, claiming that the Cumberland and Pennsylvania Railroad technically no longer had a separate existence.[17] Now, as if in contempt of its competitors, Consolidation, which had been accused of instigating the wage cut and even of dictating the wording of the posters announcing it, had restored wages and was busily shipping coal while most of the other companies were idle.[18]

Consolidation had become the sole villain of the piece. The *Alleganian*, for example, now said flatly that if the corporation had obeyed the law and reduced its rates on the Cumberland and Pennsylvania, no operator would have been forced to make demands on the miners.[19] The fact that some other companies were also still paying 65 cents a ton was ignored. They had never announced a cut, and were therefore not so unequivocally traitors to their class. The situation at the mines was chaotic. All companies which were trying to reduce wages were idle because their miners were on strike. Some companies were operating as usual at 65 cents. Others which had not announced wage cuts were not shipping coal because of their quarrel with the railroad. Still others continued to pay 65 cents a ton, but instituted economies by closing some of their mines. In reducing forces, they retained the married men, and the luckless bachelors were as much out of work as if they had been on strike.[20] However, they could derive some satisfaction from knowing that they were theoretically 65-cent men.

With some skepticism the editor of the *Alleganian* inquired, "Do the miners now at work for the companies paying sixty-five imagine that such rates will continue during the remainder of the season? If they do believe, they never labored under a greater delusion. No sooner have these companies gobbled up the big annual contracts and driven the other companies out of market, than they will reduce peremptorily." [21]

Borden Mines—which had been paying its miners 65 cents

and shipping over the Cumberland and Pennsylvania, though protesting its rates—suspended production on April 27. By May 1 only Consolidation was furnishing any business to the Chesapeake and Ohio Canal.

The *Alleganian* concentrated its editorials on the iniquities of the Consolidation Coal Company, warning that "by and by under the soft velvet glove it will show the miners the iron hand beneath."[22] Further attacks appeared throughout April. The combined efforts of the press and the aggrieved coal companies finally brought official action; by May 8 the attorney general of Maryland had begun an investigation of Consolidation, the wealthiest company in the mining region. The state did not, however, succeed in its subsequent attempt to revoke the company's charter either in the Allegany County circuit court (*State of Maryland* v. *Consolidation Coal Company*, No. 2288 Equity),[23] or in the court of appeals (*State* v. *Consolidation Coal Company*, 46 Md. 1).

Although the miners had succeeded in holding the rate to 65 cents a ton, it was at a tremendous sacrifice. Those who had struck rather than work for 55 cents were in desperate straits. Many of them had been in a precarious economic position even at the beginning of the mining season.[24] In their determination not to give way, the employed men accepted the responsibility of aiding those who were actually carrying the burden of the strike. Some money was collected for out-of-work miners in Pompey Smash at the end of May,[25] but early in June it became necessary to consider the question of giving more systematic aid to the strikers, who had asked for help.[26] A resolution passed at a meeting of employed miners called upon every man still at work to contribute five cents for each ton of coal he dug. Other mine labor was to contribute in proportion, so that everyone participating in the strike would get a share of the support.

On June 21 a number of miners conferred with the agents of certain unnamed coal companies to see what relief they might expect in the future. Work had been scarce all season, they said. "With the exception of the miners employed by the Consolidation company, the others have scarcely been able to maintain their families." [27] The agents were not sanguine. They felt that as long as Consolidation kept on getting high freight rates, the other companies would do little business. So far as they could see, the only hope for improvement lay in acceptance of pay cuts by both canal boatmen and coal workers.

The situation steadily grew worse: the Hampshire and Baltimore Company shut down its mines; the New Central Company announced that it would soon close unless its men would accept 55 cents; and rumors spread that Swanton mine would close indefinitely. However, if it was any consolation, things were even gloomier in Pennsylvania, where the latest suspension, as a result of the general economic depression, had put 30,000 men out of work.[28]

As 1876 drew to a close, the mining rate for those few still employed in Maryland remained at 65 cents a ton. In spite of great want, the men had presented a solid front of opposition to the companies, which had, in contrast, shown themselves to be greatly divided on policy. The harmonious days of the old Superintendents' Association had vanished. However, ominous signs portended that this situation would change. The Baltimore and Ohio Railroad was quietly buying the stock of the Consolidation Coal Company,[29] which was expected to join the other Maryland producers in an attempt to cut wages early in 1877.[30]

Cumberland coal shipments fell off one-third in 1876; during that same year, Clearfield shipments increased one-third.[31] The price for mining in Maryland during 1876 had been the highest paid in any eastern coal region, but the miners had not worked more than six months of the year, and only a few had had steady employment. If Consolidation introduced a cut, the other companies would probably follow. Although the operators said they would be willing to keep up the prevailing price if the Pennsylvania miners were paid at the same rate, they asserted that with the current difference in wages, they simply could not compete. The editor of the *Alleganian* urged the miners for their own good to accept a cut to 45 cents a ton in return for firm contracts from the companies for eight hours of work a day and steady employment.[32] He did not explain how the operators were to be persuaded to grant such contracts.

In the Maryland coal region, notices of wage reductions or demands for higher pay were usually made in January or February. As long as the Chesapeake and Ohio Canal remained an important element in coal transportation, its closing each winter meant that mining activity dropped almost to a mere maintenance level. During the period from late December to early April most miners expected to be idle, living on what they had managed to save during the preceding work season. Consequently for at least three months of every year the companies and their men could argue about wages or working conditions without much harm to either contestant. The miners held meetings, discussing the question until they had reached approximately unanimous agreement; in the meantime, the companies presented their point of view in statements to the press. Until late March management and labor were often so far apart that solution seemed impossible, but inevitably, as April approached, one side or the other gave way, and the trade opened with the opening of the canal. This was the normal procedure in the Cumberland region. Since 1866 it had been management which gave way; the Maryland miners had accepted only two wage cuts after the war.

Very early in 1877 it was apparent that wage determination would cease to be a winter's diversion with an almost certain outcome. Although there were no satisfactory statistics, contemporary observers reported that unemployment throughout the nation was extensive.[33] The Maryland coal region had its own peculiar unemployment problem arising from the ten-year maintenance of high wages. The Baltimore *Sun* of January 20 said, "It is claimed by some that mine labor is not as well paid anywhere else as here, and in consequence there is always an influx of men here in the spring, some of whom get places, and sometimes crowd out the regulars." Moreover, George's Creek miners employed during 1876 had worked only six months of the year,[34] and out of their limited income had supported their fellows who were striking in protest against a wage reduction.

The majority of the miners therefore reacted with considerable indignation, not to say consternation, to the news that sixty or seventy men had approached the Consolidation Coal Company's superintendent and had offered to work in the New Hope mine for fifty cents a ton.[35] The voluntary nature of their action was not clear at first, and prominent miners at Frostburg complained about the way in which Consolidation had supposedly tried to reduce wages without consulting them. The facts emerged at the fourth of a series of miners' meetings arranged to deal with this specific problem and to discuss wages in general.

The first of these meetings called by the "Miners Union" opened at the home of John Donahue at Frostburg on January 13,[36] but grew to such proportions that it had to be moved to Forester's Hall. A committee reported that the superintendent had refused them permission to go into the New Hope mine and talk to the men; so another group was appointed to approach these employees on the public highways and ask them to leave. There was to be no intimidation; a resolution adopted by the afternoon session specifically said, "it is not our intention to bulldoze or Molly Maguireize any one." [37] Further meetings held on Janurary 15 and 16 accomplished nothing more. The volunteers at New Hope (described as not miners, but "window-frame and sash makers, shoe makers, tailors and bulldozers [ruffians]") refused to leave, and a number of the old "practical miners" who had been receiving 65 cents a ton in that mine quit rather than work with the lower-paid newcomers.[38]

Whether or not the New Hope recusants were genuine miners, the damage was done. The 65-cent wage, so stoutly defended during the hard times of 1876, had been breached, and the Maryland men must decide what to do before the companies took concerted action. Already Franklin and New Central miners had heard that they would have to work for 50 cents or quit.[39]

Delegates from all mines met at Lonaconing on January 19.[40] It was obvious from the discussion that the men did not want a strike.[41] At this time "no formidable combination of the companies [had] been effected," and most of the operators had not indicated what price they would offer. With a good deal of reluctance the delegates finally proposed 55 cents a ton "under pressure of cry of lost trade on account of high mining wages." [42]

Of all the companies, only Borden took any immediate action. Superintendent A. C. Greene requested his miners to go to work for 50 cents a ton, but after conferring with their committee, agreed to pay 55 cents "for the present." [43] In spite of Greene's acceptance of the compromise, the Baltimore *Sun* reported that other superintendents were "strongly of the opinion that 50 cents will eventually be the price at all the mines."

Understandably, then, the affair of the New Hope men still rankled. Although the action of the miners' delegates had "virtually fixe[d]" wages at 55 cents for the whole region "so far as the miners can fix them," [44] the New Hope men persisted in working for 50 cents, and the bulk of labor in the region feared that the acceptance of the lower rate might become "contagious." [45] At a mass meeting in Frostburg on January 26 some individuals made threats against the miners working at New Hope.[46] In an earlier episode of violence on January 25, protesters stoned the house of a "blackleg" Franklin miner and hurt three of his children.[47] Subsequently, after being showered with stones, a number of New Hope miners in Frostburg went to the mine on January 31 and took out their tools, "declining longer work at 50 cents a ton which has to be reached through sprinkles of road gravel." [48]

By mid-February the operators had determined on their course of action. With the exception of the Maryland and Piedmont coal companies, the employers "mutually agreed" not to pay more than 50 cents a ton and further agreed to lock out their miners beginning on February 19 if the men would not accept the reduction.[49] The Baltimore and Ohio Railroad, with its strong interest in Consolidation, allegedly had a hand in the move,[50] refusing to adjust its freight rates until the miners' wages were settled,[51] and withholding cars from any coal company which threatened to withdraw from the wage agreement.[52]

Faced with the concerted action of the companies, the men began to waver in their decision. Consolidation's regular New Hope miners voted to accept the company offer and go back to work on March 13.[53] At a meeting in Lonaconing on March 16 the employees of the Maryland and George's Creek companies voted to resume on company terms, whereupon "bulldozers" from Pekin broke up the meeting.[54] On March 17 the New Central men met at Lonaconing and resolved to settle for 50 cents.[55] These meetings were not orderly. At one "there was such a scene of general confusion that further business was impossible." [56] At another, one miner was arrested on a charge of assault and battery.[57] On March 19 the Cumberland *Alleganian and Times* announced, "The strike [which it had previously called a 'lock-out'] may now be considered as effectually ended."

The Consolidation Coal Company did not reemploy its former New Hope miners who had engaged in the strike.[58] Although a labor meeting on January 26 had adopted a resolution to the effect that all members of the union would stop work if any member were victimized by his employer,[59] the organization did not attempt to oppose the post-strike discrimination.

In a comment on the events of the spring of 1877, Consolidation later reported to its stockholders, "The most important question in the past year connected with the mining department, was a reduction of miners' and laborers' wages to figures more nearly approaching those paid in other regions." [60] In justice to this company, it must be stated that coincident with the resumption of work at the lower wage, Consolidation reduced the rent of its houses at rates varying from 20 to 25 percent, and also gave its employees a lower price on coal purchased for their home use.[61]

At the time when the miners capitulated, few of the companies had contracts to supply coal.[62] Consolidation's competitors had been awaiting the decision of the court of appeals on the freight rates charged by the Cumberland and Pennsylvania Railroad, and had not actively looked for business. The decision of the court ordering a reduction of rates, coupled with the pay cut, pointed to a favorable season for the Cumberland coal trade, but nevertheless it remained dull, and a number of miners left the area to look for work elsewhere.[63] It is impossible to say whether under normal conditions the situation would have improved during the summer, for on June 19 the canal boatmen struck and blockaded the canal to prevent the passage of boats.[64] Their action curtailed coal production and threw a large number of miners out of work. The great railroad strike which began on July 16 completed the process of closing the Maryland mines.[65]

Four of the largest eastern railroads (New York Central, Erie, Pennsylvania, and Baltimore and Ohio) which had been engaged in ruinous rate wars during the depression, agreed in April 1877 to call a halt to the competition which threatened to bankrupt all of them. They would make up for whatever losses they incurred by cooperating also in cutting the wages of their employees by a flat ten percent. This unannounced reduction, coming when trainmen were employed only half-time and were complaining about the increased tonnage and length of freight trains, set off strikes on a number of lines. Trouble began on the Baltimore and Ohio at Martinsburg, West Virginia. The governor of West Virginia at once ordered out the state militia and wired President Hayes for federal troops. When peace was restored at Martinsburg, the strike spread to Maryland.

The miners met at Barton on July 20 and adopted resolutions expressing their sympathy with the workmen and employees of the Baltimore and Ohio Railroad and criticizing the governor of West Virginia for calling for federal troops.[66] A mass meeting of miners at Frostburg on July 28 (after the Pittsburgh riots) adopted a similar resolution of sympathy, but expressed disapproval of the destruction of property and other unlawful acts. John Betz of Eckhart Mines was chairman of the committee which drafted the Frostburg resolution.[67]

Maryland's Governor Carroll, unable at first to move the state militia out of Baltimore because of rioting in that city, in turn called for federal assistance. Two companies of United States troops arrived in Cumberland on July 21. Actual disorder in western Maryland was short-lived, but the Baltimore press printed sensational and unfounded stories of thousands of miners and canal men roaming the streets of Cumberland ready to meet violence with violence.[68] Meanwhile Pittsburgh experienced widespread rioting, arson, and looting, resulting in the loss of twenty-six lives in skirmishes between mobs and militia. The Cumberland and mining area newspapers reacted indignantly to the sensationalism of the big-city press. Typically, an editorial from the *Alleganian and Times* of July 26 retorted that "among the most quiet and orderly of the laboring class at present are these very miners who are so much maligned," commending them for their peaceable behavior under "most trying circumstances."

Trains were moving again on the Baltimore and Ohio by the end of July.[69] Knowing that the coal companies would be desperately eager to begin filling their contracts, the miners decided to risk another lock-out and demand a higher wage, together with "rectification of a long complained of wrong . . . relative to the amount of coal required to fill what are termed by the companies two-ton cars."[70] A mass meeting of six or seven hundred held at Knapp's Meadow on July 31 to demand a return to 65 cents a ton adjourned without reaching any decision,[71] possibly influenced by a statement from James A. Milholland, general manager of the Consolidation Coal Company, that his company would not pay more than 50 cents and in the event of a strike would close its mines.[72] Furthermore, "a strong conservative element, principally Welsh and Scotch miners," reportedly opposed the strike.[73] At the time of the meeting three mines paying 50 cents a ton were operating: Swanton, employing 70 men; Piedmont, 40; and Franklin, 150.[74]

Some newspapers which reported the meeting also carried a pronouncement from Charles F. Mayer, the new president of

Consolidation, making the usual references to Clearfield competition and denying that any "prominent companies" had ever demanded that miners furnish more coal than they were paid for.[75] A letter from a Barton miner promptly returned, "I will say then that there are no prominent companies on George's Creek, or in the whole region, as it is a well known fact that there is from three to five hundred [pounds] on each mine car over and above that which the miner is paid for." [76]

On August 2 the miners met again, with delegates present from 17 of the 20 mines in the region.[77] After a session of several hours, the 54 delegates, each representing 50 men, voted 31 to 23 to demand an increase to 55 cents a ton instead of holding out for the 65 cents originally suggested. The orderly nature of the meeting was due to the skill and humor of the chairman, Philip O'Rourke, who kept the debate away from personal accusation and acrimony.[78]

In spite of statements by company agents that they could not pay the extra five cents without losing all profit on their present contracts,[79] after two uneasy weeks of suspension the three largest companies—Consolidation, American, and Borden—gave in and announced that they would resume at 55 cents on August 18.[80] Their action led the *Alleganian and Times* to observe in an editorial on August 27, "However unpleasant it is to the coal companies, the fact remains that there is no cohesion among them—their pledges are ropes of sand. With all their wealth they have not half the stamina of their workmen." Before the surrender of the big companies, even while ostensibly subscribing to an agreement not to yield to the miners' demands, several of the smaller operators offered 55 cents as a "temporary" measure so that they might fill a few orders. The price of coal in Baltimore had risen considerably during the railroad strike.

While the miners were waiting for the companies to accede to their demands, federal troops moved in to open the canal blockade.[81] By August 28 nearly all the companies were sending out coal by both rail and canal. Unusually heavy shipments were expected to continue until the closing of the canal some time in

December, providing the miners with steady work at good wages which would enable them to pay off some of the debt incurred during the hard times of 1876 and 1877.[82]

However, Consolidation did not intend to let this state of affairs continue. Criticizing the "absence of united action on the part of all the companies [which] resulted in a resumption of work by the entire region at 55 cents," its annual report for 1877 went on to say, "The absolute necessity of reducing the cost of our coal, at the mouth of the mine, to that which prevails in the regions competing with us, is admitted by all the companies in our region, and steps are now being taken to accomplish this by a reduction of miners' wages and outside labor to an approximation of what is being paid in the competing regions."

Adopting a different approach to their economic problems, the Cumberland coal operators met in Baltimore on January 15, 1878, and decided "to invite the miners, through a delegation, to a conference with a committee of producers . . . to consider the mutual interests of the miners and the companies in the necessity which exists for a reduction in wages."[83] The joint conference was a device which newspaper editors in the area had often suggested, but although the miners had from time to time invited the employers to be present at their meetings, the only face-to-face discussion between mine management and mine labor up to this time had taken the form of brief interviews between individual superintendents and small groups of their men. The *Alleganian and Times* congratulated the corporations, and the miners themselves "recognized with great appreciation the advances made by the coal companies in thus inviting them to a conference."[84]

The conference was held on Saturday, January 26, with William Borden of the Borden Mining Company as chairman, and Henry Loveridge, president of the Maryland Coal Company, as secretary. On the management side, American, George's Creek, Borden, Consolidation, Hampshire and Baltimore, Piedmont, Swanton, Maryland, Blaen Avon, Franklin and George's Creek, and Potomac companies were represented. On the labor side, delegates were present from almost all of the mines.[85]

The operators presented their case almost entirely in terms of Clearfield competition, stating that the Cumberland trade had fallen off 42 percent since 1874, while Clearfield trade had gained 33 percent.[86] They had kept the Cumberland mines open, even though they had made no profit and at times sustained a loss, in order "to save business and furnish employment to the miners." Clearfield operators "nominally paid 40 cents per ton for mining," but "the companies all had their stores and made a profit out of their sales—the miners being required to deal there." Although Clearfield coal was competing successfully with their product at Baltimore, the Cumberland operators piously expressed a reluctance to force a reduction of wages, hoping that the men themselves would suggest a rate of 40 cents a ton.[87]

The miners' delegates were carefully attentive but insisted that they were there to listen, not to make or accept "propositions as to wages." They pointed out that their existing rate of 55 cents was also only nominal because of the overweight required. Without committing themselves, they suggested that if concessions were made as to weight, the price per ton might be adjusted.[88]

The meeting then adjourned, and the miners' representatives went home to report on the proceedings. As a result, a committee of 41 delegates representing over 1,900 men from 17 mines met at Lonaconing on January 30. Seven mines were not represented.[89] The final vote was for a reduction of only five cents, which the operators regarded as inadequate. The men discussed the question of weight at length, but decided to leave this problem to the legislature, which was studying a new mining bill.

Not all of the miners were willing to accept the decision of their committee. Those at the Hampshire and Baltimore company's Franklin and Phoenix mines signed a contract to mine coal for a year at 40 cents a ton with full weight allowance.[90] Caledonia miners accepted 45 cents a ton.[91] After holding out until the middle of March, the men of the Maryland, Swanton, and Potomac companies decided that work at 40 cents was better than none.[92] Simultaneously the New Central miners voted 110 to

38 to take 45 cents and just weight.[93] By March 21 even the 45 cents had been lost, and all mines had resumed at 40 cents a ton.[94]

This surrender was more complete than would appear on the surface. For years in periods of slack employment the Cumberland men had forestalled a wage reduction in their own region by going to work temporarily in the Clearfield and Meyersdale, Pennsylvania, mines, and accepting the lower rates there. In this way they made it possible for the miners remaining at home to hold firm to the 65 cents paid since 1866.[95] Now, because of the depression, the Cumberland miners had nowhere close at hand to go. Furthermore, the Clearfield miners were beginning to resent the gap between their wage and that in Maryland, and it was rumored that they were ready to strike for an increase.[96] If the Cumberland men looked for employment in Clearfield under these circumstances, they would be regarded as blacklegs.

The companies had thus succeeded in reducing wages to the level they desired. In addition, the problem of the "superfluous amount of labor" complained of by Consolidation was "partly overcome by the Company through reducing the number of its miners by discharging those most capable of seeking employment elsewhere; and the same result [had] been attained in the region generally, by emigration therefrom." [97] Although no figures are available to show to what extent the mining force in the region diminished in 1878, we surmise that the reduction, together with the 15-cent cut in wages, made this year one of the worst the miners had yet undergone.

It is therefore not surprising to find a resurgence of interest in some form of association to speak for united miners in their conflict with united employers. After the collapse of the Miners' National in 1876, the Maryland men had maintained an organization of sorts referred to by Cumberland newspapers early in 1877 merely as the "Miners Union." This "union" may have been the beginning of Knights of Labor activity in the area.[98]

The Noble and Holy Order of the Knights of Labor was a loosely-knit body which admitted to membership all persons who worked or ever had worked for wages (excluding specifically doctors, lawyers, bankers, and those whose living came from the sale of "intoxicating drink"). The idealistic aims of the Order were bound to appeal to men whose views were as moderate as those of the Maryland miners. Opposition to strikes, resort to arbitration, and emphasis on improving the status of the wage earner through education and legislation, were combined with the element of fraternal ritual which had long been popular in this coal field.

The K. of L. was in no sense an association of trades unions.[99] Because of its broad membership, it had in some localities mixed assemblies composed of those engaged in a number of occupations. In other areas it had district assemblies made up almost entirely of members of a single trade or craft, as was the case in the George's Creek-Frostburg section of Maryland.[100]

On August 30, 1878, Samuel H. Vaughan, a Frostburg miner, received his commission as district organizer for the Knights of Labor in western Maryland.[101] Within a few months he formed a number of local assemblies, and on November 7 secured official recognition of District Assembly No. 25, with headquarters in Frostburg.[102] Thereafter membership statistics of the George's Creek mining district appeared in the official proceedings of the Order. By January 1, 1879, District 25 had 500 members.[103]

Like many other districts in the Order, particularly among the skilled trades, District 25 regarded its mission as the promotion of the welfare of its members. While content to operate under the general supervision of the national headquarters, it involved itself only incidentally in seeking "to promote the interests of society by elevating to higher planes of thought, education and action, the masses of laborers." [104] It found much more to the point the maintenance of a living wage, the limitation of the working day, and the establishment of better working conditions. Scattered letters from officers of District 25 to T. V. Powderly, grand master workman, from 1879 to early 1882, concern mainly questions about procedure. A lone inquiry about setting up a cooperative "Farming, Mining & Manufacturing Association" appears not to have been followed by any action.[105]

The membership of D. A. 25 initially contained a large proportion of men who had been dismissed by the coal companies but who had not left the region.[106] One of the first actions of the assembly, therefore, was directed toward supporting the surplus labor in the area. In 1879, D. A. 25 adopted rules restricting each miner's production to four tons of coal a day.[107] This voluntary limitation was at first confined to Consolidation mines, and in response that company hired additional men.[108]

Having "solved" the unemployment problem by sharing the work, the union miners next turned their attention to wages. By August 1879 they felt that times had improved sufficiently to warrant raising the mining rate to 50 cents a ton. They issued their request in the form of a notice published in local newspapers, because they had found in earlier years that the practice of choosing committees to meet with the company superintendents often "proved disastrous to some one or more thus appointed." [109]

Although the notices were signed by the secretary of the miners' organization, the companies refused to pay any attention to what they termed an "anonymous" communication, and a strike which lasted for almost two months began on September first.[110] Less than a third of the mine workers in this region were then members of the Knights of Labor. As of October 1, the eight locals of District Assembly 25 had 959 members; [111] the number of adult mine workers reported for the region in the 1880 census was 3,165.[112] If the bulk of non-union members had refused to join in the strike, the Knights of Labor organization in the Maryland coal fields might well have come to an early end. However, the non-union men held mass meetings of their own, and voted to sustain the demand for 50 cents a ton.[113]

It could not have been easy to keep the George's Creek region idle for seven weeks, for the miners had needed most of their reduced income merely to exist. The union officers wrote to Powderly to inquire whether the local assemblies of his district in Scranton, Pennsylvania, might help them [114] but he replied that his locals were already assessed to the limit to support a mining strike of their own.[115] Nevertheless, the Maryland men remained firm, and the operators one by one gave in and granted the 10-cent advance. By October 28 all the companies were digging and shipping coal.[116] It is significant that toward the end of the Maryland strike the miners in Meyersdale and Clearfield, Pennsylvania, also struck and won an increase to 50 cents a ton.[117] Moreover, in many other bituminous districts throughout the country miners were making successful demands for higher wages.[118]

Despite the fact that they had received no financial aid, Maryland mine workers flocked to join the organization which had achieved so much in the first year of its existence in the region. A Cumberland newspaper estimated that at the funeral of a miner killed in an accident in December 1879 there were "fully 1,500 miners" in attendance, nearly all of them members of the Knights of Labor.[119]

By the spring of 1880 practical politicians in Allegany County sought the block of votes controlled by the union. Writing in "strict secresy" to Governor William T. Hamilton concerning the appointment of a mine inspector, a Democratic party worker said: "We have an organization called the 'Knights of Labor,' which is a sworn secret society and numbers two thousand and is still increasing, and certainly comprises *nine-tenths* of the mining population. . . . Our party without their aid will be hopelessly in a minority. While a few of the more religious Irish are not members of the order,* yet the great body of them are! and *all* are in sympathy with the order." [120]

* Not knowing exactly what their position should be, the parish priests at Frostburg and Lonaconing "merely advised" Catholics against joining the Order and requested those who were already members of the K. of L. "to promise to leave if the Church should condemn it." Henry J. Browne, *The Catholic Church and the Knights of Labor* (Washington, 1949), pp. 57–58. There is no indication that the two priests later requested their parishioners to withdraw from the labor organization. The religious issue did not arise squarely until 1885, when, at the request of Father Peter C. Manning in Lonaconing, the Archbishop of Baltimore studied the ritual and constitution of the Knights and found nothing ob-

The growing strength of the union also aroused concern in other quarters. A commission appointed by Governor John Lee Carroll to study conditions at the coal mines in 1879–80 urged that the Knights of Labor be replaced in the mining area by an open organization which would "give the strong conservative men strength to counteract the evil influence of secret associations." The commission suggested "that an act of incorporation for a Miners Association be passed . . . with full power to arrange and settle all questions between them and the coal operators, including that of the price for mining."[121] The legislature did not act on this revolutionary suggestion.

During 1880 the union gained further concessions from the companies without any great difficulty. Early in February the ten-hour day went into effect for the dumpers and drivers.[122] Beginning on February 1, the miners refused to pay more than one-half cent a ton for the sharpening of their tools.[123] The former charge of one cent a ton they considered "exorbitant," and consistently cited it as a principal grievance.

In February, also, the miners presented the operators with the choice of paying 65 cents a ton on and after February 23 or of having work stop until the increase was granted.[124] The operators yielded to this demand without protest.[125]

All things considered, the years 1876 through 1880 were among the most significant in the history of the Maryland miners. They encompassed the worst depression these men were to experience in the nineteenth century; they saw the establishment of what promised to be a potent labor organization; and they were marked by the passage of the first laws intended to protect mine workers' health and safety. These laws and their effect will be discussed in the following chapter.

---

jectionable therein. Patrick Cavan to T. V. Powderly, April 13 and June 29, 1885 (Powderly Papers, Boxes 12 and 13).

Chapter 13

Ventilation and Inspection Legislation (1876-1880)

Of the various state laws directly affecting the mining population of Maryland, by far the most important were those that provided for mine inspection, ventilation, and in general the physical protection of miners and laborers. Not only did the legislation result in improved working conditions; it also provided a continuing focus of political activity in the mining region. Passage of a ventilation and inspection law in 1876 was followed by a steady drive to strengthen it. Almost every session of the legislature from 1876 to 1910 considered proposed amendments to the mining law, and the original bill was three times repealed and reenacted with extensive revision.

Beginning in 1873 the miners became increasingly involved in politics with a view to putting their own men in the state house. Regrettably, little can be learned of the actual processes by which candidates were chosen. Some four to six weeks before the November elections in the odd-numbered years, primaries held in the various election districts selected delegates to county conventions of the major political parties, which in turn decided on candidates for the legislature and for county offices. It is impossible to determine on what basis the names of available prospects were presented to the conventions.

It is also impossible to follow the course of mining legislation precisely. The official journals of the Maryland House and Senate did not record debate, and newspaper coverage was skimpy. Exactly how the miners exerted political influence remains a mystery. It would be particularly interesting to know what sort of horse-trading resulted in the passage of laws to benefit a small Republican minority in a predominantly Democratic state. Occasionally the nature of the *quid pro quo* is suggested, as in passage of the 1878 bill discussed below. And occasionally the miners must have been forced to accept a statute repugnant to them, as was probably the case in 1880, when they salvaged their inspection law but at the same time had a coal and iron police act thrust upon them.

In 1873 the newly organized Miners and Laborers' Protective and Benevolent Association forwarded to the party conventions the names of five union members who would be acceptable candidates for the Maryland House and Senate.[1] Both Republicans and Democrats refused to place the name of a miner on the ballot, and neither party would pass a resolution instructing its candidates to support legislation for the protection of the health and lives of miners.[2] Under these circumstances the miners presented a third-party "Citizens' Ticket," which was roundly defeated in the election.[3] Nevertheless the union did not give up its attempt to secure an inspection and ventilation law for the coal mines, but late in the 1874 session of the legislature managed to have a bill introduced and passed in the House of Delegates.[4] This bill, rejected only after its third reading in the Senate,[5] came close to becoming a law.

When the legislature met in 1876, the miners joined with the rest of the Allegany County community in protesting a proposed tax of two cents a ton on coal, and held meetings for that purpose at Pompey Smash, Eckhart, Frostburg, and Lonaconing.[6] A committee consisting of Peter Cain, Francis McMahon, and Philip O'Rourke consulted on this matter with Colonel Alfred Spates, a former state senator, and advised him at the same time that the miners were "unanimously in favor of the passage of a law for the proper ventilation of mines, and for the appointment of a State inspector of mines to secure the proper ventilation and to look to the interests of miners." [7] At the tax-protest meetings in Frostburg on January 31 and Lonaconing on February 2, the miners adopted a long resolution in which they described the perils of coal mining and their complete dependence upon the employers for safe working conditions.[8]

Actually they were not asking for a great deal. Andrew Roy, mine inspector for the state of Ohio, in describing the various coal fields of the United States in the 1870's, said of the Cumberland coal region: "As the working places of the miners are high and dry, and there is no [explosive] gas to contend with, the matter of mine ventilation is simple, inexpensive and of easy control. There can be no excuse, whatever, for the presence of bad air in any of the mines, or in any of the workings of a single mine. If bad air exists, it is the result of ignorance or neglect on the part of the responsible manager of the mine." [9]

As in 1874, the miners had no representation in the legislature. The House delegation from Allegany County consisted of two farmers, a Frostburg justice of the peace, and a Cumberland banker. It is not apparent that any of them sponsored or gave enthusiastic support to the miners' request for relief. In 1876 the miners did not, as in 1874, attempt to draft their own law. The bill which William R. Getty of Garrett County introduced in the Senate was said to have been drawn up by Colonel Alfred Spates,[10] a political figure of some influence. The Cumberland *Alleganian* vigorously supported this bill, keeping close watch over its progress in the General Assembly and publishing favorable editorials during February and March. It secured an advance copy of the law from its Annapolis correspondent and invited leading miners and other interested parties to study it so that they might propose amendments if necessary.[11] The *Alleganian's* editor further advised that if advocates of the measure wished to assure its passage before adjournment, they should either send a delegation to Annapolis or submit to the legislature a petition signed by every miner in the region.[12]

The miners appointed John W. Evans of Lonaconing and Francis J. McMahon of Frostburg to go to the state capital and lobby for the ventilation bill.[13] Shortly after they arrived in Annapolis, Evans and McMahon "solicited and obtained an interview with the Finance Committee [of the Senate], and represented the case so clearly that the bill was immediately reported favorably and passed to the second reading." [14] The *Alleganian* of March 22 reported that the bill was going to its third reading in the Senate, where it seemed to have no opposition. It would then proceed to the House, where it would probably pass within a week. The only danger now threatening it was the short interval of time remaining before the end of the session. On March 24 the *Alleganian* published a special dispatch from Annapolis announcing the unanimous passage of the bill on its third reading in the House. The only amendment mentioned raised the minimum age of the mine inspector from twenty-five to thirty-five years. The law went into effect immediately after the governor signed it on March 30.

This "ACT regulating the working and proper ventilation of coal mines in Allegany and Garrett counties, and providing for the appointment of an Inspector for said mines" [15] was by no means a radical assault on established practices. It set out the qualifications and duties of the mine inspector, requiring him to report annually to the governor the condition of the mines and any accidents which had occurred, and to make suggestions for further legislation. The inspector was to visit each mine at least once a month to see that safety precautions were being taken and that the provisions of the act were being enforced. He was also to attend inquests held in cases of fatal accidents. In this respect Section 4 of the act also provided that if a mining death were fully proven to be caused "by any violation of this act, or any wilful failure to comply with its provisions," the widow or lineal heirs might bring suit to recover damages.

The companies were to furnish the inspector with up-to-date maps of their mines. Within six months after passage of the act they were to provide "a proper system of pure air ventilation to be well circulated . . . throughout the entire mine" and were also to keep out all harmful gases or impure air. Each mine was to have "a competent and practical inside overseer" appointed by the operator to be responsible for the maintenance of the ventilation system and the general safety of the men at work in the mine. If the inspector found gas or impure air resulting from "the bad condition of the ventilating apparatus," all he was empowered to do was to notify the operator to expel the noxious gases and provide proper ventilation. The mine owners were to furnish at their own expense all necessary props and timber "at the place where the miner is at work." Refusal or neglect of operators or their agents to comply with the provisions of the law was a misdemeanor punishable by a fine of not less than $100 or more than $500 "in the discretion of the court."

There is no way of ascertaining who was responsible for the provisions of Section 12, which to a large extent nullified the purposes of the act: "*And be it enacted*, That the mode of ventilation shall be the same as is now in use in the coal mines of Allegany and Garrett counties." Only five of the twenty-seven mines being worked in Allegany County in 1877 had furnaces.[16] The rest depended upon the "mode" of "natural ventilation," which was almost completely ineffective except in the most shallow workings.

This new law "affecting the coal interest and the management of the mines, was not looked upon with favor [by the companies] and to some extent, was regarded as an encroachment upon the rights of owners and producers of coal, and an improper interference with private rights and antagonistic to the interests of the trade generally." [17] The companies had not expected that the bill would pass and apparently had not bothered to oppose it actively. In his characteristically blunt fashion, A. C. Greene, superintendent of Borden Mines, wrote to an applicant for the inspectorship, "I had hoped that this wholly unnecessary Act would not have passed the Legislature. It can do no good and is likely to be productive of much mischief. We have as you well know no dangerous or explosive gases in our coal, and an Inspector would really have nothing to do." [18] In the same letter Greene expressed the opinion that "the miners would never have urged the passage of the Bill but in the expectation of getting the office among themselves here," and he added cynically, "I wish the Governor would appoint some stranger from abroad so that he might see from the outcry that would arise how little the measure is needed."

The post of mine inspector carried with it a salary of $2,000 a year to be paid by the state. The only statutory requirements for the office were that the inspector be at least thirty-five years old and "possess a competent knowledge of all the different systems of mining, and working and proper ventilation of coal mines, and the nature and constituent parts of noxious gases of mines, and of the various ways of expelling the same from said mines." [19] He was also "in no manner whatever [to] be under the employ of any of the mining companies operating mines in said counties." [20]

Naturally, the position attracted many applicants. Among these were some whose sponsors recommended them principally as loyal Democrats, Allegany County being a Republican thorn in the flesh of the Democratic state. One applicant went so far as to write to Governor Carroll, "I gave Tom Brown A good Knock down on your accounte." [21] The correspondence files of the governor of Maryland for this period show that several of the applicants were completely unqualified, and we can understand with what consternation Greene wrote from Borden Mines to Senator Thomas G. McCulloh at Annapolis:

I confess to feeling very considerable alarm when I see the kind of men who are applying for the office of Inspector of Mines under the late Act of Assembly. The provisions of the Act as to *qualifications* appear to me to be utterly ignored by many of these applicants, for it cannot possibly be said that they possess them or even approach their possession.

My opinion of the Bill is that it is wholly useless and unnecessary, and that unless it is administered with the greatest tact and judgment it will work incalculable mischief. In fact, the only thing the Inspector can do without doing damage is to do *nothing*. The moment that an ignorant pretender begins to meddle with the plans of ventilation which have been adopted at the several mines from experience of their several requirements—he makes a muddle. The difficulty with a great many of those whose names have been on the lists of applicants is that they cannot refrain from meddling. They must show their authority even if it exposes their ignorance.

"The Inspector," concluded Greene with the ineffable authority of the upper middle class, "in order that injury to all parties may be avoided, should be a gentleman—a man of sense and education—and whose character and reputation should command respect. These are qualities in addition to those required in the Bill." [22] The Baltimore *Sun* of April 1, 1876, reported that the miners, on the other hand, were unanimously opposed to the appointment of anyone but a practical miner, "a man who by simply breathing the air in a mine can instantly detect any deleterious quality in it."

Peter Cain of Pompey Smash, who was sworn in as inspector of mines for Allegany and Garrett counties before the clerk of the circuit court at Cumberland on April 27,[23] had been recommended to the governor by William R. Barnard, a member of the House of Delegates from Garrett County.[24] He was apparently not a candidate Greene approved of, for Senator McCulloh had written to the governor, "I hope some question may be put to Kain [*sic*] as I am satisfied he is in no way qualified." [25] However, the *Frostburg Mining Journal* of April 29 described the new inspector as "a sound, sensible, honorable gentleman" who would carry out his duties faithfully but would not impose unnecessary expenses upon the mine owners. The status of "gentleman" was here conferred by virtue of character, rather than of birth (as in Greene's use of the word), for Cain was a miner and had been active in promoting the miners' union in 1874.

Like his successors, Cain found only too soon that his job was a thankless one. The miners seemed unable to get along with any of the inspectors appointed by the governor, even when these functionaries came from labor ranks, and constantly questioned their vigilance or impugned their motives. The Cumberland *Al-*

leganian and Times, as well as a number of miners, criticized Cain for not recommending changes in the existing mode of ventilation,[26] and the newspaper editor further rebuked him for not mentioning that some of the mines were wet. Neither of Cain's reports for his two years in office noted "any marked neglect on the part of operators, or any serious violation of the law." [27] He did, however, recommend that the act be amended to impose a penalty for each day on which the operators refused compliance, because as the law stood the operator could postpone making improvements until the end of the season and then pay a single fine. In the meantime, to cite one possibility, the miners would have been in danger of losing their lives as the result of failure to supply props and timbers.[28]

Moreover, Cain initiated a court test of Section 9, which required the operators of each mine to furnish "a sufficient quantity of timber at the place where the miner is at work." This the George's Creek Coal Company refused to do. The mine inspector reported the refusal to the grand jury, which brought an indictment against the company. The case, *State* v. *George's Creek Coal Company*, was heard in circuit court on January 15, 1877.[29] Although the company had placed the timber for props at the foot of the plane some distance from the mouth of the mine, the court absolved it, finding not only that Section 9 interfered with the right of private contract,[30] but also that "the ventilation law was made to protect the life and limbs of the miner, and not to save him any labor, and that the carrying of the timbers from wherever laid, and the loading them on the Company's cars, were questions of labor over which the Legislature had no power." [31]

The Consolidation Coal Company very astutely precluded any difficulty with the law by issuing a set of "Rules for the Government of its Miners and Others" shortly after the passage of the ventilation bill.[32] These regulations shifted all responsibility for safety to the miners, who were put on notice that they worked at their own risk. Of their own initiative they were to stop work when the air was bad and were to report all dangerous conditions. Since in 1876 most miners looked upon Consolidation as their champion in maintaining the 65-cent wage, one imagines that the company's employees did not balk even at Rule No. 2, which read: "And in consideration of the present price paid Miners for mining coal, it is considered to cover also their service in taking props and other timber from the place of deposit outside of the mine, to the place of working as usually done heretofore." Continued work after posting of the rules was interpreted as willingness on the part of the employees to serve under the current superintendent, boss miner, and overseer, at their own risk, and in full acceptance of the rules.

In the interval between passage of the bill and the next session of the legislature, Governor John Lee Carroll came to the conclusion that the law was serving no purpose and recommended that it be repealed. Although one surmises that the companies applied some pressure, Governor Carroll's message to the legislature in 1878 attributed his decision to the deficiencies of the mine inspector's reports: [33] "The last report of the Mine Inspector does not show such an amount of efficient work as would warrant the continued payment of his salary from the Treasury of the State, and I have not been able, from the reports made to me, to see the value of a law providing such an office. If the miners have grievances they do not appear to arise from the bad ventilation of the mines, for in this respect there have been few changes made, and if there are troubles from other sources a State officer is not the person to correct them." The Cumberland *Alleganian and Times* criticized the inspector because his reports did not describe the benefits which had resulted from enforcement of the provisions of the bill, and therefore gave the impression that the legislation had been unnecessary.[34] Both the newspaper and the Allegany County delegates to the General Assembly agreed that the ventilation law should be amended, rather than repealed, so that the mines might continue to be "regulated by law under the eye and authority of an executive officer." [35]

In the elections of 1877 the miners succeeded at last in putting one of their own candidates into the House of Delegates. The *Alleganian and Times* described Patrick Carroll, a miner from Lonaconing, as "a man of considerable intelligence, and thoroughly acquainted with the needs and wants of the workingmen." [36] Born in Ireland in 1829, he came to America in 1846 and worked at railroading and farming before he settled in Lonaconing in 1863. He had held no public office previous to his election to the legislature in 1877.[37] One of his aims in entering politics was to get "a law passed to have coal measured by a sworn official at the place where it is taken out, to put a stop to the enormous frauds practiced on the poor employees by the mining companies." [38] In order to be elected, Carroll, a Democrat, must have received a substantial number of votes from his Republican fellow-workmen. The Allegany County delegation also included the editor of the *Frostburg Mining Journal*, J. B. Oder, who had supported the miners during their wage controversies in 1876 and 1877. Still a third delegate sympathetic to mining legislation was W. M. McKaig, a prominent Cumberland attorney.

Although the existing law gave the inspector limited powers concerned with the physical safety and well-being of mine employees, it made no provision for his supervising the weighing of the coal, and it in no manner interfered with the weighing policies of the companies. In his first report as mine inspector, Peter Cain suggested that if the companies provided proper scales and weighed the coal at the mouth of the mine before it was dumped into the hoppers, they would eliminate one of the miners' principal grievances.[39] In support of appropriate legislation, he argued that it would also insure that the companies would not pay more than the miner was entitled to (an unlikely possibility).

In the wage troubles during the spring of 1877 the miners claimed that they were obliged to furnish on every car sent out at least 500 to 600 pounds of coal for which they were not paid.[40] This same complaint reappeared in the second 1877 strike,[41] in the course of which the *Frostburg Mining Journal* remarked, "In the *ante-bellum* times . . . steady work at 2,240 pounds net per ton enabled the miners to make a respectable living [at a lower rate of pay]. Now, with quarter, third and half time, they get fifty cents per ton by frequently digging from a quarter to half a ton for nothing." [42] Closely associated with the failure to pay for all the coal brought out was the docking system, which deducted the price of one ton if a miner's car contained as little as ten pounds of slate or if the coal sent out was too hard or too soft.[43] Consolidation's miners at Ocean tried to fight the docking system and asked for the discharge of the dock boss, whom they found "objectionable." Told that they could quit if they did not like the boss, they did so, and the mine was closed.[44]

The bill which Patrick Carroll introduced in the Maryland legislature in January 1878 provided among other things for giving the men full and fair payment for the coal they mined.[45] After passing its first reading in the House of Delegates, it appeared in full in the *Frostburg Mining Journal* of February 2. It was intended to repeal the mining law of 1876 and reenact it with amendments to Sections 2, 3, 7, and 12. Only the significant changes proposed will be noted here. Section 2 as amended added to the qualifications required of the mine inspector: he must have had at least ten years' experience as a practical miner, and he must have lived in either Garrett or Allegany county for the five years preceding his appointment. Section 3 cut his salary from $2,000 to $1,500 a year.

Section 7 as redrawn made the ventilation requirements more specific. Carroll's bill called for a system which would provide for each person and animal working underground "a circulation of one hundred cubic feet per minute of wholesome air" at the place where the miners were at work. It empowered the mine inspector virtually to close a mine in case of noncompliance; an operator who failed to provide proper ventilation could not employ more than ten men in the mine at any one time until the situation was corrected.

Section 12 provided at length for the proper weighing of coal, authorizing the mine inspector to act as inspector of weights at the mines, to test the scales, and otherwise to ascertain whether the miners were being allowed full weight. Weighmasters under oath were to keep accurate records and were to "credit and allow . . . the full weight of coal in each mining car, to the party or parties who mined [it]." Violations of this section could result in fines of $100 to $500.

Carroll's original bill was referred to the House judiciary committee, which amended it in two respects before it was passed and sent to the Senate.[46] The penalty for failing to provide proper ventilation became merely a fine, and Section 12 defined a ton as 2,240 pounds.

On February 7 the bill went to the Senate corporations committee, where opposition was expected unless the miners, who were then resisting a wage cut, would agree to accept 40 cents a ton.[47] On February 13 the committee began hearings.[48] A representative of the Maryland Coal Company argued against the bill for two hours.* He maintained that the 1876 law was adequate respecting ventilation, and he referred to the inspector's report to prove that when black damp was found, proper remedies had removed it. Providing the ventilation required by the new bill would cost the companies too much. The Maryland Coal Company's spokesman expressed himself as being in favor of "just weight" but not full weight. He thought the miners should provide 2,340 pounds of coal before they were paid for a ton. The worker, not the company, "should make allowance for the coal that was lost by evaporation, theft, &c., on its way to market." Furthermore, this operator objected to "the weigh-boss being a sworn man." In concluding, he contended that the miners did not want the bill, that differences between the men and the companies could be settled without legislation, and described Carroll as "coming to the Maryland Legislature to revolutionize the coal trade of Allegany county." [49]

* The *Frostburg Mining Journal*, whose editor, J. B. Oder, was a member of the House at this session, says that Henry Loveridge, president of the company, argued its case against the bill, whereas the *Alleganian and Times* identifies the speaker as the vice-president of the company.

Carroll replied that ventilation reform would actually cost very little and would allow the companies to get out more coal and make greater profits. As to weight, the miners were asking only for justice and wanted the weigh boss to be a "sworn man" in order to "protect themselves from treachery."

A. C. Greene made a brief statement on February 14 and conceded that the method of working some of the mines on George's Creek was wrong and that the changes Carroll suggested would provide better ventilation.[50] Greene also said that he hoped the question of weight would be settled because his company, which gave full allowance, was being injured by the unfair practices of other operators who in essence robbed their employees and were thus able to put their coal on the market at a lower rate.

Of the large establishments, only the Maryland Coal Company made any extensive objection to the new mining bill. Neither Consolidation nor the George's Creek Coal Company (both of which had mines reported as model in 1880 by the mine inspector and the mine commissioners) bothered to put in an appearance at the Senate hearing.

On March 7 the ventilation bill as modified by the committee on corporations went to its second reading in the Senate.[51] On March 9 Patrick Carroll addressed a meeting of the Lonaconing area miners to discuss the amendments and see if they were acceptable to his constituents.[52] The Senate committee had removed the qualification that the inspector be a practical miner, and had deleted the requirement for furnishing an exact amount of pure air. It had not changed the weight provisions. The miners were to be paid for 2,240 pounds to the ton, and the weigh master was to be a sworn officer. The miners at this meeting indicated their willingness to accept the revisions. Carroll later complained that the *Alleganian and Times* had made it appear that he was bowing to company wishes in accepting the compromise bill.[53] Not endowed with prevision, he could scarcely have foreseen that mining legislation would be built on compromise for many years to come, and that his Frostburg colleague in the House of Delegates would write in the *Mining Journal* more than 20 years later, "there is not a mining law in the Maryland code that was not fathered or step-fathered by a Maryland miner, . . . who accepted at last 'the best that was to be had.' "[54]

On February 2 the editor of the *Frostburg Mining Journal* had written from Annapolis, "It is reported here that the acceptance by miners of forty cents will modify opposition to the bill in the Senate." By March 18 the bulk of the mining force had decided to return to work at 40 cents a ton.[55] News of full capitulation by the men on March 21 was followed in the March 22 issue of the *Alleganian and Times* by announcement of passage of the ventilation bill in the Senate. Without proof of the connection of these events, one can merely ponder the coincidence.

As approved by both houses on March 27, the ventilation act of 1878 (Maryland *Laws* 1878, Ch. 157) was substantially the same as the law of 1876 but with some significant additions. The qualifications for mine inspector did not include the proposed new experience and residence requirements. His salary, however, was reduced to $1,500. Ventilation specifications were unchanged. They did not mention 100 cubic feet of air per man, but merely stated that "pure air ventilation" should be provided "by such modes as are now in use in the coal mines of Allegheny [*sic*] and Garrett counties." Section 12, adopted in the form drafted by Carroll, provided new regulations about weighing and full payment. Section 13 supplemented Section 12 by requiring that the companies furnish scales for weighing the coal in the mine cars. The penalties set out in 1876 had been fines of from $100 to $500; under the new law, the range was $10 to $500.

If any justification for Sections 12 and 13 were needed, the president of the Hampshire and Baltimore Coal Company inadvertently supplied it in his annual report for 1878. In explaining to the stockholders why coal sales for that year had been "unusually unproductive of satisfactory results," he cited "the loss of long customary over-weight from miners, who by legislative enactment are now paid for each and every pound of coal put upon the mine cars." [56]

To satisfy the governor, who had recommended complete repeal of the mining bill and abolition of the office of mine inspector, the act included a provision for an impartial commission to determine whether the mining law and/or the inspector were necessary.[57] Section 15 authorized the governor to "appoint a commission of three competent persons to examine the mines of Alleghany [*sic*] and Garrett counties and report before the next session of the General Assembly [1880], with recommendations for such legislation as to said commission may appear just and necessary for the proper supervision and regulation of said mines."

A tantalizing note in the *Alleganian and Times* for March 27 placed A. C. Greene and Henry Loveridge in Annapolis, purportedly advocating repeal of the mining bill just passed. They had not furnished the press with any details of their plan, and since the session was due to adjourn in a few days, one cannot see what they hoped to accomplish. The only clue to this mystery is a last-gasp amendment in the form of a separate bill approved on April 1, which added Section 15½ to the law.[58] Section 15½ obviously was of benefit to the Borden Company, which Greene represented. It provided that any company mining coal by shaft might contract with its miners to pay by the cubic yard instead of by the ton. In this event the company would not need to provide scales or a weighmaster, but its mine cars must be measured by a "sworn measurer" and have their capacity marked upon them. Borden operated the only shaft mine in the region. If Greene and Loveridge went to Annapolis specifically to obtain this amendment, Loveridge's interest is unexplainable unless he intended to begin a shaft operation for the Maryland Company, of which he was president.

The mine inspector appointed to succeed Peter Cain was Owen Riordan of Westernport, "a civil engineer and practical miner" who had been employed in West Virginia during 1877 making a survey for English capitalists.[59] The *Alleganian and Times* described him as "affable and agreeable to all," having the confidence of the miners and possessing the ability to reason with the operators and secure changes in ventilating methods if necessary.[60] During the first month of his term he visited a large number of mines and tested the air "by a very delicate and expensive instrument [anemometer]—the first, we believe, ever introduced into this region." [61]

Neither Riordan's affability nor his scientific approach prevented his being severely criticized by the miners. Before he had been in office six months, he found it expedient to write to the editor of the *Frostburg Mining Journal:* "Please insert the following lines to correct a misunderstanding that exists in reference to my duty as Inspector of Mines. I am very much annoyed with anonymous communications purporting to come from a certain class of people. I will pay no attention to any of them hereafter. I do say to the miners that if they have any grievances to redress, let me know and I will attend to them at once, but I request the writer to sign his name to his information and I pledge myself to keep it strictly secret." [62]

Apparently no one quarreled with the inspector's introduction of the anemometer for measuring the strength of ventilation currents in the mines. However, after publication of Riordan's first report, a number of miners wrote letters challenging his understanding of the instrument and his interpretation of its readings.[63] The tenor of these letters was that he noted the volume of air at the entrance of the mines but failed to check circulation in the more remote working places. "In fact," said one of these correspondents sarcastically, "since the anemometer appeared in our mines in the hands of our present Inspector, the ventilation of our mines is so fast improving that I have no doubt I will soon hear that the miners of Georges Creek have to be anchored to a prop while engaged at work." [64]

Other letters implied that Riordan had failed to mention black damp and bad air conditions in a number of mines, and one specifically questioned working conditions in the Borden Shaft, which he had reported as satisfactory with a circulation of only

34 cubic feet of air per man. Why, then, asked this writer, had the Borden management of its own volition built a new ventilating furnace since the report was made? [65]

Hostility toward the mine inspector may have been exacerbated by the presence of the mine commissioners in the community. It was a matter of common understanding that this "impartial" group would determine the fate of the ventilation and inspection bill; the inspector's failure to point out dangerous conditions could substantially damage the miners' cause.

The governor waited for more than a year after passage of the enabling act before he appointed the commission of three competent persons to examine the mines and report to the legislature. Initially he had no intention of giving the miners a representative in the investigating body. A letter from J. B. Oder, editor of the *Frostburg Mining Journal*, reveals that Governor Carroll had selected William H. Smith and C. H. Hamill, and was looking for a mining engineer to complete the trio. Oder wrote to the governor to protest that since Smith and Hamill were ex-operators, the naming of "a scientific gentleman" as the third member "would be regarded by the great mass of our people, for the most part, miners, as one-sided, partial and discriminative against them." [66] He suggested that the appointment of John Ryan of Lonaconing would assure "a fair, honest, valuable, satisfactory investigation and report," adding, "Ryan was formerly a miner, now a popular and highly-esteemed merchant, and a very sensible, just, impartial, reliable man."

Smith, Hamill, and Ryan were formally commissioned on May 22, 1879, and began their inspections on July 25.[67] Keeping a journal of their progress, they reported each day, almost without exception, that the mines which they had visited were well ventilated and properly timbered. They encountered bad air and black damp at only two locations, in mines which were practically worked out. Equally monotonous was the response of miners when questioned, always in the presence of the master miner; the men invariably replied that they had nothing to complain of and were satisfied with existing conditions.

The companies cooperated fully in the inspection of their

properties, but were less willing to reveal the amount of business they were doing or the wages they paid. "It may be proper to remark in this connection," wrote the commissioners, "that the companies throughout the region have not manifested any disposition . . . to furnish any information outside of the working and ventilation of the mines." [68] The commission waited in vain for the wage figures promised them by the Consolidation Coal Company.[69] On the question of weight, "all information of the actual amount of coal shipped was refused." [70] Although eighteen months had elapsed since passage of the statute guaranteeing fair payment to the miners, "The Commission had reliable information that some of the Companies exacted over-weight from their miners, but in consequence of a refusal of the companies to give the information necessary to solve the question, we had to abandon it." [71]

After finishing their tour of the mines, the commissioners planned to hear the miners at Lonaconing on October 22 and Frostburg on October 23, and advertised these prospective meetings in the Cumberland and Frostburg newspapers. "For reasons unknown," said the *Frostburg Mining Journal*, the miners did not attend these hearings.[72] At Lonaconing a large group of miners left the meeting after criticizing the Commission's methods of investigation. A splinter group denied having any substantial grievances and pronounced the mining law "a failure, owing to the court having decided that section 10,* referring to the delivery of timber in the mines, interfered with the right of private contract." [73] At Frostburg on the following day, the miners failed to show up at all. Because of their reluctance to come forward, the Frostburg editor wrote, 90 percent of the miners were unheard, and it was the testimony of the 10 percent who supposedly called the law "inefficient" which was placed before the legislature.[74]

* On this point the commissioners said in their recommendation for repeal, "If the construction of that section claimed by the miners had been endorsed by the opinion of the Courts, it would have undoubtedly interfered with the proper control and working of the mines by the Companies' agents, and seriously affected the investment of capital in coal lands."

A possible explanation of the miners' unwillingness to appear at the hearings was Commissioner Smith's close ties with the George's Creek Coal and Iron Company management. Smith, who was superintendent of the company during the strike of 1854, would have been remembered by many as heading the company "militia" and personally facing the strikers "armed with a cavalry sword buckled to his side." [75] Furthermore, while the commission was at work in the mining area, Smith and his wife were guests of John Douglas, then superintendent for the George's Creek company.[76]

As a result of its six months' study, the mine commission advised Governor Carroll that it had "failed to see any good effects either to the miner or to the true interest of the State by the appointment of a mine Inspector, and can see no good to result from its continuance, and therefore recommend the repeal of the Law, except so much of Sections 12 and 13 as applies to the correct weight of coal."

The mine inspector's report published at about this same time opposed repeal of the law and abolition of the office of inspector.[77] Although he made no recommendations for improving ventilation, Riordan's comments about weighing practices justified the commission's decision to retain that part of the law. "There is circumstantial evidence enough to satisfy me that there are some of the Companies not doing full justice to the weight, and there is not power enough given to the inspector to prevent it." [78] Riordan's second report, taken in conjunction with the commission's recommendation to repeal the law, led to a protest from one miner that the trouble was not with the law but with "the non-execution of it by the Inspectors," whose reports had not presented the facts concerning ventilation in the mines.[79]

The political campaign which took place during the mine commissioners' visit was the first in which the major party leaders in Allegany County were forced to take cognizance of the power of the miners' vote. The Knights of Labor had acquired a large membership in 1879 even before the successful strike added to the union rolls, and as early as August the Democratic *Cumberland Daily Times* predicted that the Order would "act a lively part in the election" to secure candidates who would protect the rights of laboring men. The newspaper therefore urged that the Democratic party choose men who would attract K. of L. support, saying, "This vote, in some portions of the County, is quite large; and if there were no other incentive, we should select good men for the offices." [80]

In their conventions both Democrats and Republicans made it a point to give the miners fair representation on their tickets. Each party nominated mine workers to run for two of the four available seats in the House of Delegates.[81] The county vote in November was predominantly Republican, and all four of the party's candidates were elected to the House of Delegates—J. J. Bruce, a Cumberland physician, B. L. Turner, a Flintstone merchant, and Reuben Anthony and David D. Shearer, mine workers from Frostburg and Barton, respectively.[82]

When the Maryland General Assembly convened early in 1880, William T. Hamilton, who succeeded John Lee Carroll as governor, called the attention of the legislators to the mine commissioners' recommendation to dispense with the office of mine inspector, and added, "If it be not abolished, as recommended, I submit to you that the salary is too high and should be reduced." [83] In the meantime the miners and their advocates were busily circulating a petition protesting repeal of the inspection law. State senator John S. Combs, a Lonaconing merchant, presented this document, bearing more than 1,800 signatures, in the Senate on February 25. According to the *Frostburg Mining Journal* of February 28, the chief argument for retention of the law was "the incompetency of the Mines Commission and the consequently erroneous character of their report."

In spite of the petition, a bill introduced in the Senate on March 2 proposed to repeal and reenact the law of 1878.[84] Proponents of the bill failed to secure its passage. A second bill, which provided merely for reducing the mine inspector's salary, was introduced on March 4, went through its three readings, and was passed and sent to the House on March 16.[85] There it was read and referred to the committee on ways and means, which reported it favorably on March 27, almost at the end of the session.[86] The bill apparently died in the House, since the Maryland Code for 1880 indicates the mine inspector's salary as being $1,500, the amount provided by the 1878 law. The governor himself appears to have expected no repealing legislation from this session of the General Assembly, for he named a new mine inspector, Thomas Brown of Pompey Smash, late in February.[87]

In spite of its inadequacies as to ventilation, which became more pronounced when the mines penetrated deeper into the hillsides, the mining law of 1878 continued in force until 1898. Even with the powerful aid of the Knights of Labor in other parts of Maryland, the miners failed to secure the revisions that they sought.

Chapter 14

The Knights of Labor and the Great Strike of 1882

During the rapid growth and almost spectacular successes of the Knights of Labor in 1879 and 1880, no public figure in the Maryland mining area seems to have condemned the organization. On the contrary, the *Frostburg Mining Journal* of August 30, 1879, described the membership as embracing "a surprising number of the sober, solid, good men of the region." From various published sources it is possible to identify well over 250 of the members and officers of D. A. 25.[1] Many of these were indeed well-known and respected members of the community, long-time residents, and in some cases property owners. A number of persons prominent in the earlier union movement—among them Peter Cain, John Donahue, George Hoskin, James Park, and E. M. Davis—were active in the new organization. The notable exception was Thomas Brown, who as mine inspector could not appropriately be involved in K. of L. affairs.*

Even the mining companies, so far as we know, did not at first view the Knights of Labor with alarm or try to prevent their employees from joining the organization. By October 1880 District Assembly 25 had eleven locals with 1,757 members in the

* However, Brown did not hesitate to praise union miners and their aims in his official report for 1881 (pp. 23–24) and publicly approved the attempt to "equalize" Clearfield wages at the Cumberland level. See *National Labor Tribune*, July 2, 1881.

Maryland coal-mining towns. In addition, it had acquired three locals in the neighboring coal regions of Meyersdale, Pennsylvania, and Irondale and Flemington, West Virginia, and two nonmining locals at Cumberland, Maryland, giving a total district membership of 2,170.[2]

Whether or not the Knights of Labor members in the Cumberland coal region felt themselves "impregnably entrenched,"[3] by the spring of 1881 they were confident enough to reveal the essential objectives of their program for improved working conditions in the mines. On May 7 they began to publish weekly in the *Frostburg Mining Journal* "resolutions . . . adopted as a guide for the miners and laboring men of this coal region." (Appendix D.)

Under these regulations no miner was to work below ground for less than $2.50 a day. The working day was to be ten hours long, from 7 A.M. to 5 P.M. Miners were to be paid extra when operating in certain situations which would slow down their production of coal; they should not have to work more than a certain distance beyond access to a roadway or push the mining cars more than 400 feet in the rooms without extra compensation. Five cars (10 tons) were to constitute a standard day's production for two men. Dead-work was to be paid for at the rate of $2.50 a day, and timber was to be delivered at the working place instead of at the mouth of the mine. The company was to furnish the tools. If a man were discharged, some other union member was to take his place until his case was investigated.

The regulations were not one-sided. The union pledged itself not to support any driver or laborer who did not perform his duty faithfully or who tried to avoid working the hours agreed on, or (in the case of drivers) who refused to start a final trip which might carry him a little past quitting time. Furthermore, it would not countenance "any recklessness or carelessness to the destruction of life and property."

The stage was set for the great strike of 1882, because the coal companies felt that these regulations infringed the rights of management. Consolidation's historian wrote that the union "not only dictated prices for mining and labor which made the cost of production much higher than in competing regions, but it forced upon all the companies rules and regulations governing the conduct of work which were as injurious to business as the high prices set for labor." [4]

In order to understand the wage issue which also arose in the 1882 strike, it is necessary to review certain earlier events. In February 1880 the miners demanded and the companies granted an increase to 65 cents a ton. The reason for the companies' accession to this demand is not clear. Possibly they expected Clearfield and Cumberland wages to be "equalized" as they had been in 1879, in which case the Pennsylvania region would lose its competitive advantage. As it happened, Clearfield miners struck for 65 cents on February 23, but after two months went back to work without having gained their increase.[5] Their defeat was accomplished mainly by the importation of foreign labor—Belgians, Hungarians, Poles and Swedes. "The hosts of immigrants ready for temporary employment were too much and too many for the strikers." [6] The Clearfield region in 1880 had only the skeleton of an organization under the Knights of Labor: 50 members at DuBois, 187 at Houtzdale, and 39 at Osceola Mills.[7] Still mindful of the Molly Maguires, the Catholic clergy in the Pennsylvania coal fields opposed the semi-secrecy and binding oath of the Knights of Labor. Specifically, so far as Clearfield was concerned, a K. of L. member had written to Powderly at the end of 1879, "All the members of the Catholic faith have had to withdraw from the order upon account of the oath." [8]

The George's Creek men provided financial aid for Clearfield throughout that region's twelve-week strike,[9] for they realized that maintenance of their own wage depended on what happened in the competitive area. The general feeling around Cumberland at the end of the strike was that the Clearfield miners had in some way betrayed the common cause. This was particularly true in midsummer, when the Maryland companies tried to drop wages to 50 cents a ton.[10] At that time the editor of the Cumberland *Daily Times* commented: "We blame the Clearfield miners for yielding their position. They were fed by our miners; no one was threatened with starvation; the substitutes of the companies were incompetent men, and while the miners were kept from starvation, it was their duty to stand by their pledge to our region. Resumption with them at the old rate was certain to seriously affect this locality, as is evidenced by the movement of the operators to secure a decrease in the cost of mining here; and we believe eventually our miners will be the sufferers from the yielding of those in the Clearfield region." [11]

While the Clearfield strike was in progress, the Maryland companies profited from the extra business which came their way. Once Clearfield had resumed at its lower mining rate, the Cumberland operators prepared an "address to the miners," signed by the presidents of eleven of the companies, asking them to volunteer to take a cut to 50 cents a ton.[12] The men refused to make this concession while freight rates remained at their current high level.[13] Wages in the area continued at 65 cents a ton. In spite of assertions from some sources that the Cumberland coal trade was "languishing," [14] conditions in this region throughout 1880 were "exceptionally prosperous," and shipments for the year were the largest since 1875.[15]

However, it could not realistically be assumed that the Cumberland operators would indefinitely tolerate a wage difference which put them at a disadvantage *vis-à-vis* their competitors in Clearfield. The Cumberland region miners knew that eventually the wages in the two regions would be equalized, either by raising Clearfield men to 65 cents a ton (the preferred solution) or by cutting the Maryland men back to 50 cents. Early in February 1882 the executive committee of the K. of L. District 25 in Frostburg appealed to the Clearfield miners to demand 65 cents a ton,[16] pointing out that the bituminous coal trade "was never in a better condition" and that trade journals predicted prosperity for the coming season.

By the end of February Clearfield miners had not responded

to the pleas from George's Creek, and the Cumberland *Daily Times*, contemptuous of the ineffectual efforts of the labor organization, proclaimed that the Maryland coal operators found "the Knights of Labor unequal to the task of equalizing the cost of production after three years' trial at the companies' expense." [17] Predicting a bitter fight ahead, the *Times* reported a secret meeting of all the companies in Baltimore in mid-February and asked rhetorically: "Is there no meaning in this? When the Maryland [Coal Company] lamb and the Consolidation lion lie down together, is there no meaning in it?" "We believe the companies are a unit; that their plans are formed," the editor added, unaware to what an extent his pronouncement was justified.[18]

On the first of March, the companies decided to challenge the union. Acting in concert they issued a circular announcing a lower rate of wages for miners, drivers, and laborers alike, and the replacement of the ten-hour day with a working day of twelve hours, one hour being allowed for dinner.[19] The changes were to go into effect on March 15. The companies did not baldly say that they intended to get rid of the union, but their circular included a statement that they had determined to adopt "such rules and regulations as may be necessary for the proper control of their property, in which they have of late been greatly embarrassed by the arbitrary and unjust rules instituted by the labor organization of the region—regulations that have been most unjust to the companies and really advantageous to no one." [20]

Nobody in the Maryland coal region was surprised by the notification of the wage cut, which the management circular postulated as necessary in order to meet Clearfield competition. Cumberland miners were only too familiar with the situation in this part of Pennsylvania, and had no intention of being reduced to that level. Clearfield miners, said the Baltimore *Sun*, "receive nominally 50 cents per ton, but because of a rule obliging them to trade in the companies' stores on what is known as the 'pluck-me' system, in reality they get no more than 40 cents. Their condition is represented to be but little better than that of animals, and their homes are likened to hovels. To the Cumberland miner, whose home is all order and cleanliness, and whose pay gives him some of the luxuries of life, such a prospect [equalization at Clearfield rates] is not pleasing to contemplate, and they contend that they cannot live decently if forced to accept the demands of the circular." [21]

Pending a formal reply to the company circular, the officers of District 25 posted notices of their refusal to accept the wage reduction and their decision to suspend work after March 14.[22] Wage disputes were an old story. What really staggered the regional Knights of Labor was the companies' reference to "arbitrary and unjust rules." Union men were convinced that their regulations protected their interests. By adopting a ten-hour day and by allowing no man to load more than five tons a day, the union had succeeded in getting fairly steady employment throughout the year. The strength of the organization had given the men more freedom to ask for their rights without fear of being discharged. Furthermore, as one member put it, "Since the adoption of the ten-hour system and the other rules of the organization the society of miners has greatly changed for the better. They now have time to spend a few hours with their families and improve their minds by reading." [23]

In their official reply to the company circular the miners refused to concede that their rules had been "really advantageous to no one." [24] Rejecting also the proposals for lower wages and a longer working day, they suggested that between March 15 and April 1 committees from both sides meet and arbitrate their differences.

Coincident with their reply, the Knights of Labor posted official notices that the miners and laborers of the George's Creek region would stop work on March 14 "in this region *or in any other region that produces coal for the same market at a lower price than for what we are now contending;* as war has been declared on all mining labor, all should fight the battle together." In addition, "Any one working at any labor in the mining department in this *or any competing region* while this question of wages is pending, shall be considered a black-leg and an enemy to the cause of labor." (Italics supplied.) [25] George's Creek men therefore could not look for work at Elk Garden, Meyersdale, Broad Top or Clearfield—all competitors of Cumberland, and all paying only 50 cents a ton.[26]

Thus began what the Baltimore *Sun* described as a strike of 3,000 miners and 800 laborers, carpenters, and mechanics, against representatives of $75,000,000 invested in mining and shipping coal from the region.[27] As a result of their careful planning, the companies were well prepared to cope with the situation. They had a solid agreement among themselves to forfeit from one to two dollars a ton for every ton of coal mined at the price labor demanded.[28] They had provided an anti-strike fund to which the area railroads also contributed. The George's Creek and Cumberland Railroad (controlled by the American and Maryland Coal companies) and the Cumberland and Pennsylvania Railroad (wholly owned by the Consolidation Coal Company) were said to have donated $25,000 and $50,000, respectively. These railroads agreed moreover, on pain of forfeit, not to carry any coal mined at more than 50 cents a ton. The operators' compact, which had been "carefully kept from the public," [29] was intended to last only until December 31, 1882, but was still in effect in 1886.

In addition to their gentlemen's agreement, the operators were able to derive comfort from the passage in 1880 of a law permitting railroad, steamboat, canal, and coal and iron companies to request the governor of Maryland to commission persons chosen by themselves "to act as policemen for the protection of the property of said corporation . . . , and for the preservation of peace and good order on their respective premises." [30] Compensation for these special officers was to come from the corporations concerned, not from the state or county governments. Furthermore, should they need some handy formula for arrest in a mining area which had a long-standing record of nonviolence, the companies could resort to a recently enacted addition to the Code of Public General Laws concerning disturbance of the peace.[31]

Morale among the miners was buoyed up by a firm conviction that the national offices of the Knights of Labor would provide financial support. The locals of District Assembly 25 had been contributing regularly to a "defense fund" established by the 1880 general assembly of the Knights of Labor.[32] However, only 30 percent of the fund was available for strikes, the remainder being earmarked for the expenses of traveling organizers and the promotion of cooperative enterprises and education. Payment to support strikes was so hedged about with requirements for four separate attempts at arbitration (on higher and higher levels, terminating with the national executive board) that it was virtually impossible to receive help from the defense fund. In any event, broad interpretation of "education" by a former grand secretary had almost completely consumed the fund by 1882, when the Order was deluged with pleas for assistance against union-breaking attempts by employers all over the nation.[33] Powderly received more than fifty cries for help in April and May, 1882.[34] The most that the national officers could do was to permit an appeal for aid (which usually brought in nothing) or in rare instances to levy an assessment (equally ineffective because so many locals had their own battles to fight and could not help others).[35]

Soon after the strike began, the companies conferred at Baltimore with a miners' committee composed of R. W. Price, James Park, Andrew Patrick, and Samuel Vaughan,[36] insisting, however, upon meeting these men as employees, not as Knights of Labor. The names of all four of the labor representatives at this unproductive discussion subsequently appeared on the companies' blacklist.[37]

Throughout April and May the employers' organization took no action, not even notifying the occupants of company houses that they must vacate. The striking miners busied themselves with their homes and gardens, putting up fences, repairing outbuildings, painting, whitewashing, digging, planting, and leading what the Baltimore *American* described as "an arcadian existence." [38] After the Consolidation Coal Company paid off its men on March 30 for the period March 1–14 (averaging $12 per man after deductions for fuel, rent, smithing and medical care), no more mining wages came into the villages.[39]

Support from the miners of competitive bituminous regions proved unreliable. Meyersdale, with eleven mines and 1,000 employees, struck on April 1,[40] but the men returned to work at 50 cents a ton after only a few days' idleness.[41] Miners at Elk Garden, a new and very small West Virginia region, continued working at 50 cents a ton.[42] Clearfield miners made a timid request for a conference with their employers to discuss an increase of 15 cents a ton, but the operators ignored them.[43] Clearfield's District Assembly 40 had only 225 members in good standing, and did not intend to strike and meet certain defeat.[44] The Cumberland *Daily Times* sneered that the Clearfield workers had shown some sympathy for the Maryland men, but "no sooner had they received a hint that their houses would be needed in ten days after they came out of the mines, than a change came over them." [45]

Early in April, District 25 sent Robert W. Price to Pittsburgh to talk with the national executive board of the K. of L. concerning financial support of the strike. The executive board promised an appeal for funds, but Price unfortunately came away from the meeting with the mistaken impression that not an "appeal" but an "assessment" had been authorized.[46] Shortly after the Pittsburgh meeting, Robert Layton, grand secretary of the Knights of Labor, went to Cumberland to investigate the possibility of arbitration. "If I can succeed in effecting a settlement in this way," he wrote to Powderly, "I think that a precedent will have been established in this branch of our Order, that will eventually work wonders." [47] Layton met with Robert Price, James Park, "and other persons from the George's Creek region who are supposed to be members of the Knights of Labor Executive Committee" on April 19.[48] On April 28, still in Cumberland, he assured Powderly that everything was proceeding "very smoothly,"[49] but on May 2 he was back at headquarters, having achieved nothing.[50] However, as a result of Layton's visit to Maryland, "there [was] an apparent stiffening up again of the miners."[51]

In addition to its efforts toward arbitration, national headquarters stepped in to take part in organizing the Clearfield region. Since equalizaton at 65 cents a ton seemed to some K. of L. officials the best solution to the Maryland strike, Layton dispatched Myles McPadden, grand organizer, to Clearfield about April 7.[52] Here McPadden, working against intense opposition from the operators, and himself "catching regular *hell* up in Osceola,"[53] by May 17 had succeeded in bolstering the Clearfield miners' courage to the point of issuing a mildly worded request for a 15-cent increase or an agreement to submit the matter to arbitration.[54]

Unfortunately for the cause of solidarity, at about this time the officers of District Assembly 25 discovered that the financial aid they were expecting would not materialize because no actual assessment had been made. Layton, blamed for the misunderstanding, asked Powderly to confirm his interpretation of the decision made by the national executive board in April.[55] Powderly, who seems never to have understood that more was at issue here than a mere matter of wages, replied: "Appeal is the word and not assessment. It is just as I expected, the old story. *Strike at the boss and hit yourself*."[56]

Under these circumstances Layton was indeed naive if, as he said, he could not understand the refusal of D. A. 25 to furnish funds to support a possible strike in Clearfield.[57] Failing an assurance of aid from Cumberland, the Clearfield miners postponed their strike and made still another attempt to secure a conciliatory meeting with their employers.[58] Since the employers wasted no time in discharging any one suspected of union activity, suffering was widespread in the district. Powderly, when appealed to, answered peevishly that he wished the men would "have patience and become organized before launching out into strikes, but it is the old story over again. Strike in a hurry and repent at leisure." [59] Here he entirely missed the point that the Clearfield men had not struck, but were discharged precisely because they were being patient and trying to become organized.

By the end of May the Maryland strikers had much more to worry them than Clearfield's shilly-shallying. It had been generally predicted that the companies would make no move until the first of June and would then import labor if the miners had not returned to work.[60] Any suspicion that the threat of foreign labor was a bluff was dispelled by the Consolidation Coal Company's beginning construction of buildings at Eckhart to house its new employees and the special police who would guard them.[61] One building 100 by 25 feet, fitted with 82 double bunks and mattresses, was ready for the police on May 29.[62] Two buildings 150 by 25 feet, for housing a total of 400 men, were under construction, and the old machine shop at Eckhart had already been fitted up with 304 bunks and with long tables, so that it could be used as a combination dormitory and mess hall.[63] Except for police headquarters, all of the buildings to accommodate the new men at Camp Mayer (in honor of Charles F. Mayer, president of Consolidation) were enclosed within a six-foot stockade.[64]

At 5 A.M. on May 30 a police force of 125 arrived at Eckhart.[65] A justice of the peace from Cumberland swore in the men, who then went to their barracks, and donned their official uniforms—blue pants and blouses with brass buttons, slouch hats, and badges reading "State Police." Armed with revolvers and clubs, these guardians of peace and good order, not to mention property, were to receive from the Consolidation Coal Company two dollars a day and board for an unspecified term of employment.[66] Close on the heels of the special police came the first contingent of strikebreakers. On May 31 a group of 91 men and three women, fresh from Castle Garden, got off their special train at Eckhart guarded by 50 police officers who had boarded the train at Frostburg.[67] Most of the new men were Germans and Austrians, but some Poles, Hungarians and Swedes were among them. Almost as if viewing animals at the zoo, crowds of strikers and their families at Frostburg and Eckhart stared at the new arrivals and commented on their appearance, but created no disturbance. Only about a quarter of this initial group had any mining experience. All had signed a specific contract with Consolidation to mine coal at 50 cents a ton, and the company had promised to "guarantee the foreigners all the comforts needed and full protection."[68] A second group of 115 foreign miners arrived on June 7. Only 24 of these were said to be practical miners.[69]

Even when the new men actually went into the mines and began to send out coal, the mining population in general believed that they would not produce enough to repay Consolidation for what it was spending. The old residents were confident that the company would soon summon them to replace the incompetent greenhorns, but the mine inspector did not share their belief. On the contrary, he assured Governor Hamilton that "This new labor is going to be a success beyond a doubt if the experiment is pushed and there seems to be no doubt of that in mind judging from the action of the Coal Companies."[70]

Tensions of various kinds began to mount almost immediately. Swarms of miners watched the new men go into the mines, and an extra squad of police had to be sent to the mine opening at Eckhart on the first day of work.[71] Dissatisfaction also arose within Camp Mayer, and a few of the Germans left after discovering that they were being used as strikebreakers.[72] The Catholic and Lutheran clergymen of Frostburg objected to the company's isolating the new men from the solace of religion. They wanted the occupants of Camp Mayer to be free to attend churches in Frostburg, Eckhart, and vicinity.[73] Some of the police, saying that they were union men, refused to act against the miners and returned to Baltimore. Others got tired of Eckhart life and went back to the city.[74] Still others broke the monotony by drinking and were discharged.[75] Between replacements and resignations, the personnel of the guard force was constantly changing.[76]

Uneasiness among the strikers increased after some occupants of company houses received notices to vacate. Of the estimated 300 to 400 company houses in the region,[77] the Consolidation Coal Company was said to own "about two hundred cottages around Ocean, Hoffman, New Hope, Pompey Smash, Washington, Frost, Allegheny, and Eckhart mines."[78] On May 31 the constable and his assistants served notice on 50 occupants of Consolidation's houses at Ocean mine that they must be off the premises in thirty days, that is, by July first.[79] According to rumor, just before the deadline, the men would "be given the option of immediately going to work on the companies' terms, in which event they [would] be allowed to remain, *with a few exceptions*." (Italics supplied.) Consolidation had similar notices served on all old employees occupying its houses at Eckhart.[80] The New Central Company at Lonaconing on May 29 informed an unspecified number of families that they must vacate within 30 days.[81] The George's Creek Coal and Iron Company also ordered its tenants out by July 1.[82]

No startling developments occurred in June. Trainmen on the Eckhart branch of the railroad, acting in sympathy with the miners, refused to haul coal dug by the foreigners, and police ran the trains.[83] "By mistake * it was made to appear that Mine Inspector Brown would aid in instructing the new labor."[84] The strikers smuggled into Consolidation's stockade handbills written in German, explaining the labor trouble and asking the new men to quit.[85] The union politely requested mine bosses and weighmasters at Eckhart to cease work "now . . . that capital dares to degrade labor to the condition of European serfs."[86] Teamsters refused to haul provisions from Eckhart to the stockade, and supplies "had to be rolled up the hill into camp by the men."[87] The police force was armed with breech-loading rifles.[88] A few rowdies threw stones, but hurt no one.[89] The Frostburg clergy were told that it would not be "prudent"

* This misunderstanding was undoubtedly the beginning of the disagreement between Inspector Brown and the union which eventually led to his resignation and his scathing denunciation of the Knights of Labor in his official report for 1883.

for the new men to leave the enclosure, but priests and ministers were welcome to visit and hold services at the camp.[90]

On this last point Henry J. Browne in his *Catholic Church and the Knights of Labor* reveals an episode which demonstrates how the attitude of the church to the K. of L. differed in Maryland and in Pennsylvania. During the 1882 strike Consolidation's president, Charles F. Mayer, complained to Archbishop (later Cardinal) Gibbons that Valentine F. Schmitt, pastor of St. Michael's Catholic Church in Frostburg, was "not only a strong sympathizer, but an encourager of the Knights of Labor." [91] Mayer asked that Gibbons appoint some other priest to serve the foreign miners at Eckhart.

Father Schmitt informed the archbishop that he had been impartial, but added that in his own mind he had "a leaning toward the poor miners" and found it hard to see his people "victimized and driven from their houses and homes." Fearing violence, neither Schmitt nor the Protestant ministers would cooperate with Mayer by holding services within the alien labor camp. Browne concluded that Archbishop Gibbons' failure to remove Schmitt from his Frostburg parish suggested a "latent sympathy toward organized labor." [92]

In general, the region remained quiet, greatly to the annoyance of those visiting journalists who had looked for blood and violence when the new men were introduced. All *Frank Leslie's Illustrated Newspaper* could produce to accompany the sketches its artist had prepared on the scene was an unfounded allegation that the tyrannical Knights of Labor were forcing all shopkeepers to close at 7:30 * in the evening.[93] The miners played a waiting game, giving the companies time to tire of their expensive and reputedly inefficient labor. Many left the region to look for jobs elsewhere as mechanics and laborers.[94] Between 500 and 600 worked on the extension of the West Virginia Central and Pittsburgh Railroad, earning from $1.25 to $1.40 a day.[95] Some of those who remained at home gained "a fair livelihood"

* Actually storekeepers were experimenting with early closing at the request of their clerks. *Frostburg Mining Journal*, June 3, 1882.

by gathering herbs in the mountains.[96] But many more families had no income at all, and the rumors of large sums of relief money coming in from K. of L. assemblies outside the district were entirely without basis. District Assembly 25 once more inquired of Powderly why it was that grand secretary Layton had prevented their getting an assessment. Powderly absolved Layton, but he again revealed his misunderstanding of the Maryland situation when he wrote: "I am opposed to strikes and if your D. A. had kept at work until *now* and put the money spent in the strike in a fund you could have purchased a mine of your own and be independent of bosses, but as you have struck, stick to it. I send you five dollars . . . as my contribution to you." [97] However, in this same letter Powderly promised to grant an assessment if the executive board agreed.

It is ironic that a more accurate appraisal of the feeling in D. A. 25 came not from a union official but from a mere bystander, a "prominent and influential gentleman in Cumberland," who remarked shrewdly that he was sorry to see the miners "turned out of their homes and sent adrift for a sentimental principle." He felt that the men would give in on wages, might even give in on the ten-hour day, but would not surrender their organization. "Once assured that they will be permitted to hold their membership in the Knights of Labor," he said, "they will flock like sheep to work." [98]

On June 24 the Clearfield miners decided to strike in support of the George's Creek region.[99] Organizer McPadden urged Powderly to collect a strike tax, saying he believed that locals in Pittsburgh "could spare $500 easily." [100] "I assure you," he wrote, "there is suffering and want in Cumberland Region and if some means is not devised to help these men it will be a lasting disgrace on our Order." The national office of the K. of L. finally levied an assessment in July,[101] but its effectiveness can be judged from the response of two of the Pittsburgh districts. The master workman of District Assembly 3 said that the assessment was unjust because the executive board well knew that D. A. 3 itself needed money to sustain the strikes going on there.[102]

The secretary of District Assembly 9 also protested. Its members had been on strike for four months, had not received any assistance from the national office, and could not be expected to help Maryland while their own families were suffering.[103]

Miners' committees from the Cumberland and Clearfield regions met at Bedford, Pennsylvania, on June 29 to discuss plans for the joint strike effort.[104] Here they requested Powderly to send Myles McPadden back to Clearfield to hold the men in line.[105] Powderly's reply is not contained among his papers, but we know that McPadden returned to Clearfield County, where he was arrested at Morrisdale on charges of "intimidation and threat" and again at Houtzdale on charges of "conspiracy and unlawful assemblage." [106]

The two districts also officially requested the national office of the Knights of Labor to act on their behalf in an attempt at arbitration. Consequently, on June 30, grand secretary Layton informed S. S. Lee, chairman of the Maryland operators' association, that "upon notification from the operators interested of a desire to arbitrate or settle the differences in an honorable manner, the executive board pledge their influence to effect an honorable settlement." [107]

Lee replied on July 5, "We are at all times ready and willing to confer with the men in our employ, but we do not realize that there is any need of an intermediary." Furthermore, he continued, "I am bound in frankness, to say to you, that the condition of things in our region has been, and still is such, that it is impossible for the companies to accept anything but what they have asked for." [108] Following this rebuff, national headquarters let matters stand until later in the month.

After July first many developments in the George's Creek region suggested that the operators were getting ready to take a firmer stand. A spokesman for the George's Creek Coal and Iron Company announced that as soon as the 30 or 40 company houses could be vacated, the company would bring in new labor. Before the strike it had employed 270 men at Lonaconing.[109] The Maryland Coal Company, also at Lonaconing, filed bills of complaint for the eviction of seven of its old miners, giving them six days to show cause why they should not vacate.[110] This company was the only one which had "executed leases with its miners providing for tenancy at will, each and every tenant agreeing to leave on thirty days' notice." [111] Consolidation did not press its tenants, depending upon facilities at Camp Mayer to accommodate the small detachments of new laborers who arrived almost daily to supplement the original force. By July 15 Consolidation was employing about 300 of these men,[112] and was planning to reopen Hoffman mine, which had places for from 150 to 200 miners, even though it was partially flooded.[113] Hoffman workers could enter from the Eckhart mine opening, thus eliminating the problem of providing a special escort for them.

Up to this time, only Consolidation had asked the governor for special police, and the authority of the officers appointed had been limited to protecting the property and preserving the peace on the premises of that corporation and its ancillary carrier, the Cumberland and Pennsylvania Railroad. Early in July revised commissions permitted the special police to have jurisdiction on the property of all the coal companies.[114] One hundred fifteen of the new authorizations were issued on July 10,[115] and their holders were sworn in on July 14.[116] Ten extra police, urgently requested by Charles F. Mayer received appointments soon after the new policy went into effect.[117] Consolidation no longer footed the entire bill; assessment for support of the force was to be in proportion to each company's normal output of coal.[118]

With the opening of the Hoffman mine, the previously quiet region began to boil with excitement. Crowds were sometimes so thick that mine bosses and others living at Frostburg had to have police escorts to their homes.[119] The more conservative elements among the strikers tried to restrain outbreaks of hooting and jeering, and were to some extent successful. The harassment of mine inspector Brown in particular explains his biased report of the strike. From the moment that he inspected Eckhart Mine (as required by law) before the new men went in, rumors flew that he was instructing the new miners in the art of digging coal; that he was recruiting foreign labor in New York; and that he was employed by Consolidation to help break the strike. He denied all of these canards most vigorously.[120] Indicative of the feeling toward Brown was the mock funeral tendered him to the strains of "In the Swede [*sic*] Bye and Bye." [121] Brown wrote to the governor that "a reign of terror has been inaugurated," and asked for "such advice and assistance as you may be able to give." [122]

The stepping up of nonviolent demonstrations coincided also with the failure of the strike at Clearfield. McPadden complained to Powderly on July 20 that there had been no response to the assessment, "and without money [the strike] will be a failure. We cannot hold the Clearfield men out. The clergy there are very bitter. Rev. Fathers Maher and Lynch of Houtzdale and Osceola respectively denounced me on last Sunday in a most violent manner & encouraged the men to go to work." [123] The following day he again wrote to Powderly that in view of the lack of funds he could foresee nothing but defeat for both Clearfield and Maryland.[124]

It is difficult to understand Powderly's uncharitable reply. For months the Clearfield operators had been pursuing a policy of eviction, dismissal, withholding of wages, and the importation of strikebreakers.[125] Yet all Powderly found to say to McPadden was: "I felt when you started to the Clearfield region that nothing could be accomplished there. The men of that region are as craven hearted a set with few exceptions as I ever heard of." [126] At the end of July, Cornelius Cotter, an officer of D. A. 40 who had been arrested at the same time as McPadden, advised Powderly that the Clearfield strike had been called off.[127]

At the close of the Clearfield strike, McPadden and James Campbell, a member of the grand executive board of the K. of L., decided to go to Cumberland and try again to bring about a negotiated settlement.[128] Campbell sent a telegram to Stephen S. Lee, who received all communications for the employers, on July 27, asking him if he would be willing to meet a committee representative of the miners. Lee indicated that the companies were willing to meet their men, but stated emphatically that they would "decline conferring with officials of any kind or any outside parties."[129] Nevertheless, Campbell and McPadden proceeded to Frostburg to try to arrange a meeting of operators and men.

After this the situation in Maryland deteriorated rapidly. The New Central Coal Company at Lonaconing began building barracks for about a hundred new men with the intention of opening its Koontz mine, and special police were sent from Camp Mayer to guard the mine property.[130] These preparations set off large and noisy demonstrations in Lonaconing, marked particularly by episodes of "bull banding," serenading nonstrikers (often maintenance personnel) with drums and tin pans. The participants in the outbreaks were more likely to be women, children, and teen-aged boys than adult strikers.[131] Among those singled out for attention were the storekeeper who sold groceries to the barracks, the baker who supplied bread for the police, the hardware dealer who furnished the stove for cooking at police headquarters, and the engineman at the Koontz mine. Shortly after these incidents the special police refused to stay at Lonaconing, saying the miners had ordered them to leave.[132] The commanding officer sent no replacements, leaving the sheriff to maintain order and protect New Central's property. There was some talk of calling in the militia,[133] and it seems that this possibility was seriously considered at Annapolis. Correspondence between Governor Hamilton and C. J. M. Gwinn, attorney general of Maryland, indicates that the latter drew up a proclamation ready for use in any "emergency growing out of the labor troubles."[134] No emergency arose. The sheriff arrested a few persons in connection with the bull-banding, and swore in deputies to prevent interference with New Central's preparations. The state's attorney for Allegany County issued a circular explaining what constituted unlawful assembly and disturbance of the peace, and the Lonaconing area quieted down, again at the urging of conservative elements among the miners.[135] However, fear of an attack on Camp Mayer led to the placing of a Gatling gun at the entrance to the stockade.[136]

On August 8, hoping to secure a conference with the operators, employees of the various mines named delegates to meet with company officials.[137] McPadden, who was still on the scene, communicated to Powderly at this juncture his belief that if the men would "Consede the price" the operators would give way in everything else.[138] To avoid "bloody war" on George's Creek, where he described the members of D. A. 25 as being "armed to the teeth," * McPadden thought it advisable for them to "meet the operators, compromise, save the organization and go to work for 50¢ a ton." "But," he complained, "The D. A. officers are blind or fools. They will hear to nothing nor compromise nothing."

What McPadden interpreted as blindness on the part of the district officers was in fact an open-eyed appreciation of the realities. The companies had no intention of giving way in "everything else." This was made unmistakably clear when they finally granted the miners a conference on August 15.[139] The miners had previously met at Frostburg to decide on what basis they would settle, and had agreed on 55 cents a ton, no blacklist, a ten-hour day, and payment every two weeks. The operators' attitude was uncompromising: wages must come down to the Clearfield level; the men could make up the difference by discarding the 5-tons-a-day limit. The companies would not depart from the conditions set out in their March circular, and would make no promises as to reemployment. Consolidation specifically stated that it would be unjust to dismiss its imported labor. After five hours of discussion, the conference adjourned without coming to any decision.

Even before the conference, many influences were at work to pull the Lonaconing miners away from those at Frostburg, who would be the principal sufferers if the strike were abandoned.

* All of McPadden's letters to Powderly tend to be dramatic. Whether or not "armed to the teeth," the George's Creek men did not engage in violence in this strike.

The New Central Coal Company posted handbills in Lonaconing advertising for 300 miners to work at 50 cents a ton and warning, "Our old men are urged not to let their places be filled by new men, or they will be permanently lost." [140] The president of the George's Creek Coal and Iron Company had made it clear in interviews with some of his old employees that he would in no event make concessions, but would resume when the men were willing to meet his terms.[141] Furthermore, the George's Creek Company prepared to serve notices to quit on the balance of its tenants,[142] and let it be known that it would not reemploy anyone who sought to avoid eviction by going to law.[143]

While the lower section of George's Creek was being exposed to this kind of intimidation, Consolidation tightened the screws on its old miners near Frostburg by announcing that the Hoffman mine was in full production with a hundred new men and that Ocean mine would open next.[144]

The inevitable break came when new labor arrived at New Central's Koontz mine. At that time a group of miners of the George's Creek Coal and Iron Company told their superintendent that they were ready to go to work.[145] By evening of the following day, about 76 of the company's normal quota of close to 300 had asked for places, and as the news spread, the New Central men also flocked to their superintendent's office.[146]

The labor organization worked frantically to postpone a piecemeal surrender. Notices displayed at Lonaconing read: "Attention! Be it known that any man who will go to work Monday morning or before the undersigned order it will be branded as a BLACKLEG. Now, men, be careful not to bring on yourselves and children something that may never be forgotten. K. of L. August 19, 1882." [147] Under one of these notices some one had scribbled, "We're going to work; go to h-ll." [148] On the morning of August 21, when the George's Creek Company planned to resume, a crowd of about 2,000 assembled and by sheer bulk prevented the Lonaconing men from reaching the mine.[149] A much smaller crowd assembled for the same purpose on August 23. No violence broke out, and it would seem that the sheriff expected none. Although he made a point of being on the scene, and went so far as to swear in a posse, paradoxically enough he drew his deputies from among the ranks of the strikers.[150]

The strike was officially ended on August 24. After the companies refused any further conference, the executive board of the district assembly of the Knights of Labor posted a notice calling the members back to work.[151] The miners gained nothing from the prolonged suspension which the Baltimore *Sun* of August 23 estimated had caused a total money loss of about $1,500,000. They went back on the terms originally offered by the companies at the beginning of March: the 12-hour day, wages of 50 cents a ton, and company control of operations. In addition, many returning to work for Consolidation allegedly were forced to sign yellow-dog or iron-clad contracts in order to get employment.[152]

At the Consolidation Coal Company's mines about 800 old miners were initially without work.[153] Foreign labor had filled some 400 places, and the company did not propose to resume immediately on its former scale. The Lonaconing companies had "blacklists" of a smaller size.[154]

As was customary in this region, most of those left jobless at the end of the strike were fairly quickly reabsorbed into the working force.* Only a few were blacklisted in the true sense. By October 6, the number of those "uniformly blacklisted throughout the region" had declined to 234.[155] The bulk of these were from Consolidation mines. The names of some of these "uniformly blacklisted" reappeared as mine employees in newspaper accounts of subsequent management-labor conferences. A short while after D. A. 25 published its roster of unem-

* As a Lonaconing miner described the process of reemployment which was typical also after subsequent strikes: "The different mine superintendents here are exchanging victims. For example, Mr. Jones refuses to employ Mr. Williams in his mine for certain causes, and then employs Mr. Baxter, who has been victimized by Mr. Jackson for a similar cause." Letter from John W. Evans, *National Labor Tribune*, September 30, 1882.

ployed (see Appendix E), the number of foreign miners had dwindled to "about seventy," [156] providing openings for at least 300 men.

Certain differences between the Clearfield and Cumberland strikes have been intimated; they should be stated more definitely before closing this chapter. McPadden, Layton, Campbell and other "outside" figures moved freely in the George's Creek region without interference by the law. Arrests in the Maryland region were few and for minor offenses—bull-banding, unlawful assembly, and the like—and were promptly tried before the local magistrates instead of being continued for months as were the conspiracy cases in Clearfield.[157] In the matter of eviction, only ten days' notice was necessary in the company-owned Clearfield area.[158] Thirty days' notice was given in Maryland, but by various delaying tactics on the part of the miners the cases were postponed over and over, and only about a half dozen ever came to trial.[159] One result of the strike was that the Maryland companies prepared new leases in which the tenants agreed in writing to leave on due notice.[160] Clearfield miners had to deal with company stores, and so had no source of credit to maintain them over a long strike. In contrast, merchants in the Maryland mining area were generally sympathetic to the strikers (in fact, many belonged to the K. of L.),[161] and extended credit as long as they could. As an example, Walter Edwards of Frostburg, a member of L. A. 718, carried fellow-members on his books during the strike to the amount of $12,000; similarly, former state senator John S. Combs, a Lonaconing merchant, provided the strikers in his town with "work, money and bread." [162]

This widely-publicized Maryland coal strike revealed the inherent weaknesses of the Knights of Labor central organization. Principal among these, at least from the point of view of the strikers, was the lack of financial support. On earlier occasions the region's miners had sustained themselves throughout strikes without outside assistance, but circumstances had been different. The companies, because of their extreme rivalry, had not opposed a united front to their employees as in 1882. More important, no portion of miners' wages had been paid to an all-encompassing defense fund which supposedly would return the contribution in case of need. The Knights of Labor defense fund, even if it had not been diverted to other purposes, could not possibly have provided support for the multitude of labor disputes which arose throughout the 1880's. Local and district assemblies were left to conduct and finance their own small wars and to help one another if they felt so inclined.

The question of helping one another was a particulary touchy one in George's Creek, which had traditionally shied clear of involvement in any Pennsylvania trouble. It was a great step away from parochialism for these men to support the Clearfield strikers in 1880. To be asked again to provide funds for Clearfield in 1882, while the national organization did nothing, was disillusioning for the members of D. A. 25, and explains in large measure their later reluctance to join the U.M.W.A.'s general bituminous strike in 1894.

A second serious weakness was the failure of the national office to appreciate the problems of the district organizations. The Powderly correspondence contains evidence of the pathetic struggles being carried on in the coal fields against union-breaking practices, yet neither Powderly nor his officers demonstrated any real sympathy for these attempts to build up or to preserve the labor organization. Powderly himself almost automatically equated "strike" with "wages," as we have seen in the case of his misunderstanding of the Maryland and Clearfield situations. His naivete in suggesting that the Maryland strikers should have purchased a mine is almost beyond belief.

The insistence on "one big union" worked to the disadvantage of groups like the bituminous miners who very clearly needed a national organization of their own, with leaders who would be conversant with the problems of the soft coal trade as well as of the mine workers. The widespread location of bituminous deposits brought about extreme competition for regional markets, and most employers, in trying to cut costs, began by attacking wages. They also ruthlessly suppressed miners' attempts to organize. Unless a strike could be maintained within a whole competitive district, workers in one field suffered while their non-striking neighbors profited through the increase of business. Powderly did not see this, even though organizer McPadden repeatedly urged: "Organize all these districts solid and when they move let it be with a united district and if they must strike stop every mine that sends a pound of coal to the eastern Market." [163]

Possibly the most regrettable result of the Maryland strike was the cleavage between the Consolidation Coal Company miners in and around Frostburg and the employees of other companies further down the creek near Lonaconing. The stampede to work of the George's Creek Coal and Iron and New Central men was taken as a betrayal of trust. The division continued, and throughout the remainder of the nineteenth century prevented any lasting or effective organization under the various labor unions which succeeded the Knights of Labor.

Chapter 15

Decade of Uncertainty (1882-1892)

At the end of the great strike of 1882 there was every reason to assume, as did the *Coal Trade Journal*, that "the miners' organization along Georges Creek is defunct, so far as any power of controlling the affairs of the companies is concerned." [1] The effect of concentrated company opposition to the union during and after the 1882 strike was catastrophic (Table 13). The organization did not entirely disappear from the Maryland coal region, but its membership there shrank from 1,910 in mid-1882 to 497 in the summer of 1883.

In an interview with the Baltimore *American* the president of the Consolidation Coal Company made it clear that the companies would not tolerate "the least interference by the Knights of Labor or anyone else," and that they would abolish the five-tons-a-day limit established by the Knights.[2] Whether or not Consolidation did exact a promise to resign from the Order as the price for reemployment, the significant loss in K. of L. membership was in those locals made up chiefly of Consolidation miners: Frostburg, Ocean, Pompey Smash, Eckhart, and Allegany.[3] At Barton, Lonaconing, and Westernport the companies apparently did not share the deep-seated antipathy of Consolidation toward the Knights of Labor. This difference in attitude, however slight, accentuated the division between the Consolidation and the lower-creek miners. However, the latter did not

Table 13. Knights of Labor membership by locals in Maryland mining region, on dates indicated, 1879–1885

Local	Oct. 1 1879	Oct. 1 1880	Oct. 1 1881	July 1 1882	July 1 1883	July 1 1884	July 1 1885
718 Frostburg	264	253	209	312	42	28	35
767 Barton	95	171	111	148	17	17	24
775 Ocean	60	66	79	80	21	20	29
848 Lonaconing	197	446	359	585	201	84	120
853 Pompey Smash	69	47	88	94	25	39	28
1205 Westernport	166	408	253	288	93	34	28
1311 Eckhart	52	87	97	116	—	—	—
1319 Alleghany	56	73	106	104	40	19	46
1476 Mt. Savage	—	80	46	—	—	—	—
1513 Frostburg	—	107	153	161	45	20	—
1581 Frostburg	—	19	23	22	13	—	—
	959	1,757	1,524	1,910	497	261	310

Source: Knights of Labor, Proceedings of the General Assembly, 1879–1885.

enjoy any special concessions concerning wages or hours; the mining rate for the entire region was 50 cents a ton, and twelve hours were required of all mine workers.

In contrast with Clearfield, where the district organization of the Knights collapsed during 1883,[4] the Order in western Maryland had at least a good basis for revival when and if favorable circumstances should occur. Although the miners who returned to work after the strike were said to be "not very attentive to their lodge meetings," [5] nine local assemblies remained active (the smallest having thirteen members), and district officers did not concede defeat. They secured an organizer's commission for William B. Baird, one of the victimized brothers from Frostburg, and would have succeeded in making him a general organizer for the bituminous fields if Powderly had not felt obligated to retain McPadden "now that the operators are persecuting him." [6] Baird himself was philosophical about post-strike conditions, writing to Powderly that he would try to organize the seventy German strikebreakers still remaining at Eckhart.[7]

While it seemed possible that the Knights of Labor might recover their strength in the Maryland mining district, the grand secretary picked up a rumor of unwelcome competition. Early in January 1883 he wrote to Powderly that the miners of Maryland, Pennsylvania, and Ohio were "agitating a National organization." He added with a note of pettiness: "Do you think Maryland could be induced to take part in such an enterprise? I still hold a few hundred dollars [in relief funds] for that region and will continue to do so until I *know*." [8]

William B. Baird signed the address to bituminous miners of Pennsylvania, Ohio, and Maryland proposing a national miners' organization and calling for a convention at Pittsburgh on May 15, 1883.[9] He was not present at the convention, but P. A. Crowe, another blacklisted miner from George's Creek, served on a number of committees and was elected secretary of the new Amalgamated Association of Miners of the United States.[10] This organization was destroyed by the Ohio strikes of 1884–85.[11]

Some Maryland miners, therefore, as early as 1883 were presumably agitating for a national union separate from the Order. Others who were devoted to Powderly and sentimentally attached to the Knights of Labor initiated the movement for what eventually became National Trades Assembly 135, a miners' national within the Order. At the first regular session of the K. of L. general assembly following the unsuccessful Maryland strike, Christopher O'Brien, delegate from D. A. 25, presented a resolution calling for the appointment of a practical miner as general organizer of all miners in the United States.[12] The resolution was passed, and soon after the convention, the secretary of D. A. 25 asked Powderly to commission Rees D. Rees, one of the leaders of the 1882 strike, as a "general organizer of miners." [13] Powderly's reply is not contained in his papers, but he apparently did not grant the request. Rees was one of the few genuinely blacklisted and forced to leave Maryland.[14]

Partly because of a brief national slump in business, the year following the great strike was not a good one for this mining region. Employment was slack; between September 1882 and June 1883 the miners worked only three full months.[15] The mine inspector reported that only a portion of the trade lost to competing regions had been regained, causing some mines to remain closed throughout the season and others to be worked part-time.[16] There was a covert defiance of the companies' decree against limiting the daily output per man. In the autumn of 1882 the *Coal Trade Journal* reported a movement afoot among the miners to limit production to five tons a day and thus spread the work to some of the unemployed,[17] but in spite of this measure many mining families were in desperate need by midsummer.[18]

Dissatisfaction with the 12-hour day was general. Hitherto strict working hours had been imposed only upon those paid by the day; they did not affect miners, who were paid by the ton. During the regime of the Knights of Labor, "when two men working together had loaded five cars [10 tons], their day's work was complete, no matter if it had been accomplished in three hours, and they left the mine." [19] Under the company "manifesto" of 1882 the men were not to be allowed to restrict their labor to five tons a day. Presumably the intention was to keep them in the mines for 12 hours so that they would have to produce more. The inspector commented that the "magnanimous spirit displayed by the employers" in "acceding one hour" during the winter months "had the effect of wiping out a great deal of the bitterness" after the strike.[20] However, he went on to say, "I am sorry to be compelled to state that at a number of the mines the dinner hour is not adhered to—only half of this time being allowed. This breach of contract is causing great complaint throughout the coal field, and no doubt will cause trouble in the near future."

Trouble was a luxury which the miners could not afford at this point. Instead of striking, they went to the legislature in 1884 to secure a local ten-hour law for labor in and about the mines of Allegany and Garrett counties. Although the length of the working day throughout most of the state was ten hours, the miners' measure met with unexpected opposition and passed the House of Delegates with a bare majority.[21] As finally approved on April 8, the statute included no penalties for violation. It provided only that labor employed in the mines should work ten hours a day beginning at 7 o'clock in the morning, and did not prohibit the working of extra hours should a miner so desire and so contract with his employer.[22]

The companies proceeded as if the law had not been passed. Except in the winter months they exacted 12 hours a day, with an hour off for dinner.[23] Not all operators adhered to the 7 A.M. beginning hour. Some were open from 6 A.M. to 6 P.M. in the summer and from 6:30 to 5:30 in the winter.[24] The law therefore brought no improvement, for, as one miner said, "that has been the way we have been working for the last few years." [25] Men who lived far from the mines in order to obtain cheaper rents added considerably to their day by the time consumed in walking to their jobs; "as early as four o'clock in the morning the dancing lights of their lamps can be seen ascending the hills as the men trudge to their labor." [26]

What, actually, did the operators gain by what seems to have been a punitive ruling on the number of hours miners should remain underground? If their intention was to get more than five tons a day out of each man, they most signally failed; in 1884 daily production per man dropped to three and one-half tons.[27] It is obvious that the miners were sharing the work and resisting attempts to clear out surplus labor. Only by taking on extra men could the companies have shipped the record tonnage of 1884. In this year the region sent out 700,000 tons more than it had in 1881, and for the first time surpassed production in 1873.[28]

Even with their self-imposed limitation, Maryland miners managed to survive in 1884, and were better off than those in some other locations. Miners at Palatine, West Virginia, a newly-developing competitive area, made only $1.00 a day in that year.[29] Knights of Labor statistics showed that miners at Frostburg and Lonaconing earned $1.50 a day,[30] and the Maryland Bureau of Industrial Statistics credited them with $1.75.[31] The daily wage of $1.50 at Clearfield [32] represented greater effort, for the rate there had dropped to 40 cents a ton on April 1.[33] So far as miners' annual earnings were concerned, the average in western Maryland in 1884 was $445;[34] in the Pennsylvania bituminous fields the average was $369.20.[35]

Soon after the Clearfield operators announced their wage cut, speculation arose as to its effect on the George's Creek region. Early in June, coal trade gossip hinted that the Maryland companies, acting in concert, intended to equalize Cumberland wages with Clearfield wages.[36] However, Maryland operators waited until the end of the season to post notice of a reduction to 40 cents a ton, effective on the first of December.[37] As usual, the operators pleaded the need to meet Clearfield competition, although Clearfield production had actually lagged behind that of the Maryland coal field during 1884.[38] Having heard this excuse for years, the men replied, "The argument presented in reference to the competition between the George's Creek and Clearfield regions, in our opinion is threadbare." [39]

On July 1, 1884, Knights of Labor membership in the surviving locals of D. A. 25 had dropped to 261.[40] Nevertheless, the mine workers issued a protest to their employers beginning, "We, the Miners of the George's Creek Region, through the Knights of Labor represented, beg leave to make the following reply to the manifesto recently submitted by you." [41] This was scarcely the way to approach company presidents who had steadfastly refused to recognize the Order as an interested party in labor disputes. The companies ignored both the answer prepared by the men and the suggestion of arbitration which they put forward. On December 1 work went on as usual in those mines which were still open, and the wage reduction was accepted "with great discontent." [42] It is possible that the operators had more than Clearfield competition in mind and used the 10-cent reduction to end voluntary sharing of the work. The *Engineering and Mining Journal* remarked apropos of the Maryland wage cut, "The struggle is simply one to settle the question whether, as formerly, the Knights of Labor or the Companies are to run the business. . . . The Companies are acting as a unit, and are making a contest . . . for the recognition of a principle not sufficiently established during the last strike." [43]

The men did not represent themselves as Knights of Labor in 1885 when they sent out a circular requesting the employers to restore the 50-cent mining rate and to comply with the law requiring a ten-hour day in the mines.[44] Signing themselves "Your Employees," the petitioners asserted that it was impossible for them to sustain their families at the existing wage rate.

Some company representatives described this leaflet as the work of agitators, or "persons seeking to make political capital of the state of affairs in the region."[45] Reliable sources predicted that the operators would ignore the wage demand.[46] As to hours, the grand jury for Allegany County had reported to the judges of the circuit court that although it appeared from the evidence that nearly all the coal companies in the county required their employees to work eleven hours, the ten-hour law "wholly fails in any of its provisions to afford the relief for this grievance."[47] *

The officers of D. A. 25 may not have known that the employees were going to issue a circular, but they wasted no time in coming to the defense of the document and its sponsors.[48] After inquiring carefully as to the origin of the wage request, the executive board of D. A. 25 announced that it came from miners, not politicians. Furthermore, the executive board agreed that wages in the region were too low. Scoffing at the operators' statement that the miners would not strike because their union was broken up, it declared: "We admit that the miners' union did get a good shaking up. . . . But the miners' organization did not fall with the strike. It still lives and is very healthy."

The board expressed a hope that there would be no strike, and offered to arbitrate the controversy with the employers. At the same time, it rather inconsistently boasted that it could call a strike at any time it wished, and advised miners who were not members of the Order to stop making appeals in print, and to "act." Although there had been a general business depression in

* This defect was "corrected" in 1886 by an amendment providing that any violation of the ten-hour law should be deemed a misdemeanor punishable by a fine of not more than $50. Maryland *Laws* 1886, Ch. 303.

the United States since 1883, the K. of L. district officers at Frostburg accused the companies of going beyond normal operating economies; they had defied the law in regard to hours and weight, and had not made proper reductions "in rent, blacksmithing, etc., in proportion to reduction of wages."

The employees who signed the circular paid no attention to the exhortations of the district executive board, but after May 1 addressed another communication to the operators, asking for a committee to meet them and discuss wages and other matters.[49] Their spokesman was Hector Cochrane, who had been a member of the Maryland House of Delegates in 1884. S. S. Lee replied that representatives of the operators' association would willingly meet a committee of miners from the companies in the region "provided the members of that committee are not Knights of Labor. We wish it distinctly understood that we will not directly or indirectly hold any conference with men identified with that organization, which has caused so much trouble in our region, both to employers and employees." [50] Cochrane promptly replied that his group would "not be influenced by any organization or Knights of Labor, but appointed by the miners at an open meeting." [51]

The operators' postponement of the meeting for a day gave the conference an inauspicious start, for the miners' committee members who were on hand on May 15 lost two days' work instead of one. Although the discussion was conducted with great decorum, and the men were allowed to present their arguments at length,* the companies in the end made no concession. C. F. Mayer, president of Consolidation, speaking on behalf of all management, said the men should realize that the corporation presidents had their interests at heart and never discussed any labor subject "without a careful consideration of its bearing

* One delegate's plea for higher wages was taken almost verbatim from the Knights of Labor statement of April 28 quoted in part above. At least two of the men on the committee had been active Knights of Labor in 1882, for their names are on the blacklist published by D. A. 25 after the strike in that year. Cochrane himself was a member of the K. of L. in 1884.

upon our men, and an earnest desire to make the decision just and fair to them."[52] Consolidation alone had lost over 200,000 tons in orders by not reducing its wages to the Clearfield level in April 1884. None of the Maryland operators cut wages until December 1884, "after a loss to the companies," because they had made previous contracts with the expectation of paying 50 cents a ton and did not think it fair to the men to make a reduction until those contracts were filled.

The men were under a misapprehension, said Mayer, if they thought that the Clearfield miners worked shorter hours. The ten-hour system "did not prevail in Clearfield in actual practice," and many mines worked "11 hours and more when necessity required it." Moreover, Clearfield could send out coal all year round, but Maryland producers, depending on ocean transportation out of Baltimore, found many of the ports to which they shipped closed by ice early in the winter. It was therefore necessary to do as much business as possible in the summer, even if this involved long working hours.

At the beginning of the conference, Hector Cochrane, speaking for the miners, made it clear that they were not threatening to strike.[53] And although out-talked on every point they raised, the men did not strike, accepting the verities of economics as expounded by the operators. The chief of the Maryland Bureau of Industrial Statistics commented in his first official report, "When I visited the coal fields in 1885, there was an exceedingly feeble body of the K. of L., and not the least intention to strike manifested. The memory of the great strike of 1882 was continually referred to in conversation, and the misery and poverty produced in the district was vividly described."[54]

For the George's Creek men, the failure to wring any concession from the operators in the 1885 conference was merely the latest disappointment in a whole series of reverses and rebuffs which began with the strike of 1882. The legislature had given them an hours law without teeth[55] and a mining school without adequate funds,[56] and had failed to pass a revised ventilation bill both in 1882 and 1884. The law-makers had also ignored a petition from the mining region asking for repeal of the coal and iron police act.[57]

As if the legislative fiasco were not enough to bear, the miners had been confronted with two wage reductions in a two-year period in spite of the fact that some of the companies pleading poverty had paid dividends.[58] The executive board of D. A. 25 estimated that at 40 cents a ton the average miner would earn about $320 in 1885. He would pay $50 a year for rent, $12 for the doctor ("whether needed or not"), and roughly $8 for smithing, leaving him $250, "or an average of about 69 cents per day with which to feed and clothe his family." [59] Incomes in 1885 were actually a little higher than this estimate. The Bureau of Industrial Statistics reported that the average paid miners for the year was $356.[60] But even at the higher figure a miner had only 95 cents a day with which to pay for clothing, shelter, food, medical care, and the expenses incident to his work—blacksmithing at one cent a ton and the oil and cotton for his lamp. A gallon of oil cost 55 cents and lasted only half a month.[61]

William Baird, district organizer of the K. of L., writing to Powderly from Frostburg, described the state of affairs in D. A. 25: "There is a tendancy to reorganize here at present. But the financial condition of the people will not admit of it. The average wage earned here last year was less than one dollar per day. . . . I reorganized L. A. 2061 located at Meyersdale Pa. it cost me $3.00 to go there & they could not pay it, & that is the condition of the majority of our people." [62]

Labor in Somerset County, Pennsylvania, was much worse off than that along George's Creek or in Frostburg. The average daily wage of a miner working full time in Somerset County in 1885 was about 78 cents. Meyersdale, located in this county, fared better, for a man there could make about $1.40 a day "at present rates with very hard work. Present rates are 25 cents per ton of 2,240 pounds. Allowing a miner to dig and load six tons—which is a good day's work—would give him $1.50. His expenses would be 3 cents for smithing [less than in George's

Creek], 4 cents for oil, and 10 cents for powder, and to keep up wear and tear of tools, boots, and mining clothes would make a considerable item." [63] The use of powder, forbidden in George's Creek, enabled Meyersdale men to get out more coal in a day.

In view of conditions prevailing in the bituminous fields, it is not surprising that as early as September 1885 a new miners' organization, the National Federation of Miners and Mine Laborers, had attracted sufficient members to set up its District 3 covering the Irwin, Clearfield, Garrett, and Meyersdale sections of Pennsylvania; Elk Garden, West Virginia; and George's Creek, Maryland.[64] The Federation had been organized at Indianapolis in July without the participation of Maryland delegates.[65] Many K. of L. members apparently did not view the Federation as a rival organization, and in the eastern soft coal regions membership in the Knights of Labor and the Federation overlapped. John R. Paisley, of Clearfield, a member of the Federation's executive board for District 3, was at the same time active in organizing for the K. of L., as indicated in Powderly's correspondence. C. V. Ogden, one of the Maryland delegates to a Federation convention early in 1886, was also listed as district worthy foreman on the letterhead of District Assembly 25, Knights of Labor. William Baird, as master workman and organizer for the Knights of Labor in western Maryland, proceeded to drum up business for the Federation at the same time that he was trying to revive the lapsed locals of the K. of L. in his district. His activities caused some consternation in Local 1205 (Westernport), which asked Powderly for a ruling on Baird's decision that the district assembly could "force any Local Assembly or individual member thereoff into any organization—(Federation of M. & M. L. the organization in question)—not under the jurisdiction of the K. of L. when it was a benefit to the D. A." [66] Powderly, who did not like trades unions, scrawled on the bottom of the local's communication: "The case you refer to me is the most peculiar one I have ever had referred to me. Neither D. A. or L. A. or a member of the Order can be forced into any other order. It should be our aim to build up the K. of L. If we [do] that we will have enough to attend to without making any effort to drive our members into any other organization."

It was not necessary to drive K. of L. members into the Federation. The introduction of the joint conference between operators and miners at Pittsburgh in December 1885 [67] was an innovation which seemed promising to workmen in many coal regions. Early in 1886, District 3 attempted to bring the joint conference idea to George's Creek and its competitive fields. At a meeting in Cumberland on February 18, miners' and laborers' delegates adopted a scale of prices per ton for mining coal in the various regions after March 8: [68] Clearfield, 50 cents; Irwin, 60 cents; Garrett, 40 cents; Meyersdale, 40 cents; George's Creek, 50 cents; Elk Garden, 50 cents; and Pocahontas, 45 cents. The scale represented an across the board increase of 10 cents a ton. To justify the new prices the executive committee of District 3 issued a circular to the various company presidents, asserting: "At the present prices your employees cannot, or at least do not, receive sufficient compensation for their labor to enable them to live as becomes Christians and civilized beings, possessed with a desire to feed, clothe and support their families in a respectable manner." [69]

The Federation's circular also pointedly addressed the operators on the need "to prevent the fatal consequences of ruinous competition, the result of which falls heavily on the laborer." It urged the companies to form a combination with broader purposes than strikebreaking. By agreeing to raise wages on a uniform scale, they could also uniformly raise the market price of coal, thus losing nothing in the way of profits. "By amalgamating the workers of various competing districts," wrote the union officials, "we set you an example which we trust you will not be slow to follow."

The communication closed with a request to the company presidents to meet the union officers at Cumberland on March 1 if they were "of the opinion that the scale prepared above is not uniform or equal."

The scale adopted at Cumberland was to be submitted for approval at the national convention of the Federation in Columbus, Ohio, on February, 23.[70] Maryland miners sent representatives to the national convention, but the Maryland companies did not participate.[71] Because of their absence, "The Columbus convention met and adjourned, but Maryland appeared to have no share in the disposition of affairs." [72]

The executive board of District 3 met again on March 2 and issued an address to the mine workers within its jurisdiction saying that the failure of the operators to meet them at Cumberland on March 1 was an admission "that the scale was uniform." [73] All miners who did not receive an advance to the union scale were to stop work on March 8. The Cumberland companies, acting together under the same agreement as in 1882, at once issued a statement that competition from other regions had dropped coal prices to such a level that an advance of wages was "absolutely impossible." [74] The suspension called by the Federation began on March 8 and lasted for ten weeks in the Maryland region. The miners here were the first in District 3 to seek settlement with their employers.[75] Although it lacked dramatic incidents, this strike is of particular interest because it was the first officially ordered district-wide stoppage in the seaboard coal fields.

Regardless of the role played by the Federation, there is no question that Knights of Labor participated in the strike in Clearfield and in Maryland, and in the latter region the strikers were described by the Baltimore *Sun* as being Knights of Labor under District Assembly 25.[76] In April Knights of Labor members from western Maryland solicited strike support in Baltimore, where they received funds from other labor unions as well as from K. of L. assemblies.[77] Shortly thereafter, the treasurer of D. A. 25 distributed relief money "to the several lodges of the order in this region." [78] Furthermore, the general assembly of the K. of L. in 1886 granted $4,000 "to relieve the wants of the victimized brothers and their families in . . . D. A. 25.[79] In view of Powderly's hostility to the Federation, it is unlikely that he would have allowed such a grant unless Knights of Labor were significantly involved. One must therefore discount statements that after the strike of 1882 the K. of L. died out in this coal region and was superseded by the Federation of Miners and Mine Laborers.[80]

The Federation for a short period took precedence over the K. of L. in this part of Maryland but was no more able than its predecessor to combat the organized operators. The very scope of the organization in District 3, extending throughout the competitive bituminous regions supplying the eastern seaboard, gave the operators in any single field an opportunity to play on the suspicions of their men. Maryland miners were warned against Pennsylvania organizers as early as 1873, and during the 1882 strike the Cumberland *Times* had asserted that Clearfield miners would "never" aid George's Creek men.[81] In 1886 the Cumberland operators professed to believe that the delegates to the Federation did not properly represent their employees and had no authority to ask for an advance in wages.[82] Accordingly they intended to pay no attention to the Federation circular.

Another result of the interstate and interregional nature of the organization was "the most disastrous evidences of weakness and want of harmony among the workmen." [83] The executive board clearly had not contemplated what would happen if some companies granted the advance and their men dug coal while others in the district remained idle. This happened first at Meyersdale, where the operators granted a raise to 40 cents a ton.[84] The Meyersdale men resumed with the understanding that they would donate the amount of their raise for strike relief.[85] Next three small operators in Clearfield advanced their rate, and their men went to work.[86] At Broad Top some employers gave the advance and "agreed to collect from their men ten cents a ton for the support of those still on strike." [87] In their first official communication to their men after the onset of the strike, the large Clearfield corporations pointed out that Pocahontas miners "have been at work regularly and without intermission at the old wages" and that Virginia shippers were taking business away from Pennsylvania.[88] In Maryland, at a mass meeting at Lonaconing on April 14, the men voted to ask miners receiving the new rate at Meyersdale and Clearfield either to hold their output to what it was before the strike or to stop work.[89] In another resolution they demanded "total suspension of all mines that ship coal to the Eastern markets, whether they are paying the advance or not." [90] In other words, so much coal was going to market from within District 3 that the so-called suspension was a farce. Whether digging at the old rate, as in Pocahontas, or the new rate, as in Meyersdale and parts of Clearfield, the employed miners were damaging the cause of the strikers, and a few fortunate operators had as much business as they could handle.

When it became apparent that the Federation had by no means cut off the supply of bituminous coal to the eastern seaboard, some of the George's Creek men expressed dissatisfaction with the management of the strike and the personnel of the executive board of District 3. Nevertheless, at a mass meeting in Lonaconing late in April, the majority voted to endorse the district president and committee and to remain away from work.[91] Several days after this mass meeting, the Reverend Father Schmitt, pastor of the Catholic church in Frostburg, made public a letter of appeal to the coal operators, asking them, in view of the suffering in the region, to arrange a conference with the miners.[92] According to the Cumberland *Daily Times* of May 4, Father Schmitt also advised all members of his congregation to go back to their jobs at the old price.

C. F. Mayer's letter in reply to Father Schmitt appeared in full in the *Daily Times* of May 6. Making the most of the outside leadership of the strike, Mayer wrote that "unworthy men" had influenced local miners to demand an advance in wages. He said that Maryland miners received higher wages than those in any other region, and were paid for all of the coal they mined; operators in other fields demanded overweight. Along George's Creek the men received their wages regularly and in cash, "not in part or in whole in goods, as is done in some places." Furthermore, the district was "the acknowledged garden spot of all the soft coal regions of the Atlantic seaboard basin." Its miners worked in the best vein known. Work was apportioned fairly, and the men's earnings "would in many cases have been larger but for their own urgency to have friends or companions permitted to share the work with them."

As to Father Schmitt's criticism of the companies for not meeting their employees, Mayer responded that they would not confer with a Federation committee composed mostly of members from competing regions, but would willingly talk with any "proper representative body of their men." He stated that only a small minority of George's Creek men were Federation members.

The letter convinced many in the Frostburg area that they did not want to continue the suspension. Miners from Consolidation, Borden Shaft, and Blaen Avon had initially instructed their delegates to vote against the strike,[93] and had reluctantly gone along with the majority. Now, plagued by continuing suspicion of the outside leadership, and agreeing with Mayer that their situation was far from intolerable, Consolidation miners at Hoffman were the first in the region to indicate that they wanted to break with the Federation and send their employer "a delegation that the company will hear."[94] As a result of the Hoffman miners' efforts, and in spite of union protest against interference from the clergy and others, the operators granted a conference on May 15.[95] The companies apparently did not specify that the employees' delegation must consist of men without labor organization ties. The committee named in the Cumberland *Daily Times* of May 15 included Garrett Burns and John Stoddart, blacklisted members of the K. of L. after the 1882 strike; C. V. Ogden and Thomas Murphy, both currently officers in D. A. 25 of the Knights of Labor; and Isaac Anderton, a member of the Federation committee which distributed relief money in the Frostburg area. Ogden and Stoddart were also prominent in local Federation activities. Possibly the omission of union members from the competitive regions satisfied the operators that this delegation was a "proper representative body of their men."

The discussion proceeded with great politeness. Hector Cochrane, speaking for the miners, opened by thanking the companies "for their presence and cheerful attention to our solicitation." [96] In closing, S. S. Lee, speaking for the operators, said that the company presidents appreciated the reasonable attitude of the men and "deeply regretted that they could not give the advance asked." During the interval between these expressions of courtesy, the operators disposed of the wage raise as economically unfeasible, and explained in logical fashion why they could not grant reductions in smithing charges, semi-monthly pay, and a 7 A.M. beginning for the working day.

After the conference, miners' delegates met at Lonaconing and by a close vote of 32 to 31 decided to return to work.[97] Consolidation's Eckhart men did not wait for the decision, but applied for their tools and went into the mine. The rest of the Consolidation men at Ocean, Hoffman and Pompey Smash were quick to follow their example.[98] The situation of 1882 had been reversed; this time it was the Frostburg men who "stampeded" for resumption, leaving the Lonaconing strikers without adequate support. Lonaconing remained briefly the center of agitation for continued suspension. The executive board of District 3 met there on May 21, with delegates from Clearfield, Broad Top, Irwin, Meyersdale, Elk Garden, Newburg, Fairmont, and Clarksburg "to inquire into the causes that were brought to bear upon the men influencing them to resume work, and see if they cannot be persuaded to hold out a little longer." [99]

As in 1882, work began with a diminished labor force after the strike. During the conference of May 15, C. F. Mayer had declared: "The [low] price paid per man is due to the fact that there are too many men in the region. In 1882, when we resumed, we put in men enough, we thought, to make good wages. Since that time I have, on every trip, been applied to to put men to work. That resulted in having 150 men more than we

[Consolidation] need. . . . The excess of men is the real difficulty."[100] As of May 21 the operators had announced reductions totaling 372 in the following mines:

Consolidation mines (202)		*Other mines (170)*			
Allegany	50	Borden Shaft	24	New Detmold	30
Hoffman	75	Blaen Avon	10	Kingsland	35
New Hope	36	Miller	15	Jackson	15
Eckhart	11	'Coney	2	Hampshire	4
Ocean	30	Koontz	35		

Although the company presidents stated that they had not dropped any men "for spite,"[101] Consolidation, for example, with an excess of only 150 men according to Mayer, cut its force by 202. This action gives credence to a miner's statement published in the Cumberland *Daily Times* of May 22: "I can produce men who asked what they were blacklisted for and the reason given was that they had advocated the ten cents advance, and one was told that all the agitators of this strike would be weeded out."

Bitterness about the reduction in force led to a vote at Barton for resumption of the strike if the employers refused to take back the discharged men.[102] The Frostburg men, too, became agitated when confronted with 202 lost jobs at Consolidation mines and 34 more at Borden Shaft and Blaen Avon, and called a meeting to discuss the blacklist and the prospect of striking again to sustain Clearfield and Elk Garden.[103]

Talk of resuming the strike ended abruptly with settlement in Clearfield through "mutual concession on the part of both operators and miners" on May 26.[104] Clearfield miners agreed to return to work at 40 cents a ton on condition that they be guaranteed just weight, a checkweighman on every tipple, abolition of the store order system, cash payment on the fifteenth of every month, and the privilege of making purchases where they pleased. Maryland miners had enjoyed these concessions (with the exception of checkweighmen) for years.

In Cumberland, the company presidents remained adamant about the number of men they would employ, insisting that they did so for the benefit of the remainder, who could then earn good wages. To ease the situation, several operators "agreed to contribute money to pay the fare of the miners and their families to other coal regions." [105] Some of the men tried unsuccessfully to persuade those who had jobs to share the work with the unemployed.[106] This had been a long-established practice among these miners, but earnings figures for 1886 given by the Maryland Bureau of Industrial Statistics lead us to assume that there was less sharing of work than in 1885, when the average annual wage for a miner was $356. Although the Bureau remarked that the strike materially lowered incomes for the year,[107] in 1886 the average miner earned $381.24 [108] despite a drop in production of roughly 276,000 tons.[109]

After these passages of woe, it is a relief to arrive at a period of prosperity and good feeling. In the spring of 1887 the Maryland operators and their major competitors formed a "combination of the steam coal producing regions of the Atlantic seaboard . . . to harmonize and protect this large interest, which for several years has been in so deplorable a condition." [110] The Seaboard Steam Coal Association parceled out tonnage among seven districts, the largest allotments going to George's Creek and Clearfield.[111] Participants in the "soft coal pool" agreed to carry out the business-sharing arrangement for at least five years.[112] The immediate result of the combination was a rise in the price of coal, enabling the producers in both Clearfield and Cumberland to grant wage increases. After March 1 the mining rate in Maryland became 50 cents a ton, and the wages of other mine workers were restored to the 1884 level.[113] At Clearfield also the price for mining after March 1 went to 50 cents a ton. Clearfield men gave up their semi-monthly pay period but retained the other concessions granted after the strike of 1886, understanding, however, that their agreement with the companies would last only as long as the 50-cent rate and the monthly payment system continued in Cumberland.[114]

The unsolicited pay raise gave point to the operators' often-repeated contention that the uncontrolled competition among the seaboard fields had held wages down. The very fact that the increase was given voluntarily created an entirely new atmosphere in the mining region. The mine inspector made particular mention of "the kind feelings and amicable relations that exist here now between employer and employee," going on to remark that "while labor troubles have distracted business in many coal districts in other states, our people have been blessed with peace, contentment and prosperity." [115] The payroll for George's Creek miners in 1887 was almost $1,500,000, "or a fraction over six hundred dollars *per capita*." [116] As early as June the editor of the *Frostburg Mining Journal* remarked on the increased prosperity of the region,[117] and his observations were confirmed in a letter from "A Consolidation Miner" who wrote of the additional business evident in the bustle of Saturday night shopping expeditions, the improved appearance of the children at Sunday school, and the increase in church donations.[118]

Drawing its conclusions from the mine inspectors' reports for both states, the Frostburg newspaper stated that during 1887 the average miner in Maryland made more than twice as much as the average miner in the Pennsylvania bituminous fields.[119] To date this was "the halcyon year of the coal trade," [120] marked by an increased production of over three-fourths of a million tons [121] and the employment of several hundred more miners than in 1886.[122]

In 1888 the companies hired 275 additional miners, and the mine inspector felt it incumbent upon him "to call attention to the great evil of employing too many miners—a policy which, in my opinion, has more to do with promoting labor troubles in and about the mines than all other things combined." [123] This warning was gratuitous, for he concluded his report by saying, "Peace and prosperity prevail throughout this region." [124]

It was evident that the miners no longer had an agreement to restrict daily production, and a man's income depended to a large extent on what he wanted to produce (assuming equal turns and an adequate supply of cars). "You will be surprised," wrote the mine inspector, ". . . with but one uniform price paid per ton for mining in this region, at the difference in the earnings of the miners in and at the different mines."[125] The eight largest employers paid "per capita" amounts as follows in 1888:[126]

Company	*No. miners employed*	*Amount paid miners "per capita"*
George's Creek Coal & Iron	282	$799.53
American	200	735.58
Potomac	166	631.23
Maryland	292	598.26
Consolidation	882	594.75
Borden	200	544.58
New Central	165	526.44
Maryland Union	155	352.53

Miners and laborers at a few of the George's Creek mines discussed restricting the output of coal to five tons a day per man and adopting a nine-hour working day.[127] The movement proved abortive, undoubtedly because the miners, "excepting those of one company, *did not average five tons per day*" in 1887, and it would be ridiculous for them to urge restriction to a figure they had not achieved during the best year of the region's existence.[128]

Despite a drop in production, Maryland miners in 1889 were by far the best paid in the bituminous regions of the seaboard-supplying states. The average day's pay in Maryland was $2.45; in Pennsylvania, $1.93; in Virginia, $1.53; and in West Virginia, $1.86. On the basis of the average number of working days in 1889, Maryland miners received an annual average of $497.35; Pennsylvania miners, $405.30; Virginia miners, $436.05; and West Virginia miners, $414.78.[129]

In 1890 bituminous operators "all round" had a prosperous year, even exporting coal to Europe because of labor troubles in Wales.[130] Production in the Cumberland region for the first time in its history exceeded 4,000,000 tons.[131] By 1891 the number of miners employed had risen to 3,141,[132] and the average annual wage for a miner in Maryland was $635.[133] The "relations of capital to labor" were "pleasantly maintained," and the chief of the Maryland Bureau of Industrial Statistics was encouraged to remark: "I believe that the folly of strikes and lockouts was so forcibly impressed upon both miners and operators during the long and bitter struggle from March 15th to August 25th, 1882, and again in the spring of 1886, that it is probable no such disastrous methods will again be adopted, until all other means of adjusting labor difficulties shall have been tried and found to absolutely fail." [134]

The pooling agreement expired in 1892, and companies outside of Maryland set low prices for their coal.[135] Because the George's Creek companies refused to cut the standard price of their product or to reduce the rate per ton paid to their miners, their shipments fell off by about one-third of a million tons.[136] Nevertheless, the atmosphere of well-being continued throughout the year, and the Bureau of Industrial Statistics was able to report: "The general feeling existing between employers and employees throughout the region has been of the kindliest nature. No material differences have arisen, and all have worked harmoniously." [137]

During most of this period of relative prosperity the Maryland mine workers were affiliated with the Knights of Labor National Trades Assembly 135, the miners' association authorized by the Knights at their general assembly in Hamilton, Ontario, in October 1885.[138] After the unsuccessful strike of 1886, the Cumberland region had little to do with the Federation of Miners and Mine Laborers, and D. A. 25 with its twelve locals and 584 members continued to act as the employees' organization for the remainder of that year.[139] However, Maryland did send delegates to St. Louis for the convention of "coal mining Knights of Labor" which set up plans for National District Assembly 135.[140] Soon after the St. Louis meeting, the *Journal of United Labor*, official organ of the K. of L., spread the word that mining locals should attach themselves to the new national body.[141]

Early in January 1887 a convention of miners from the regions shipping to the eastern markets met in Cumberland and organized sub-district 2 of National Trades (or District) Assembly 135. M. F. Flannigan, formerly secretary of D. A. 25, was elected master workman of the sub-district, which reportedly had 5,000 members.[142] The mining locals of D. A. 25 gave up their charters and merged with N. T. A. 135 "in order to better strengthen their craft." [143] D. A. 25 thereafter became a non-mining assembly with headquarters in Cumberland. The Maryland miners continued to think of themselves as Knights of Labor throughout a two-year struggle between the K. of L. and its rivals for control of the coal miners of the United States.[144]

The United Mine Workers of America, organized in January 1890,[145] sought to establish labor harmony by combining N. T. A. 135, Knights of Labor, with the National Progressive Union (successor to the National Federation of Miners and Mine Laborers). Maryland miners took no part in setting up the UMWA,[146] and did not send a representative to its first annual convention in February 1891.[147] The Cumberland region was designated as District 16 of the new union, which had "the blessing of both Gompers and Powderly" and was affiliated with both the American Federation of Labor and the Knights of Labor.[148]

On March 4, 1891, President John Rae of the United Mine Workers issued a call for a convention of miners and operators producing for the eastern seaboard trade to "fix rates for mining, correct inequalities and establish such conditions as will do away with needless agitation and establish greater harmony between employer and employed." [149] When the meeting opened at Cumberland on March 18, twenty-five members were present from Clearfield, Meyersdale, and other Pennsylvania and West

Virginia regions, "but the George's Creek miners were conspicuous for their absence." [150] P. H. Penna, official UMW organizer from Indiana and later vice-president of the UMWA, accounted for the nonattendance of George's Creek men by saying that they were preoccupied with politics and refused to involve themselves with the union.[151]

Among the delegates were two former K. of L. organizers from Clearfield—John White and John R. Paisley. A new and significant figure on the scene was William B. Wilson from Blossburg, Pennsylvania. Only one operator from any of the competitive regions bothered to send a representative. This was the Berwind-White Coal Company, the largest and most important in the Clearfield region. Finding himself alone on the management side, the Berwind-White delegate asked to be excused. The UMW president, who attended the meeting as a delegate representing Clearfield, described the absence of the other operators as "an organized movement to ignore the miners."

The George's Creek operators sent word to the convention that they would not attend because they had no inequalities to adjust, and knew of no agitation among their employees.[152] They asserted that they were already paying a higher wage than any other competitive field, and their men were well satisfied.[153]

The refusal of the George's Creek miners to send delegates surprised both workers and operators in other states.[154] However, J. A. Kinsloe, editor of a labor newspaper published at Phillipsburg, Pennsylvania, explained that "the condition of Cumberland miners as regards earnings, is very much superior to that of any other region in this competitive district." [155] In Kinsloe's opinion it was "the low average earnings in the Pennsylvania fields, and the doing of a large amount of dead work, such as setting timbers, etc., for nothing, which . . . caused the agitation in the coal fields . . . for a year past." On the day following the convention the Cumberland *Daily Times* published an editorial which commended the miners for "holding aloof from the conference of the delegates from competitive soft coal districts," saying: "The Pennsylvania and other miners who have grievances should adjust them at home, with the operators locally interested. It was a mistake to suppose that men elsewhere not affected by the trouble of Pennsylvania would take upon them the responsibility of interference."[156] The persistence of this narrow point of view in Maryland was to prove fatal to later UMW organizing efforts.

Chapter 16

An Episode of Sympathy: 1894

In the spring and summer of 1893 the United States experienced a financial panic marked by the failure of hundreds of banks, the insolvency of business enterprises numbering into the tens of thousands, and the bankruptcy of the country's leading railroads. Without contracts from factories and transportation interests, coal fields in many parts of the nation were hard hit.

The Cumberland region, however, seems to have been a fortunate exception. It shipped roughly 4⅓ million tons in 1893, an increase of more than 300,000 tons over production in 1892.[1] Early in 1894 the George's Creek Coal and Iron Company paid a semi-annual dividend of three dollars a share; [2] the Consolidation Coal Company declared a dividend of two dollars a share; [3] and the Maryland Coal Company's annual report showed a net profit of $94,728, "out of which was paid two dividends . . . of 5 per cent, amounting to $94,180." [4]

Under these circumstances, one is inclined to view with a good deal of skepticism the arguments that the Maryland coal company presidents advanced during a conference with their men on March 22, 1894. For the first time since 1878 the operators invited their employees to send a committee to meet them to discuss wages for the coming season.[5] "The presidents were very distinct and emphatic in their statements that forty cents for mining and a corresponding reduction in all other labor was the utmost they could do; that it was entirely useless to consider anything but that, and much as they regretted this necessity, they were bound to frankly tell the men what the true situation was, and that this was in consequence their ultimatum."[6] The "true situation," according to C. F. Mayer, who presided at the meeting, was that business was generally depressed; the market price of coal had fallen; the tariff on coal was probably going to be reduced; and increased competition from Nova Scotia coal was imminent.[7]

Although the operators thus pleaded that they could not sell their coal at a price high enough to warrant paying more than 40 cents a ton, the *Coal Trade Journal*, edited from the management point of view, asserted on January 31: "The cost per ton paid for mining is on a very low basis . . . , the mere matter of wages per ton for mining being but a small fraction of the very moderate cost of delivery at the point of consumption." As to that resurrected bugaboo, Nova Scotia coal, the Baltimore *Sun* remarked in an editorial on May 11: "The frequent loading of vessels in Baltimore with cargoes of coal for Canada should quiet the apprehension of the Maryland miners and mine-owners. These citizens have been in a state of disquiet ever since it was proposed to place coal on the free list, and the more recent proposition to place a tariff duty of 40 cents a ton has only measurably restored quiet. Notwithstanding the fact that coal is still under the McKinley bill protection of 75 cents a ton, the mine-owners have seized the occasion to reduce wages. This reduction is not only supposed to prejudice the miners against the democratic party, but it saves a large amount of money to the mine-owners."

During the conference the men asked for reductions in house rent, smithing, fuel, and doctors' fees; and the company presidents stated that these charges would "be put at what they were when the rate of mining was forty cents" on earlier occasions. In accepting the wage cut in 1894, the miners indicated that they expected the companies to carry out this promise and to restore the 50-cent mining rate as soon as business improved.[8] However, at least one of the miners' delegates expressed doubt that the operators would make concessions. He said that he had worked in the region for many years "at prices ranging from $1 a ton on down" and that during all that time, rent, smithing, and doctors' fees had remained stationary.[9] The presidents' statement was in fact misleading, to say the least. When the companies reduced wages to 40 cents in 1884, they did not adjust the "off-takes," and as we saw in Chapter 15, miners' committees meeting with the company representatives in 1885 and 1886 cited this failure as one of their major grievances.[10] In 1894, regardless of what they had implied at the conference, the companies retained the blacksmithing charge of one cent per ton.[11]

Although there may have been a few hundred members of the Knights of Labor, the Maryland mine workers (numbering more than 4,000) were, for all practical purposes, unorganized. But even before the companies announced the pay cut, the editor of the Lonaconing *Review** surmised that most miners in the region favored organization. He himself believed that a union, no matter how small or how weak, would be "the nucleus from which may spring at the most unexpected moment a sturdy movement to do manly battle for the elevation of the workers." [12]

The unexpected moment came in less than a month. During March the national officers of the United Mine Workers of America laid the groundwork for a strike in all the soft-coal regions.[13] In many coal fields outside of Maryland (and particularly in Ohio), a "vast army of half idle mine workers" was in desperate straits.[14] President McBride of the UMW hoped that a suspension of operations would bring about depletion of surplus stocks of coal, allowing operators subsequently to set higher prices and pay higher wages.[15] Therefore, he called a general strike for "higher wages and improved conditions" to begin on April 21.[16] Concurrent with the announcement of the forthcoming national suspension, M. F. Flannigan and other members of "the local organization of the National Union" in western Maryland

* The Lonaconing *Review* was purchased in 1884 by Henry W. Schaidt and James A. Maxwell. Schaidt and his brother Harman (editor of the paper) had been officers of L. A. 848 of the Knights of Labor, and both had been blacklisted after the strike of 1882.

requested that a UMW organizer be sent to the region.[17] William B. Wilson was dispatched to Frostburg, where he began working as an organizer and lecturer on April 19.[18]

Wilson, a Pennsylvania miner who had been active in the labor movement since 1877, was a member of the executive board of the UMWA.[19] Almost at once he encountered resistance to his organizing efforts, particularly having to contend with insinuations that he was trying to foster Pennsylvania coal interests.[20] He himself told of the "pronounced opposition of the great bulk of the newspapers, which were dominated by the coal operators." Articles purporting to be written by miners charged him with being a coal operator from the Clearfield region. This allegation, said Wilson, "was a lie out of whole cloth. . . . It might have had the effect of discrediting me with the miners, which the coal companies desired, but for the fact that I was able to show the hard, bony callouses on my hands that could only come from the long-continued use of the pick." [21]

When the Maryland miners struck in 1882, they were already firmly organized under the Knights of Labor, and had almost unanimously agreed that they had reason to act. In 1894 the UMW at first found it difficult to build up an organization. The men were unwilling to join a strike in which they had no personal stake and in which the only initial issue was that of sympathy with miners in other soft coal regions. Notices calling on the Cumberland mine workers to quit at noon on April 21 were torn down almost as soon as they were posted.[22] The Maryland men knew from long experience that they benefited materially during miners' strikes in other regions. Furthermore, for eight years they had had no real organization, and had become accustomed to approaching their employers through delegations from individual mines. The operators, too, were in the habit of dealing only with their own employees and would have nothing to do with the employees of other companies. By the end of April 160,000 miners throughout the country were on strike,[23] but Wilson and the other organizers for the UMWA had little success in getting the Maryland men to suspend operations or to join the union until the end of the first week in May.[24] After a second week of concentrated organizing effort, including many secret meetings, the union claimed a thousand members in the region.[25] The *Journal of the Knights of Labor* commented on May 10: "To the astonishment of [UMW] President McBride, the miners of western Maryland have laid down their picks."

The greatest obstacle in the way of the union organizers was the reluctance of the men of the Consolidation Coal Company to have anything to do with the strike movement.[26] Consolidation workers had bitter memories of 1882 and were unwilling to risk so much again. Furthermore, shortly after that debacle, the company began to acquire a reputation as the best employer in the district.[27] In 1894 it was "undoubtedly true that any miner in the region would resign his present place were he assured of employment at the mines of this company. The mine superintendent is spoken of as fair, impartial and considerate of the men's feelings, there is no 'pluck me' store, the mines are ventilated in the best method at present known, the men employed there are a carefully selected and especially intelligent class,* and in many other ways this company's mines present inducements that appeal with great force to the steady, reliable miner." [28] The men of Consolidation's Eckhart, Allegany, and Hoffman mines held meetings and resolved "that having given their promise to work at the forty cent rate they were in honor bound to do so." [29] Without their participation, the strike could not succeed, and the mining population would be split into irreconcilable factions.

Those miners who struck early in May gave no reason.[30] The Baltimore *Sun* said on May 10, "They are silent as to what they have suspended for, and this is owing to the fact that they are going to leave the settlement of terms to the general conference to be held in Cleveland, Ohio, the 15th of this month, where

* However, without specifying which company he meant, "George's Creek Miner" wrote in the *United Mine Workers' Journal* of May 18, 1893: "If a man is not a company suck or a particular friend of some official he will have to leave without work. This is a plain black-legging region if ever there was such a thing. If you stand in with some of these mine bosses you can do as you please at some of the mines."

they expect the operators of this region to meet the delegates sent from here." The *Journal of the Knights of Labor* took it for granted that the mining rate was the issue, and remarked, "Even sleepy old Maryland has responded to the cry for living wages, after having suffered another reduction to 40 cents a ton a few weeks ago." [31] A Consolidation official undoubtedly reflected the views of most of the operators when he characterized the walkout as unreasonable and unjustifiable because no grievance had been alleged, and no increase of wages demanded. He felt that the strike was "a desperate attempt" on the part of the union to get a foothold in the region, and bluntly announced that his company, some of whose Ocean miners had struck, would make no concessions and would not treat with union representatives.[32]

Although other companies continued to lose men, Consolidation, except for its Ocean mine, kept on working at nearly full strength.[33] By May 11 all mines in the area were on strike except the Consolidation's Eckhart, Allegany, and Hoffman in the vicinity of Frostburg. The entire body of strikers, with reinforcements from Elk Garden, West Virginia, converged on Frostburg to make a concentrated effort to have these mines suspend.[34]

The mass movement of strikers against the Consolidation miners on May 12 failed, chiefly because of the unexpected action of the wives, daughters, and mothers of the Eckhart employees.[35] The women accompanied their men to the mouth of the mine and threatened to go underground and dig coal themselves if the men were compelled to quit. The Baltimore *Sun* felt that if it had not been for this action, the Eckhart miners would have given in. "The women are much more fearless than the men. The strikers talk boldly, but the working miners have nothing to say. The women are outspoken. They take an aggressive stand, and their husbands are bound to sustain them. If the strikers lay hands on them, there will be war." [36]

Somewhat abashed by the publicity, one of the Eckhart women wrote to the editor of the *Frostburg Mining Journal:* "Some of us fear that it does not look ladylike to the outside world for us to get mixed up in a brawl with men, but it really could not be helped. The company treats us well and gratifies all our wishes that they can (I mean especially those of us who live in Consolidation houses,) and we think it is too bad that our men should be driven from their employment to please people away off somewhere else." [37]

Expecting another great push at Eckhart on May 14, the sheriff increased his posse to 100.[38] At Lonaconing the superintendents of the American, Maryland, New Central and George's Creek Coal and Iron companies applied for injunctions "restraining the United Mine Workers of America from trespassing on the Companies' property and from interfering with their employes," and the American Coal Company notified the sheriff that it would hold Allegany County responsible for any damage done to company property by "mob, strikers or others." [39] Many of the union organizers were absent at this time, having gone to Cleveland for the national conference which was expected to settle the strike, but injunctions obtained by the George's Creek Coal and Iron Company were served against at least two UMW officials at Frostburg on May 12.[40]

The Cleveland conference ended on May 17 with no agreement to settle the strike.[41] Most companies (including those from Maryland) refused to meet with the union delegates to secure a living wage for the men and a higher price for coal with a fair profit to the mine owners. The absence of the management representatives made settlement impossible, and UMW president McBride said that the miners would continue the strike until the operators agreed on a wage scale acceptable to the union.[42] By this time operators in the Pennsylvania anthracite region and the nonstriking West Virginia mines were pouring their product into the market and defeating the UMW's program of planned scarcity. Violence erupted along the railroad lines which carried coal west to the former customers of the depressed mining areas.

When news of the conference failure reached Maryland, the strikers renewed their efforts to convert the Consolidation men. Overt hostility to nonstrikers took a more threatening turn. Up to this time the strikers had limited their "intimidation" to arguments, taunts, and liberal use of the insulting term "blackleg." Although McBride had urged delegates at the Cleveland meeting to use only peaceable means of persuasion, on May 17 houses of Consolidation miners were stoned, shots were fired, and some women and children were injured by flying glass.[43] On the morning of May 19 strikers began to accost men going to work at Eckhart, being careful to stay on the public roads and away from company property.[44] The sheriff, who had about 200 deputies on hand, made several arrests on charges filed by Eckhart employees. Meanwhile the Eckhart women continued their harassing tactics, although their sisters at Carlos, being of a different mind, stoned miners who were going home after a day underground.[45] In the face of stronger demonstrations by strikers, the Frostburg city council increased its police force and requested saloon-keepers to close at 9 P.M. until the region had quieted down.[46]

Concerned about the continuous interference with its miners, th Consolidation Coal Company secured court orders against W. B. Wilson and about 80 others, enjoining them from continuing their "unlawful assemblies"

> on or near to the said mines aforesaid, or on or near the paths and approaches leading to the mines of the said Consolidation Coal Company, and from continuing to assemble there with intent to forcibly prevent the miners working for the said Consolidation Coal Company in its said mines from going to work therein, and from then and there using threats, menaces, shouts, show of force and offers of violence to interfere with, prevent and stop the miners working in the mines of the Consolidation Coal Company from continuing their daily labor therein, and from using any of the aforesaid means to prevent the said Consolidation Coal Company, through the service and labor of said miners working for the Consolidation Coal Company, from continuing its business of mining and shipping coal from its said mines, until the further order of this Court.[47]

On May 23, more than a month after they had stopped work, the Maryland strikers issued a statement of the terms on which they would return. At a mass meeting in Midland they adopted a resolution calling for 50 cents a ton, no blacklist, the reemployment of all former workers, abolition of company stores and the docking system, establishment of the fair turn, and permission to employ checkweighmen. They also appointed a committee to ask the operators to meet them in Cumberland on Saturday, May 26.[48] It is not clear whether the companies were actually approached, but it is certain that no meeting took place in Cumberland on May 26, when the attention of the region focused on the Allegany County circuit court. Wilson, who had addressed the Midland meeting, had been summoned with three others to answer a charge of contempt. "It was alleged that in the face of the injunction these men had continued to intimidate and turn back the men" at Consolidation's Allegany mine on May 22.[49]

Probably the most significant aspect of the 1894 strike was the extent to which the companies called upon the courts and invoked legal processes. Because a Maryland statute of 1884 had made it more difficult to obtain an indictment for conspiracy,* management seized upon the injunction as a weapon against organized labor. The trial of Wilson and his co-defendants was the first in Allegany County for alleged violation of an injunction.[50]

Wilson admitted that he had addressed a meeting near Allegany mine on May 22 and that he had "stated in substance that if the strike failed the men at work in the Consolidation mines would be largely responsible for it." [51] However, he denied having anything to do with the subsequent so-called intimidating behavior of the strikers toward the working Allegany miners.[52]

* Chapter 266 of the Laws of 1884 amended the Code of Public General Laws relating to crime and punishment by adding the following section entitled "Conspiracy": "SEC. A. An agreement or combination by two or more persons to do, or procure to be done, any act in contemplation or furtherance of a trade dispute between employers and workmen shall not be punishable as a conspiracy, if such act committed by one person would not be punishable as an offence; nothing in this section shall affect the law relating to riot, unlawful assembly, breach of the peace, or any offence against any person or against property."

In summing up, counsel for the defense argued that Wilson's co-defendants had not taken part in an unlawful assembly but had been engaged in a lawful and proper movement to persuade the Allegany miners to quit work in order to better their condition. On behalf of Wilson, counsel asserted that his client had not violated the injunction but had been brought into court on the strength of his activities before the injunction was issued. Counsel for the complaining Consolidation Coal Company held that Wilson's co-defendants had clearly violated the injunction and charged that Wilson himself was the original cause of the labor trouble in the mines. Argument was concluded on May 29.[53]

Judges Boyd and Hoffman discharged Wilson because the evidence failed to show that he had been guilty of contempt. They did, however, express surprise that "non-residents have been permitted to come into [this region] and cause a large majority of our own miners to be confronted with the prospects of suffering and trouble that generally accompany idleness or suspension of work." "It is clearly shown," they continued, "that the respondent, Wm. B. Wilson, is largely responsible for the present condition of affairs, and it is equally true that whatever may be his accomplishments many of our miners are his equals in intelligence and know far better than he does what is for their interest. He is responsible, in part at least, for the loss of wages already sustained and for such lawlessness as has already occurred in this county. The evidence shows acts done or incited by him that would have unquestionably made him amenable to the law and liable to punishment if they had occurred subsequent to the service of the injunction upon him." [54]

Concerning the meeting at Allegany which led to the contempt trial, the court said: "No one will deny the right of miners or others to meet together in a lawful way and discuss their grievances, whether real or imaginary, in a proper way. But holding a meeting so near the place where these men worked, and with the avowed intention of reaching the men in that mine, together with the use of such language as is admitted by Wilson, has the appearance, at least, of intimidation and threats."[55]

Although the court found Wilson's co-defendants guilty of interfering with men going to work at Allegany mine, it released them on payment of costs. The two judges concluded their opinion with a warning: "These respondents were probably not informed as to how far they could go without violating the injunction, but the defendants in this case and those similarly situated can now inform themselves, and they may rest assured this court will not permit its mandates to be trifled with as long as they are in force."[56]

While the Wilson trial was in progress, the presidents of the American, George's Creek, and New Central companies posted notices ordering their miners to remove their tools by May 29. These companies intended to close their mines unless enough men were willing to work at 40 cents to make operation practical.[57] At this stage of the strike one operator said to a Baltimore *Sun* reporter, "The miners have placed us in a position where we could not grant their demands even if the market would afford it."[58] On May 29 other companies ordered strikers to take out their tools, but the bulk of Consolidation men continued to dig coal at Eckhart, Allegany and Hoffman, enabling the company to ship 3,000 tons a day.[59] A few men were also at work in the Union mine of the Black, Sheridan and Wilson Company.

Consolidation's Ocean miners, idle since early in May, requested a meeting with the company president, who refused to confer with men who were no longer in the service of the company and therefore had no claim on him. Mayer said that the mining rate had been fixed at forty cents a ton because of the depressed conditions throughout the country. The situation had not changed. He had made contracts on the basis of forty cents, and would quote prices for the balance of the season on the same basis.* "The Consolidation Coal

* However, an official of an unnamed coal company, "in speaking of the contracts made by the companies to supply coal for the year . . . said that each contract not only had a clause releasing the company from

Company has not made a single dollar in consequence of the temporary advance [in the price of coal], and has lost money in some cases where we deemed it wise to give our customers anthracite or foreign coal at higher prices to hold their trade." [60]

As it became increasingly apparent that the Consolidation men were not going to be moved by any appeals, a break in the strike at Lonaconing seemed possible, and the superintendents of the George's Creek and Maryland companies had their mine horses shod in anticipation of resumption.[61] To forestall surrender in the mines along George's Creek, the strikers made an all-out attempt on June 4 and 5 to stop operations at Hoffman and Eckhart.[62]

On June 4, in a letter headed "Strictly Confidential," president C. F. Mayer of Consolidation wrote to the former governor of West Virginia, A. B. Fleming, who was closely connected with the Fairmont coal interests, to warn him that "if the more desperate element [of the UMW] gets control they may send a body of men from Ohio into your region to drive out your men. . . . While this is a desperate move, it is not very unlikely to be undertaken. It is already measurably being done in our region with our own men, namely; the Consolidation Coal Company's. About 140 strikers from Davis' Elk Garden property in West Virginia and from Meyersdale, Pa., came over last night into our region and prevented the bulk of my Hoffman mine from going to work this morning. There are about 140 of these foreigners in the streets of Frostburg, and we apprehend trouble when our men return from work this evening. If, in a region like ours, which has practically been in the hands of the Sheriff, with a good strong posse for more than a month, such an effort has been made, you will realize that it is not unlikely that it might be attempted with you. Any special trains you may want to concentrate the men you need, when called for will be given to you at once." [63]

On June 5, aided by the foreigners from Elk Garden and

---

obligation in case of strikes, but a provision for the increase of price to be paid for the coal should wages be advanced." Baltimore *Sun*, May 14, 1894.

Meyersdale, about 300 Maryland strikers met at the entrance of the Eckhart mine to prevent the men from going in. They pushed past the sheriff and his eighty deputies and drove the would-be workers back.* When the sheriff cautioned them against interference, "He was told by one or two of the strikers to go to a place where coal is not needed to make things hot, and that he had no business around the mines anyway." [64]

Immediately after this demonstration, the sheriff and the state's attorney telegraphed to the governor that they could no longer preserve peace and order and requested him to dispatch troops.[65] On the afternoon of June 5 he sent the fourth and fifth infantry regiments to Allegany County. That same evening the Eckhart miners joined the strike, influenced in part by indignation over the call for the militia.[66] This left about 150 working in a region normally employing about 3,200 (plus other labor).

Soldiers had never before been used to keep order in the Maryland mining region; in 1877 the military had been in the county to subdue railroad and canal strikers. Considering the relatively mild nature of the incidents which preceded the call for aid, it is open to question whether the sheriff's request was justified. The first serious episode—dynamiting of a miner's house—occurred only an hour before the arrival of the troops.[67] However, in view of what was going on elsewhere, the sheriff and the state's attorney may have prevented serious disorder by acting when they did. By June 8, in the coal strike alone, armed forces had been called out in Iowa (where Pinkertons also were used), Colorado, Illinois, Indiana, Ohio, West Virginia, Pennsylvania, and Maryland.[68] The state of Indiana sent fifteen companies of militia and a squad of artillery to deal

* A number of persons arrested in connection with this episode came to trial in November. It was alleged that they had acted "in a riotous, tumultuous and disorderly manner to the consternation and alarm of the citizens," and that they had stopped persons on their way to work by "threats, intimidation, and the use of violent language." Women witnesses testified to experiencing a "feeling of dread and apprehension." Cumberland *Weekly Civilian*, November 16, 1894.

with striking coal miners; West Virginia had eleven military companies on the line of the B & O to protect coal train movements; and in Ohio, where strikers were blowing up railroad bridges, 1,700 troops were on duty.[69] Therefore superintendent Randolph of the Consolidation Coal Company may have had reason for saying that the militia arrived in George's Creek just in time to prevent bloodshed.[70] A good deal of blood had flowed in Pennsylvania in clashes between strikers and deputy sheriffs during which some fifteen or twenty persons were killed and more than fifty wounded.[71]

According to the order issued by the state adjutant general, the militia in Allegany County was merely to help the civil authorities keep order; the sheriff and his deputies retained the power of arrest. Nevertheless, the strikers were generally indignant at the presence of the troops, insisting, "We do not intimidate; we simply argue the case and explain the situation to the men who insist on working."[72] The *Sun's* Frostburg correspondent, however, commented, "The explanation and arguments . . . are not generally appreciated, and one of the men at work said the polite requests not to work have the flavor of the favor asked by the polite highwayman when he asks a man to oblige him by handing over his valuables. It would not be healthy to refuse."[73]

On June 6 word came from Cleveland that a meeting of UMW delegates had decided to allow miners in each region to settle their differences with their own operators.[74] William B. Wilson, who had been present, explained on his return to Frostburg that the delegates had finally concluded that they could not arrange a national conference with the bituminous operators and must therefore try to negotiate separate agreements in marketing areas or districts. Company officials in most regions would be asked to confer with union members, but the George's Creek operators, because of their strong objections to the union, would be asked to meet their men not as organization representatives but merely as employees.[75]

By this time the partial suspension had cost Maryland participants an estimated $132,000 in wages, which the *Coal Trade Journal* described as "a pretty good price to pay for sympathy," inquiring, "When did Ohio and Clearfield do as much for Cumberland?" [76] Although a joint committee of miners and operators agreed on a compromise wage scale for Pittsburgh, the Hocking Valley, Indiana, and Illinois, this decision would not affect seaboard fields.[77] Maryland strikers said they had already tried three times to confer with their employers, but without success.[78] The companies felt that their settlement with the men had been made in March, when the 40-cent rate was offered and accepted. A typical management response to the UMW's "Columbus compromise" came from Consolidation's superintendent, who told the Baltimore *Sun:* "To grant the increase would be giving in to the organization, and that could not be considered. It would bring no end of trouble." [79]

On June 13, miners of the George's Creek Coal and Iron Company, the Maryland Coal Company, and the New Central Coal Company, appointed delegates to ask for a conference at Cumberland on June 15.[80] The operators took the position that since they had discharged their employees, they had none to meet, and in no event would they confer with a committee of members of the United Mine Workers.[81] However, the peace feelers of these miners at Lonaconing led to rumors of a break in the strike. Demonstrations by strikers in this portion of the mining region resulted in the serious beating of one miner, and troops moved into town to protect the few employees of the George's Creek Coal and Iron Company at the hours when they entered and left the mine.[82] The women of Lonaconing "were quite liberal with their jeers and epithets" directed toward the soldiers.[83] During the Lonaconing disturbances, organizer Wilson and D. E. Dick, "a prominent local agitator," were arrested on charges of criminal conspiracy and held in $500 bail for appearance at the October term of court.[84]

At Frostburg, where the troops were first stationed, no disorder occurred during the first ten days of the occupation. But on the evening of June 15, as workers came out of the Union mine, they encountered "about sixty women, armed with tin pans, buckets, base-ball bats and babies." The women called the miners "blacklegs" and "sneaks" who were taking the bread out of the babies' mouths. For a while it looked as if the men would undergo rough handling, but deputy sheriffs broke up the crowd.[85]

With the guarantee of protection from both civil and military authorities, those Consolidation men who had been forced to strike "through intimidation and threats" began to return to work.[86] No company showed any signs of yielding to the terms demanded by the miners on May 23, and none would promise full reemployment without a blacklist. Consequently, except for Hoffman, Allegany, and Eckhart mines, the region remained idle until late June in what the *Sun's* Lonaconing reporter described as the "most peculiar" strike in the history of the Maryland coal field.[87] News of resumption at Elk Garden was disconcerting [88] because the West Virginia miners had been among the most stalwart advocates of the suspension. A group of about 175 Maryland strikers who went to "argue the case" with the men going into the Elk Garden mines found the West Virginia authorities prepared to resist their interference.[89]

The long-expected break at Lonaconing came on June 27. On June 26 the George's Creek Coal and Iron Company put a crew to work cutting timber on a large tract of land near one of its mines. Rumors spread at once that it was clearing land to put up buildings for imported labor.[90] The company, which did not deny the rumors, began hauling lumber to the cleared site and at the same time told a number of employees living in its houses that they must either go to work or vacate.[91] Next day the men began to report at the mine. By June 28 the strike was over,[92] and by June 30 the troops had left.[93]

There was not even the face-saving dignity of a pre-resumption conference. After the rush of George's Creek Company men to get their tools, the men of the Maryland, American, and New Central companies, also at Lonaconing, decided to abandon the strike, and sent delegates to notify their superintendents. In regard to the blacklist, these companies asserted that they "would retain their privilege to employ whomsoever they pleased." [94] The men most active in the strike did not get work, and the Baltimore *Sun* of June 29 stated, "It has been made a condition of employment in some cases that withdrawal cards from the Local Union United Mine Workers be shown to the superintendent." At least 400 men failed to obtain jobs immediately after the strike.[95] Consolidation at first refused to open its Ocean mine, whose men had been among the first to quit, but within two weeks had resumed there with about one-third of its usual complement.[96]

C. F. Mayer wrote to ex-governor Fleming at Fairmont, West Virginia, on July 7: "You are probably aware that quite a number of the most turbulent men in the Georges Creek Region, largely composed of the young men who thought they would not be given work, have left the Region. I am told that quite a number of these have gone to Fairmont to work for your different companies there. They are not a very desirable set of men to have in your service, and I therefore advise you of what I hear, in order that you may be on the lookout for them. They are the class of men who had far better go to Ohio and countries of that kind, for they are sure to give trouble if they should get among your people." [97]

Consolidation's superintendent Randolph denied that motives of revenge dictated his company's employment policy, but said the management had selected workers with whom the company could be on "friendly terms." [98] At Lonaconing, too, the *Sun's* correspondent reported shortly after the strike, "Perfect harmony and good-will seem to prevail on all sides." [99]

However harmonious may have been the relations of the Frostburg and Lonaconing companies with their men, it is obvious that conditions at the extreme lower end of the creek were not entirely satisfactory. A disgruntled Westernport miner wrote to the *United Mine Workers' Journal* in midsummer: "We are getting the usual amount of attentions from the bosses that generally falls to the lot of unorganized and defeated miners. All the petty tyrannies and abuses that ignorant, half-paid bosses can devise they are heaping upon us. In the mines that are working there is not the semblance of a square turn kept. The good and obedient miners get all the cars and the best places to load them in, while the bad boys that give back-talk are put in deficient places, and not even allowed to shoot their coal.* We get 40 cents per ton and 2,240 pounds for a ton, saying nothing about the fun we have in laying our own turns and switches. Oh no, you don't get paid for any dead work here. And if you do any kicking the bosses will tell you to take your sticks out, as there are over 500 blacklisted men patroling this region we have to keep very quiet." [100]

In the middle of August, Consolidation rewarded its faithful by extending "to all men who worked 24 days or more between May 12th and June 26th, 1894, and to all others who entered the service of the company on or before June 16, 1894, and worked continuously thereafter . . . rent free and coal free from July 1st, 1894, to April 1st, 1895." [101]

After the resumption, certain matters still remained to be settled at law. Men arrested in June awaited trial at the October term of the Allegany County circuit court. These included a number charged with unlawful assembly, as well as William B. Wilson and others charged with conspiracy. Several persons were found guilty of unlawful assembly, but the court reserved sentence, saying: "I propose to allow these parties to go free, upon their own recognizance, until the next April term of court. Some of them, if not all, are married men and have families, and have been without work for a long time, and to impose the payment of the costs of the case upon them at this time would be merely subjecting them to confinement in the jail for a period of not less than sixty days. That would be a hardship upon their families, who would be thrown upon the community for support during the winter." [102]

* This remark seems to indicate that the writer of the letter was a newcomer to this region, where shooting coal was not customary until after 1900.

In the case of *State* v. *William B. Wilson*, the warrant was alleged to be missing, and the state's attorney asked for continuance to allow the sheriff to search for it.[103] The Wilson case is particularly interesting because its central figure later became the first Secretary of Labor for the United States, serving under Woodrow Wilson. Early in his career in the cabinet, Secretary Wilson found it expedient to make a public statement concerning his involvement in this coal strike. This testimony, coming from a responsible government official, sheds some revealing light on company tactics in Maryland.[104]

Wilson said that his arrest for conspiracy was contrived in such a way that for at least twenty-four hours none of his friends knew that he was in jail at Cumberland. At the end of the strike he had returned to Pennsylvania, but was unable to find employment except in the winter time, and then only in small mines. For two years he was under bond waiting for his trial on the conspiracy charge. The case never came to trial.*

According to Wilson's account, "The Coal Companies used the usual 'big business' methods to break the strikes; thugs were employed to follow at my heels wherever I went. They went so far as to employ one man, paying him $500, to kidnap me and hold me a prisoner in an abandoned farm house, in order that they might be able to announce to the miners that I had sold out and left the region, hoping thereby to discourage the strikers and break the strike. The plan did not succeed solely because one of the men who was asked to help carry it into effect sent me a warning to be on my guard."

After the strike of 1894, union activity in the Maryland coal field practically ceased until 1899, when organizer William Warner began work in the region.[105] The United Mine Workers' organizing effort in 1894 did not unite the region, but on the contrary divided it. The division unfortunately persisted at least into the first decade of the twentieth century.

* This is similar to McPadden's experience in the strike of 1882.

Chapter 17

The Grasping Corporations (1895-1900)

The failure of the general strike in 1894 almost destroyed the United Mine Workers. Its national membership declined to 8,000, and its treasury was almost depleted. Regions such as Maryland, involved only by sympathy, failed to send delegates to the organization's annual convention in 1895.[1] The Maryland operators' massive resort to the courts, coupled with their appeal for military assistance, indicated that they had not regarded the union lightly, but had considered it a real threat to their oligarchy. After 1894, by playing upon the region's distrust of out-of-state agitators, by taking advantage of an anti-union faction among their men, and later by resorting to aggressive action, the Cumberland coal companies consistently pursued a policy of squelching further union movements. Moreover, by refusing even to experiment with the joint-conference procedure which had succeeded in establishing a pay scale and preventing much labor trouble in central Pennsylvania, they eliminated Clearfield as a basis of comparison for George's Creek mining rates and working conditions, and began to cite competition from the nonunion fields of Pocahontas, Fairmont, and Somerset as their principal reason for not raising wages. In all of these regions large coal combinations took over most of the business at the beginning of the twentieth century.

After 1900 Maryland was somewhat of an anomaly in the coal-mining world. So small in area and in the size of its labor force as to be almost insignificant beside its giant neighbors, Pennsylvania and West Virginia, Maryland continued with pick mining in an era of machines, producing a specialized product for a limited market. Indeed, it is doubtful whether the principal companies could have survived on their George's Creek output alone. The Big Vein had been exploited in a scandalously wasteful manner, and many mines were already worked out. Although improved methods made it possible to recover a good deal of the "lost" coal, the more farsighted corporations looked to the day of eventual depletion and began to acquire property elsewhere. The George's Creek Coal and Iron Company found it advantageous to extend its operations to West Virginia, and Consolidation began its expansion by purchasing a large tract of coal land in Somerset County, Pennsylvania.[2]

Whether by accident or by design of the operators, the grievances of the Maryland miners during the period 1895–1900 were less serious than those of fellow-workmen in the competitive seaboard fields. As we have seen, the mining villages were not company towns in the usual sense. Although working conditions in some mines left a good deal to be desired, the largest companies, Consolidation in particular, continued to improve conditions underground by the installation of more efficient ventilating fans, pumps, and haulage equipment. Work was steady, and, in the light of publicity given to wages in West Virginia, for example, Maryland miners could congratulate themselves on being relatively well paid. Ignoring the union scale, the employers periodically granted voluntary increases. This policy was customary in nonunion fields, where operators would give a small advance in the mining rate to counter a larger raise negotiated by the union in a region nearby.[3]

Were it not for their attitude toward organized labor, the Maryland operators might have been described as benevolent paternalists. As it was, confirmed unionists viewed the mulishness of the company presidents as despotism, and one staunch organization man declaimed during the strike of 1900:

The grasping corporations had the audacity to say,
If you all deny the union and foreswear your liberty,
We'll give you all a chance to live and die in slavery.[4]

However, for several years after the strike of 1894, workers in the Maryland field had no particular incentive for uniting in their own interest. In April 1896 the employers, enjoying a foretaste of McKinley prosperity, raised the mining rate to 45 cents a ton.[5] Beachley cites this increase and the adoption of the semi-monthly pay system as "typical instances of the consideration which Consolidation has striven to maintain for the welfare of its employes." [6] (The semi-monthly pay system which went into effect in 1896 was required by law and had been forestalled by the Maryland operators for decades.) Although the miners had no protective association, Beachley implies a working agreement among the companies, for he refers to the increase as being given by Consolidation "acting in concert with all the other companies in the Georges Creek, Cumberland, and Clearfield Regions."

The five-cent advance did not restore wages to the 1893 level, but it provided better living than at Clearfield, where the concession of 45 cents was accompanied by reduction to half time after April 1.[7] George's Creek miners were also measurably better off than those at Fairmont, West Virginia, where the Monongah Coal & Coke Company paid 30 cents a ton for screened coal.[8] Even when wages at Fairmont went up two and a half cents a ton in the spring of 1896,[9] the resultant 32.5 cents was not calculated to arouse jealousy in the Cumberland region. The mine inspector reported harmonious relations between operators and employees in Maryland, and added: "The fact that the operators are continually besieged for work by miners from other regions seems to show beyond question that employment here is attractive and desirable. This is further attested by the corresponding fact that of the number who go from here to other regions for work, a large majority return after a short experience." [10]

In the second half of 1897 the UMW, rejuvenated by funds from the American Federation of Labor, called a strike in the soft coal fields of Ohio, Indiana, Illinois, and parts of Pennsylvania, West Virginia, and Kentucky, to enforce the union wage scale for those states.[11] This strike did not affect most of the seaboard producers. The Fairmont region of West Virginia, paying wages varying from 27 to 40 cents a ton (weighted average 34.5 cents a ton),[12] was involved because it shipped a large portion of its product westward and should theoretically have paid the scale set by the UMW for the midwestern states. With substantial aid from the A. F. of L., the Mine Workers carried the twelve-week strike in the central competitive field to a victory which added considerably to the prestige and membership of the UMWA. The operators agreed to a wage increase, an eight-hour day, abolition of company stores, recognition of the union, and annual conferences with the UMW representatives.[13]

Maryland producers, working full time throughout the strike, had not been afraid that their miners would quit, "as they realize that business conditions have not as yet improved sufficient to justify an advance in their wages." [14] Soon after the strike began, the Cumberland region reported the largest tonnage in its history, with coal trains running night and day.[15] Under these circumstances the Maryland workers were not disposed either to organize or to strike, although they did contribute money for the support of the strikers.[16]

Thanks to the war with Spain, Consolidation men worked nearly the year around in 1898.[17] The corporation reported large government orders of George's Creek coal for navy use in the Pacific as well as in the Atlantic. As a result of the activity of the coal trade, "the condition of the whole George's Creek region has been unusually prosperous, nearly all the companies having increased their outputs. The growth of business has not only given the miners more regular employment, and consequently improved their condition, but has tended to enlarge the population of the region." [18]

During the war Consolidation installed a mining machine at Ocean,[19] thus introducing "mechanical mining methods into a field heretofore exclusively hand worked." [20] By the end of 1899 Consolidation and the Sinclair Mining Company had eight machines in operation, but mechanization did not here pose the threat to mine labor that it did in two competitive regions. During the period covered by this narrative, machine production in Maryland was a minor part of the total. But at Elk Lick in Somerset County, Pennsylvania, the introduction of machine undercutting established a mining rate of 20 cents a gross ton for that type of work, and a miner from the region was quoted as saying that even the 20-cent rate was "fictitious." Five cents were taken off for overweight and five cents for the company store, leaving the miner 10 cents a ton for powder, oil, and subsistence.[21]

The *Frostburg Mining Journal* reported that the Clearfield operators were "rapidly drifting into machine mining. They have been paying 45 cents per ton for pick mining and 22½ cents for machine work—in some cases 25 and 27 cents." [22] Late in May, Berwind-White, the largest Clearfield Company, announced a voluntary raise to 50 cents a ton for pick mining.[23]

When the Maryland operators cut wages to 40 cents a ton in 1894, they had promised restoration of the old rate as soon as business conditions permitted. Five cents had been given back in 1896, but the men felt in 1899, in view of the exceptional prosperity of the country, that the return of the remaining five cents had been overly delayed. The *Coal Trade Journal* during the first half of the year consistently pointed to a continuing demand for soft coal on the eastern seaboard. When Berwind-White announced the increase at Clearfield, the Cumberland miners confidently expected a corresponding raise. At the end of June a number of persons "on the inside" predicted that the operators would give a voluntary increase of five cents, which would come as a "surprise" on the next payroll.[24]

However, by the end of July the companies had offered no increase and after several meetings the miners addressed a letter to the superintendents asking for an advance of ten cents a ton and requesting a conference if for some reason the raise could not be granted.[25] Instead of arranging a conference, the companies posted a circular at the various mines pleading their inability to pay higher wages because of the difficulty in competing with the cheap labor in the Virginia mines and warning that any increase in the price of Maryland coal at tidewater would divert their trade "into other regions where methods are resorted to for reducing costs which are unheard of in the Georges Creek region; we mean, of course, the store system, and payments on a basis of mine car figures, instead of on a basis of actual weight." [26]

In 1899 the Pocahontas region of Virginia was paying its miners at the rate of 60 cents per car. Some cars held two tons, some three, making the mining rate 30 cents or 20 cents per ton, depending on the size of the car. Moreover, in Pocahontas, the *Frostburg Mining Journal* said, "the real company store prevails, each company having a provision store at the mines, at which the men are required to deal, and 'provisions are very high.' " [27]

The operators' communication to their men concluded: "The prosperity which now prevails generally throughout the country has, unfortunately, not yet enabled us to advance the price of coal. It will, no doubt, in time reach this commodity. And our employees may confidently rely upon our assurance that when it does they shall share it." [28] Maryland's first Republican Senator, George L. Wellington, a native of Cumberland, later commented from Washington on "the overweening desire of the companies to hold for themselves more than the just share of profits accruing from the trade." [29] Pointing out that through his efforts the tariff on coal was increased in 1897, Senator Wellington said: "and as business increased and the price of coal ascended it was in my judgment, no more than right and proper that the miners who had been reduced during the continuance of the Wilson-Gorman bill [which reduced the 75-cents-a-ton tariff on coal to 40 cents a ton] should have been benefited to some degree by reason of the imposition of the protective tariff and the consequent return of prosperity and higher prices for our product. Such, however, was not the case. I looked forward in 1898 and 1899 for a fair increase by the companies to the miners, but it did not come."

The failure of the operators to arrange a conference aroused more bitterness than the failure to grant an increase in pay. The immediate reaction of the men was to invite the UMW to send an organizer to the region. On August 28 William Warner arrived from Pittsburgh, and on August 31 miners met at Lonaconing "to perfect their organization." [30] State Senator David E. Dick presided over the meeting, which Warner addressed. At subsequent meetings Warner insisted that his purpose was to organize the men, not to agitate for a strike.[31]

"Men of the character and habits of the Maryland miners are usually somewhat difficult to approach," wrote the editor of the *United Mine Worker's Journal*, "and generally take ample time to investigate a thing before passing their judgment upon it." [32] Nevertheless, within three months after beginning his work, Warner had organized enough locals to warrant the formation of a district comprising the George's Creek and Meyersdale regions.[33] District 16 of the UMW came into being at a meeting in Lonaconing December 12 to 14, 1899.[34] In addition to electing officers and adopting a constitution, the delegates addressed a circular to the presidents of the various coal companies in the George's Creek and Meyersdale regions asking for a conference at Cumberland on December 28 to consider an advance in the price of mining and other labor; an equal distribution of cars in the mines; and permission to employ checkweighmen.[35] By December 14, the date of this second request for a conference, an estimated one-fourth of the miners and mine laborers of the region were members of the UMW.[36]

The operators ignored the request for a conference, instead issuing a circular promising a pay raise after April 1, 1900, when their contracts for supplying coal would expire. They did not specify the amount of the raise.[37]

Indignant because of their second rebuff in a few months'

time, the miners who had assembled in Cumberland on December 28 called attention to their compliance with the operators' request for a conference in 1894 when a wage cut was contemplated. At the 1894 conference the companies had expressed their willingness to meet their employees at any time. The miners also took issue with an "inconsistent and unfair" statement which the operators gave to the press, making it appear that the companies failed to meet them because they refused to recognize the miners' organization. Not only was it true that the operators had failed to confer with their men several months earlier, when there was no labor organization, but it was also true that the companies themselves had a "thorough organization" and therefore "could not justly take exception to organization among [their] employees."[38] Listing the grievances which they wished to discuss, a joint committee of George's Creek and Meyersdale union members again invited the operators to a conference, requesting a reply by January 31, 1900.[39] Again the operators ignored the request.

The reasonableness of the miners' position is perhaps best attested by the moderation in the attitude of the *Frostburg Mining Journal*, scathingly anti-union during the 1894 strike. Early in 1900 the editor of this newspaper commented, "The *Journal* fears that the company policy of ignoring 'organization' is not *at this stage* the wisest," going on to say that the United Mine Workers' activity during the preceding six months had been "unprecedentedly free from intemperate conduct or talk."[40]

When the United Mine Workers met for their national convention in January 1900, they agreed on tentative wage scales to be submitted later to operators throughout the country.[41] The agreement did not specifically include Maryland, but because of close ties between the Clearfield and Cumberland wages in the past, it was assumed that the central Pennsylvania scale (60 cents a ton for pick mining) would extend to George's Creek. Delegates from District 16 and from the central Pennsylvania field planned to meet on March 14 to prepare a uniform scale for the two districts.[42] Almost three weeks before that meeting was held, the Maryland companies announced that after April 1 they would pay 55 cents a ton for pick mining, the highest rate offered in the region for eighteen years.[43]

George's Creek sent nine observers without vote to attend the convention of the central Pennsylvania miners at Altoona on March 14. The Pennsylvania operators agreed to a 60-cent wage, and the Maryland delegates decided to return home and ask for the same rate.[44] They also announced that they would settle the wage difference "without allowing the miners' contentions to degenerate into a strike." [45]

On March 19, after the Maryland delegates had returned from Altoona, the executive committee of the district union mailed to the operators a circular requesting them to reconsider the 10-cent advance and make it 15 cents a ton. For eighteen years, they said, the mining rate in Maryland had been the same as that in Clearfield. Since June 1899 Maryland men had worked for five cents a ton less than miners received in Clearfield. Now that the Pennsylvania miners were to get 60 cents a ton, the George's Creek men believed that they were entitled to the same rate. Pointing out their "patience and forbearance," they asked for an early reply or for a conference on March 31.[46]

It was of the utmost importance to these men that their employers recognize them—if not as union members, simply as human beings. There was little talk of striking, and if the companies had seen fit to meet with their miners, the wage question would probably have been satisfactorily adjusted. The Cumberland *Evening Times* remarked on March 29: "It is fair to state that the miners, as miners and as employes have been improperly denied the conference they asked with their employers. Whether members of the United Mine Workers or not, they most certainly should have been received and considered as Allegany miners by the operators of Allegany mines. The companies err here and they must sooner or later correct their error if they expect the support of the general and conservative public."

When the strike finally came, it was triggered by an event completely separate from the wage issue. As early as March 15 the miners and mine laborers of George's Creek had planned to join in celebrating the United Mine Workers' national eight-hour holiday on March 31. The Maryland observance was to take the form of a great mass meeting at Lonaconing, where Samuel Gompers, president of the AFL, president John Mitchell of the Mine Workers, and other UMW officers would speak.[47] The companies knew in advance that many of their men would be absent on the day of the celebration, and there was no reason to expect any objection. The operators themselves had only recently closed the mines on one occasion before the election of 1899, thereby, according to organizer William Warner, "forcing their employes to take a holiday to go and hear an address from Governor Roosevelt, of New York, and Governor Lowndes, of Maryland, who was then a candidate for reelection, and who . . . is largely interested in coal mines of George's Creek." [48]

Therefore businessmen as well as miners received with shocked surprise the news that superintendent B. S. Randolph of Consolidation had disciplined 95 of his men who attended the Lonaconing meeting. According to Randolph's own statement, he dismissed 20 because they had prevented others from working, and suspended 75 for periods of ten to twenty days, "not for attending the meeting, but for being absent from work in a body contrary to rules." [49] After futile attempts to have the men reinstated, 1,200 Consolidation miners went on strike on April 7.[50]

In the meantime, the businessmen of Midland and Frostburg formed a committee to communicate with the company president, arrange a conference on the issue of reinstatement, and stave off an almost inevitable region-wide sympathetic strike.[51] Before the businessmen completed their negotiations, the officers of UMW District 16 ordered a general suspension on April 12 to continue until the companies granted a mining rate of 60 cents a ton *or* met their employees in joint conference.[52] The official circular issued by the district officers stated: [53] "Inasmuch as our employers have refused to meet us in joint conference in response to this, our sixth appeal of April 2, we therefore now declare a general suspension of all mines in District No. 16, United Mine Workers of America, at midnight, Wednesday, April 11, 1900, until 60 cents per ton of 2,240 pounds is paid, or until our employers meet us in joint conference and settle the questions at issue." *

Meeting the press after announcement of the strike, organizer William Warner estimated that only about 500 miners in the region were nonunion.[54] He said also that reinstatement of the Consolidation men would not bring settlement, "but a joint conference of miners and operators could very easily establish peace, and, in all probability would do so if the operators would only consent to meet their employees." [55] In answer to the question whether other regions would strike in sympathy, Warner replied: "No. Contracts have been made with operators in other States for a period of one year from April 1 and they must be kept, and it is a part of their contract with other operators that this region shall, if possible, pay 60 cents." [56]

Warner consistently denied that recognition of the union was an issue.[57] Certainly the miners made every attempt to secure unobjectionable committees to meet the companies as employees, not as union men.[58] However, the Bureau of Industrial Statistics sensed a desire for "recognition by the employers for the men collectively, with a view to a further effort to organize the coal miners of the United States. This demand for recognition, while not made a part of the request for the increase [in wages], was evidently a part of the policy of the organization, as well as the leaders, whether expressed in plain terms or not." [59] The operators ignored the union's wishes. When the company presidents *collectively* decided that the time was auspicious, they *individually* met their employees and told them under what circumstances they could return to work.

* In order to avoid tedium, the narrative will pick out only the main points of the 1900 strike, which in detail strongly resembled the strike of 1894. Those interested in day-to-day coverage will find it in the Baltimore *Sun*. The strike was also the subject of a special study published in the annual report of the Maryland Bureau of Industrial Statistics for 1900.

The Consolidation Coal Company's report to its stockholders at the end of 1900 made it appear that the strike resulted entirely from a demand for higher wages, ignoring the official conclusion of the Bureau of Industrial Statistics that "it was possible, at any time after the closing of the mines, for the operators to have ended the strike by publicly signifying to their employees that they would confer with them as to their grievances." [60] Instead of making this simple concession, the companies persisted in a course of action which prolonged the dispute until August.

Consolidation's own version of the affair denied that the company took any "active measures . . . at any time during the strike for the purpose of bringing about a termination. The resumption of work was the voluntary act of the miners." [61] This statement was not entirely candid. Consolidation joined the George's Creek Coal and Iron Company, the American Coal Company, the Barton and George's Creek Valley Coal Company, and the New York Mining Company in securing sweeping injunctions against the principal UMW organizers and about 600 miners.[62]

Apart from this open active measure, Consolidation allegedly engaged in more subtle methods of discouraging the strikers. The employees of Ocean Mine No. 1 charged that "certain officials of the Consolidation Coal Company" attempted to call meetings of small numbers of the company's men "for the purpose of causing dissension in our ranks." They also expressed their "indignation at the bull-dozing tactics of Mine Foreman P. A. McMahon for his threats of violence toward our fellow-workmen." [63] Former mine inspector Frank J. McMahon, "an employee of the Union Mining Company . . . much disliked by the miners," also attempted to break the strike at Ocean.[64]

Further criticism of unspecified companies' unethical practices came later, when miners in a mass meeting at Knapp's Meadow on May 31 adopted resolutions denouncing the action of "disorderly characters who are trying to create disturbances, which lead to riot and bloodshed," and calling the attention of the public and the peaceable and law-abiding citizens "to the fact that these disorderly characters are receiving encouragement from agents of the coal companies and money to get drunk with from some unknown source."[65] The use of money was quite apparent, said Warner a few weeks later, "as the agents of the companies bowl up some individuals with drink and in that way create disturbances which at times become a little exciting."[66]

But by far the greatest harassment of the striking miners came through the companies' resort to law. Between May 18 and 23 various operators asked for and secured injunctions which forbade "collecting, congregating or loitering or continuing your unlawful assemblage at or near the said company's mines, or on the public highway conducting to or on or near the road, paths and approaches leading to the said companies mines," and also prohibited the use of "threats, menaces, outcries, clamors, shoutings, shows of force or offers of violence" for the purpose of keeping men from going to work.[67] All of the union officials and miners so enjoined were summoned to appear before the circuit court on June 4 to show cause why the injunctions should not be made permanent.[68] On May 24 the organization officials retained as counsel Jasper N. Willison and David J. Lewis to fight the injunction suits.[69] Lewis, who in his early youth had been a miner and a member of the Knights of Labor,[70] was beginning a long and distinguished career as a champion of the workingman.

So far as can be determined from newspaper reports, no arrests resulted from violation of the injunctions. Willison and Lewis, by postponing their answer to the companies' suits, left the injunctions' validity open to question until the strike had almost come to an end. The document they filed on July 28 charged that the coal companies had "unlawfully combined" to refuse to raise wages to the Clearfield level.[71] This being the case, said the miners' counsel, the companies "had no right to ask the aid of a court of equity to further [their] unlawful designs." The answer also charged that the Baltimore and Ohio and the Cumberland and Pennsylvania railroads, by threatening to discontinue the rebates they had given some companies (contrary to law), had forced those companies to join with Consolidation in refusing either to grant a wage increase or to meet their miners in conference. In short, the coal corporations and the carriers were alleged to be engaging in "an unlawful conspiracy in restraint of trade and labor."

As the strike dragged on, and the men saw their savings disappearing, a great deal of bitterness grew up between those who wanted to return to work and those who wanted to hold out. Numerous instances of assault and interference culminated in "a hot time in the old town" of Lonaconing on June 22.[72] Because of arrests for assault and rioting at an anti-strike meeting on that date, the judges of the circuit court reconvened the April grand jury for the first time in Allegany County's history.[73]

Although public sympathy was more strongly in favor of the miners than it had been during the strike of 1894, the attitude of the court had discernibly toughened. Even when troops had been brought in to preserve order, the Allegany County judges had not thought it necessary to reconvene the grand jury or to call a special session of the court to deal with alleged violations. Furthermore, they had been lenient in imposing sentences.

In his charge to the grand jury in the summer of 1900, Judge A. Hunter Boyd confirmed the right of the miners to strike, provided they did not violate the law: "If any of the miners choose to lay down their tools and refuse to work until their grievances, real or imaginary, are corrected, it is a right they possess under the law. If they see fit to organize for what they deem to be their interests, as long as they do no act prohibited by law, it is not for us to question their right to do so." At this point he interpolated a personal opinion, expressing amazement "that men of intelligence will use their hard-earned money already laid up or further involve themselves in support of those in ease and comfort whose salaries or compensation must largely depend upon the continuance of differences between employer and employee."[74] Judge Boyd also instructed the jury that men who wanted to work had the right to do so without interference and might call upon the law for protection in exercising that right.

The grand jury indicted twenty-nine persons, including national organizers William Warner and Thomas Haggerty, and district president Allen Barber.[75] The indictments charged rioting, unlawful assembly, unlawful assembly with intent to assault, and riot with intent to assault.[76] Warner was also charged with criminal libel because of certain statements published in the *United Mine Workers' Journal.*

The calling of a special term of court to try cases arising out of a strike set another judicial precedent for Allegany County.[77] Warner commented on the precipitancy of the proceedings: "The court will convene in special session to convict and imprison at once. The fact that the next regular session of the grand jury and court is not until October proves that justice is not intended; but laboring under the idea that with me out of the way the strike will end, they are badly mistaken." (Punctuation supplied.) [78]

Warner's suspicion of the motives behind the arrests and subsequent trials seems well founded. The most casual reading of the newspaper accounts reveals farcical elements worthy of the talents of W. S. Gilbert.

The jury was empaneled only after many challenges,[79] but even so, Warner was able to base his appeal in part on the choice of jurors.[80]

The witnesses were all members of the crowd unlawfully assembled, and were in that respect as guilty as their unfortunate fellows singled out for indictment.[81] "Nearly all these witnesses endeavored to exonerate themselves from any connection with the alleged rioting, giving their testimony in a rather reluctant manner." [82]

Testimony for the state was mainly to the effect that a disorderly crowd gathered in front of the meeting place and "hooted and yelled 'blackleg!' until the people of Lonaconing were terrorized." [83] David J. Lewis, attorney for all the defendants, argued that since the meeting had been advertised as an anti-strike gathering, it naturally attracted a crowd of onlookers who were present out of sheer curiosity and had no intention of rioting. The disturbance was created when a drunken man accidentally went into the meeting hall.[84] Warner, who was on the stand for two hours, denied shouting or interfering, and said that the crowd was not dangerous. A city councilman testified that it was Warner who got the crowd to disperse.[85]

Nevertheless, the jury found Warner and eleven others guilty of unlawful assembly, and later convicted still more on charges of riot, assault and unlawful assembly earlier in the strike.[86] The sentences imposed on August 3 and 4 were unprecedentedly harsh, and Judge Boyd warned that he would be even more severe in the future if the present cases did not have the desired educational effect. Warner was given six months in the house of correction, and promptly appealed. Other sentences ranged from small fines and short terms in jail to incarceration in the house of correction for five months.[87] In sentencing Warner, the judge referred to him as the principal cause of the trouble, not because he was a member of a union or an organizer, but because he failed to use his influence to break up the crowd at Lonaconing before trouble started.[88]

By the time the trials ended, the strike was over. "It was not a question of going to work, but a rush to see who could get work, and it was estimated that fully three thousand men were at work throughout the region by the 4th of August." [89]

Leading spirits in Allegany County immediately began to move for release of the strikers who were already in jail. Thomas Brown (mine inspector during the strike of 1882) began the agitation by writing to Governor John Walter Smith, asking him to free the men, who were mostly young and supposedly had learned their lesson.[90] Eventually a petition signed by more than 7,000 residents of Allegany County (including most of the jurors who had tried the case) produced the desired result. The governor pardoned the seven who were already serving sentences.[91] To the great indignation of the normally Republican mining population, the Baltimore *Evening News* insinuated that the pardons were given in exchange for a promise to vote Democratic in the forthcoming election.[92]

Warner's appeal was heard in December, and the judgment of the lower court was affirmed.[93] The union organizer and the others who had appealed with him began their terms in the house of correction on December 23.[94]

The commitments of William Warner and his fellow appellants were dated August 4, the day of sentencing, and were not corrected to allow for the delay involved in the appeal. The prison clerk said that he had called attention to the defect, but the attorney general told him to keep the prisoners until he got some order from the court to discharge them. The attorney general denied making this statement. The judge of the court of appeals said he had no knowledge of this point, which had not been at issue before him. The assistant state's attorney for Allegany County said that he didn't even know that the case had come up before the higher court. The miners concerned knew that their commitment ran from August 4 and notified a union organizer of that fact. "Of course, they did not volunteer to the court information which might keep them in custody." [95]

Warner and five others were released in February 1901 after serving only forty-five days. The Baltimore *Sun* commented: "Six men who, because of somebody's blunder, had been kept in the House of Correction after the expiration of their terms of imprisonment, were released yesterday on habeas corpus by Judge Stockbridge in the City Court." With commutation for good behavior, Warner should have been released on January 4; the others, earlier. Nothing in the commitments showed that the men had not been incarcerated since August 4.[96]

Before dropping the subject of the 1900 strike we should at least briefly consider in what ways the labor organization contributed to its failure. With the most worthy intentions, the union leaders made mistakes in judgment which cost them dearly. Their major errors were interference with free speech, dependence on out-of-state organizers, and a policy of providing work in neighboring states at less than the rate demanded by the strikers in Maryland.

It is evident from newspaper accounts and from organizer Warner's letters to the *United Mine Workers' Journal* during the eight months preceding the strike that many of the George's Creek miners joined the UMW with the understanding that they would not soon be called on to leave their jobs in support of union principles. Reluctance to strike was particularly noticeable in Lonaconing, "the stronghold of the organization," where large numbers of miners were ready to go to work as early as May 7.[97] On various occasions Warner complained of soft spots in union ranks: "I regret to say that some of the most weak men and those giving us the most trouble are members of the union."[98] And again: "We have many traitors in our camps. These have been greater enemies to the miners of George's Creek than the coal operators."[99]

In the absence of complete accord on the maintenance of the strike, Warner felt uneasy when independent groups of miners met or attempted to meet to discuss possible resumption. He objected strongly to meetings not called by the organization, and deplored the fact that miners and their wives listened to and participated in strikebreaking talk.[100] Obviously he did not favor the presentation of both sides of the question, in all probability because the anti-strike side offered great potentials for the "claptrap of coal company hirelings."[101] His attempts to silence opposition opinion or make it appear the work of "would-be scabs" disturbed even the pro-union Lonaconing *Review*, whose editor insisted that the expression of individual views must not be repressed.[102] The grand jury's report commented on the fact that some persons had questioned the right of the anti-strike element to hold meetings.[103]

The lessons of the past failed to influence the activities of the UMW in the Maryland strike of 1900. Suspicious of the leadership of outsiders, the miners had easily been separated from the National Federation of Miners and Mine Laborers in 1886 by the operators' propaganda directed against out-of-state organizers. In 1891 the George's Creek men had refused to be involved in an attempt of the UMW to adopt a wage scale for the seaboard regions, particularly the troubled Pennsylvania areas. And in 1894 the coal companies, the courts, the press, and a considerable number of nonunion miners had been quick to seize upon the out-of-state leadership as a means of discrediting the union movement.

In 1900 the organization sent "its most skilled agitators" into Maryland,[104] but these were all from competitive fields. William Warner, Thomas Haggerty, William B. Wilson, and Mother Jones came from Pennsylvania; national board member Fred Dilcher was from Ohio; and Henry Stevenson was from Kanawha, West Virginia. At the outset of the strike the operators said they would not confer with outsiders who came into the region to make trouble.[105] The Cumberland correspondent of the Baltimore *Sun* stressed the fact that the men who led the strike were from regions where men were still working and taking the trade away from George's Creek.[106] The *Frostburg Mining Journal* of June 23 discussed the Pennsylvania origins of the strike in an editorial entitled "An Unsatisfactory Statement." The *Coal Trade Journal* of June 20 remarked that the Maryland miners should not have listened "to the specious arguments of agitators from the West who know nothing of local conditions."

In its "Deductions from the Coal Strike" the Maryland Bureau of Industrial Statistics found: "There was some good reason in the argument of the operators refusing to recognize Warner, Dilcher and others, who came from other States and assumed, through the organization, to deal with questions appertaining solely to Maryland miners. This could have been averted . . . by the organization, had it adhered to the policy of 'home rule.' "[107]

Still another of the Bureau's deductions read: "The weakness of the position of the United Mine Workers' organization in this strike was in the fact that, while demanding sixty cents for mining coal in the George's Creek region, they permitted their members and men from Maryland to work in the Myersdale [*sic*] region, in Pennsylvania, for fifty-five cents all during the time the Maryland mines were closed. Whatever the private reasons entertained by the leaders for this course may have been, the apparent result to the Maryland community was that the organization was demanding one thing in Maryland and conceding another within a stone's throw, in Pennsylvania. Indeed, both regions were included in District 16 of the United Mine Workers' organization, and both regions were supposed to work under the same wage scale and under the same conditions." [108]

In the early spring of 1900 the Meyersdale miners had been receiving only 40 cents a ton and had struck for a five-cent increase to bring them up to the existing rate of 45 cents in George's Creek.[109] It is true that the Altoona convention had called for 60 cents at Meyersdale as well as on George's Creek. But the Meyersdale operators did not disdain to meet with their men (granting the conference requested and refused in Maryland) and agreed on a rate of 55 cents a gross ton for pick mining pending settlement of the Maryland dispute.[110] Moreover, the Meyersdale operators promised that they would not discriminate against the men who had recently been on strike.

About the middle of April, George's Creek miners began to leave to work elsewhere. The first group went to Elk Garden, West Virginia, where they received 45 to 50 cents a ton.[111] The real exodus did not start until after an abortive attempt to break the strike at Lonaconing early in May. Thereafter men moved out by the hundreds, securing work in various Pennsylvania or West Virginia mines. It was estimated that by May 28 approximately a thousand had left.[112] About 500 of these went to Meyersdale. Some in other areas were paid 60 cents a ton, but they worked under conditions which were not as good as those to which they had been accustomed. A Lonaconing boy wrote home to his mother from somewhere in Pennsylvania: "I arrived here safe and got work digging coal. Dear mother, I am working in water up to my knees. Write me as soon as they go to work on the Creek. We get 60 cts per ton here. . . . the work is hard and the board is not good. Dear mother, telegraph me when to come back." [113]

Union officials posted notices at Lonaconing in mid-June saying that they had secured employment for 500 men at 60 cents a ton in central Pennsylvania with transportation furnished.[114] Superintendent Randolph of Consolidation also found jobs for nearly 400 men in other fields, planning to reemploy them in Maryland as soon as the strike ended.[115]

The inconsistency of their migrating to work in a less favorable environment at a rate which they had refused in Maryland struck the George's Creek miners fairly early. The extraordinary efforts of union organizers to hold out for a general conference between company presidents and employees met with constant pressure to settle with individual executives for a promise of 55 cents and no blacklist. Eventually the latter point of view prevailed. After individual company conferences the men returned to work at 55 cents, without, however, assurance that there would be work for all.

Immediately after resumption, the union announced that the operators had a "blacklist" of 800 men.[116] Now, whatever else the companies may have done, openly or covertly, to discourage the organization, they did not have a genuine blacklist of this size. The miners tended to apply the term "blacklisted" not in its strict sense, but loosely, to anyone who did not secure employment as soon as the mines reopened. This had been true in earlier strikes, but in the present case the misapplication is so flagrant as to warrant some discussion.

Organizers had alleged from time to time throughout the process of enlisting the George's Creek men in 1899 and 1900 that fear of the "terrible blacklist" made them hesitate to join a union. Resolutions adopted on August 6, 1900, blamed earlier use of the blacklist "for the fear, and disposition to court favor at the cost of their freedom of thought and action that exists in the hearts of numbers of miners in this region, to whose action our defeat in the recent struggle can be largely attributed." [117]

No one reading accounts of the strikes of 1886 and 1894 can fail to be impressed by the willingness of the Maryland miners to stand up and be counted. They signed their names to petitions and resolutions; they identified themselves publicly and unmistakably with union movements. Some of these men were blacklisted in the true sense and were forever after denied employment in George's Creek and adjoining regions. But the significant point is that by far the greatest number of the local leaders did not lose their jobs because of their labor activities. They may have been discharged by one employer, but they were sooner or later hired by another in the region. The lists of union officers and employee committee members published in newspapers throughout the period make this fact clear. The same names recur in 1886, 1894, and 1900.

Organizer Warner began in May 1900 to use fear of the blacklist as a means of preventing the men from breaking away from the strike and returning to work in small groups at certain mines in and near Lonaconing.[118] After the attempted break at Lonaconing, prospective negotiations with company officials hinged on acceptance of 55 cents a ton with reinstatement for all.[119] The attitude of the Consolidation Coal Company was said to be decisive on the matter of reemployment. Superintendent Randolph had as early as April 19 issued a circular to his mine bosses ordering them to consider all applicants on an equal footing when the mines reopened, the period of disciplinary suspension having ended with the calling of the strike.[120] On June 12 the Baltimore *Sun* printed an interview with Consolidation's president, C. K. Lord, who said: "If a majority of the men will present themselves at the mines they can go to work. The men who remained at work on March 31, when the meeting was held at Lonaconing, will, of course, be given the preference, and those who broke the rules by leaving their work without permission will have to wait. It is not possible to close mines for eight weeks and then open them with work for the same number of men as was had when the mines were running regularly." The company's Eckhart mine, for example, had serious flooding.

When Lord received a delegation of his employees on June 23, he refused to discuss any increase in wages, but said that all who applied for tools would get them. However, the company could not put everyone to work at once.[121]

On July 19 the president of the American Coal Company met with a committee of his miners. Although he could not guarantee reinstatement for all, he thought that few would be left out once business had picked up again.[122] Subsequently officials of the New Central, Lonaconing, Maryland, and George's Creek Coal and Iron Companies conferred with their men, but the press did not report any discussion of reinstatement.[123] Officers of the Barton and George's Creek Valley, Union, Potomac, and New York coal companies promised to treat their men fairly upon their return to work.[124]

The Bureau of Industrial Statistics reported that the total number of mine workers before the strike was 4,747.[125] On August 4, about 3,000 miners were already at work. Approximately a thousand had left the region, but they were expected back momentarily.[126] Because of rumored discontent at Lonaconing, the sheriff went there on August 5 expecting trouble. He reported the number of blacklisted (presumably in the true sense) as about fifteen or sixteen at Lonaconing and possibly a few at Frostburg who had used abusive language toward the superintendents.[127]

The 800 allegedly blacklisted as of August 7, only a few days after resumption, included those genuinely blacklisted, those who had not yet applied for work, and those who had been given tools but had not yet been assigned places. Some of the Consolidation mines were still flooded and not yet in working condition.[128] Superintendent Randolph denied that he had a blacklist.[129] When organizer Warner called for a mass meeting and parade at Lonaconing on August 14, he was able to muster only 154 in a procession of the unemployed.[130] The *Frostburg Mining Journal* estimated that the number still out of work at the end of August was between 200 and 300, mostly young men.[131] At about this same time, Consolidation began to reemploy more men.[132] According to the *Proceedings* of the twelfth annual convention of the UMW in January 1901, only one hundred Maryland miners were still without jobs.[133]

By official account the true blacklist was small. The Bureau of Industrial Statistics reported the "Number of Old Employees Not Employed After Strike" as 68. Of these, 40 were former employees of the Maryland Coal Company, and 28 were former employees of the American Coal Company. However, the bureau also listed 48 "Men Employed After Strike Not Employed Before"—at the Big Vein, Union, New York, and Lonaconing coal companies.[134] Thus men rejected by one company may have been hired by another.

Undoubtedly, no Maryland miner was unemployed for any length of time. In September of 1900 a strike in the anthracite region gave additional business to the soft-coal fields, and throughout autumn conditions in the Maryland region were "probably never better." [135] Consolidation was "rushing the work at every one of their big mines to furnish coal to the United States Navy for the coaling stations," and was about to employ several hundred additional men to take care of a new opening in its Ocean mines and double shifts at Ocean No. 1 and Ocean No. 7.[136] During the early 1900's the labor force slowly increased. The special report on mines and quarries prepared by the United States Bureau of the Census in 1902 gave the number of wage earners in the Maryland mines as 4,936.

Chapter 18

Reform through Legislation (1888-1910)

The preceding two chapters dwelt on the more dramatic conflicts which resulted from the miners' attempts to organize. We must not, however, overlook the fact that coincident with their drive for unionization, the men of this coal region carried on a persistent campaign for legislation which would force the operators to improve working conditions. Although William Warner had charged that "often, at election time, . . . the miner is made to feel how he must vote, even if it be against his conscience and his intention as a citizen," [1] the evidence leads us to believe that the workers exercised their franchise freely. The companies might refuse to recognize the union; they might discriminate against union members; but they could not control the elections. Session after session the men of George's Creek sent to Annapolis representatives who would espouse their cause. Often their political leaders and the men whom they most respected were members or officers of the Knights of Labor and later of the United Mine Workers. Thus the two main streams of labor activity in western Maryland—organization and politics—frequently merged.

The miners most particularly wished to amend the inspection and ventilation laws, which had not been changed since 1878. Success had been within their grasp in 1886, when a law incorporating many of the reforms they desired was passed by both houses. But the engrossed bill disappeared under mysterious circumstances before it was signed by the governor.[2]

Although the larger corporations were introducing fans, the smaller companies could not or would not bear the expense of such innovations. The miners wrote to the Maryland Bureau of Industrial Statistics and Information in 1887: "In a state such as Maryland, containing a great mining territory, an act to compel the proper ventilation of coal mines is not only a right to which labor is entitled, but is a measure of relief commanding the sympathy of all classes of citizens. No one who has visited the men at their labor . . . confined in close and narrow headings, men, boys and mules, all working together, breathing the stagnant, unwholesome air . . . would deny them this relief." [3]

The miners also hoped to change the method of appointment of the mine inspector. The first few inspectors named by the governors had been experienced men, but when the office later became political, appointments were sometimes made without regard to the qualifications of the appointees. C. H. Hamill, the incumbent in 1887, was knowledgeable enough, but his ties were with management. Furthermore, he had been a member of the mines commission which in 1880 recommended repeal of the mining law. Therefore when the miners proposed to resubmit the bill passed in 1886, he gave them no support, but said, "It does not seem just that this industry, that has done, and is doing so much for the state and its citizens, should be vexed at every session of the legislature with proposals looking to the amelioration of whimsical grievances." [4]

State senator A. B. McKaig of Allegany County, author of the "stolen" mining bill, reintroduced it in 1888. It failed to pass the Senate because, according to the *Frostburg Mining Journal*, it would have legislated out of office the newly-appointed mine inspector, R. H. Browning, a senator from Garrett County.[5] (He was also, by way of occupation, a farmer, who owed his appointment as mine inspector, said the *Journal* subsequently, to "one of those unique developments of machine politics.") [6]

Shaken, but by no means discouraged, the mine workers in 1889 put up three candidates for the House of Delegates and elected two of them.[7] When the legislature convened in 1890, the miners' representatives again proposed substantial changes in the existing law: improved ventilation, a new method of appointing the inspector, and provision for checkweighmen.[8] The testimony of various company superintendents against the proposed statute outweighed that of the labor delegation sent to lobby for it,[9] and the bill "died with the Legislature." [10]

In 1886 and 1888, and possibly in 1890, mining legislation had support from Knights of Labor assemblies in other parts of the state.[11] However, when the local organization of the K. of L. dwindled in western Maryland, the miners no longer had a strong sponsor for their reforms. Nor were they helped by their own political complexion. When Daniel Young (later president of UMW's District 16) brought a ventilation bill to the predominantly Democratic legislature in 1892, that body almost at once rejected it. In Young's own words, "It was legislation asked for by a republican from a republican county, therefore it must die the death of all important legislation of the county of Allegany." [12]

In 1896 a new inspection and ventilation law almost became operative. The miners had two strong representatives in the House of Delegates—D. E. Dick and James Campbell—who introduced a bill calling for the election of the mine inspector by popular vote.[13] In other respects the bill resembled the drafts of 1886, 1888, 1890, and 1892. Representatives of the George's Creek Coal and Iron Company, the Consolidation Coal Company, and other operators appeared before the House committee on inspections to oppose it.[14] During the hearing Dick asserted that it was necessary to elect the inspector because the coal companies had in the past influenced the governor's appointments to the office. The president of the George's Creek Coal and Iron Company denied this. The companies objected at length to the provision for checkweighmen. In all, they proposed twelve amendments, seven of which were incorporated in the bill with the consent of Dick and Campbell.[15] Thus amended, it passed the House and the Senate.[16]

However, instead of becoming a law, the bill was killed by a pocket veto. Governor Lloyd Lowndes failed to sign it within the statutory time, saying that the attorney general had advised him of its doubtful constitutionality.[17] Governor Lowndes was a member of the coal-mining firm of Black, Sheridan and Wilson, operating in the George's Creek Valley. On April 13, D. E. Dick and James Campbell issued a statement which read in part: "Governor Lowndes has complied with the wishes of the coal operators and killed the mine inspection and company store bills,* under the most false and absurd pretence that ever came to the notice of man. Those bills were drawn up and closely examined by the best legal talent in this State, and when Governor Lowndes tries to skulk behind such a flimsy plea, it is only indicative of his lack of feeling for the laboring man." [18]

The issue of the veto arose again in the political campaign of 1901, when the Lonaconing *Review* said of former governor Lowndes: "His disregard for [the miners'] interests was too clearly manifest. Himself a coal operator and largely identified with coal mining interests in this county, he suffered his individual welfare to defeat the interests and justice due thousands who had hitherto been loyal to his political aspirations and who were instrumental in rolling up the magnificent majority given him by his own county of Allegany." [19] †

Campbell returned to the Maryland House in 1897, and Dick was elected to the state Senate.[20] Although the mine inspector closed his report for 1897 by saying that he did not find it necessary "to suggest any further legislation regarding the mines," [21] Senator Dick introduced a new bill on January 18, 1898,[22] which was not startlingly different from the old law. The principal innovations provided for the election of the mine inspector, the employment of checkweighmen by the miners at their own expense, and examination of the mine inspector by the grand jury at each term of court with a view to uncovering violations of the law.[23]

* A bill to abolish company stores in the mining region passed both House and Senate in 1896 but failed to obtain the governor's signature.

† Lowndes was a Republican.

Early in February, Dick and Campbell conferred in Baltimore with H. Crawford Black, John Sheridan, and Governor Lowndes (all of the firm of Black, Sheridan and Wilson) and president C. K. Lord of Consolidation. These gentlemen indicated that they objected to the bill and would ask for a hearing before the appropriate committee of the legislature.[24] Apparently the operators agreed to a hearing in the House, for the bill passed the Senate without amendment and without a dissenting vote.[25] The companies testified before the House committee on inspections on February 17.[26] Most of the discussion concerned the method of choosing the inspector, and in the end Dick and Campbell yielded to pressure and abandoned the provision for election. As Campbell explained in answer to criticism later in the year, "I was and still am thoroughly convinced by the strong antagonism manifested by the operators that had we insisted upon the election of the mine inspector, the whole mining bill would have again been vetoed." [27]

A redrawn bill "providing for the appointment of a mine inspector for Allegany and Garrett counties, and regulating the working and ventilation of coal mines in said counties" was substituted for the measure previously approved by the Senate.[28] Since the revised version satisfied the coal companies, it passed both houses and was signed by the governor.[29]

Comparison of the new inspection law [30] with its predecessor [31] reveals that the miners did not gain a great deal. An addition to the qualifications of the inspector required that he should have had "practical experience as a working miner" in the Maryland region for at least five years immediately preceding his appointment. His duties did not change substantially. Two new sections concerning weight definitely favored the miners: they would receive credit for fractions of a ton of coal loaded on the mining cars, and they might employ checkweighmen at their own expense. Extensive provisions related the duty of operators and owners and all persons employed in the mines "to observe all practical care, caution and prudence" to protect their own lives, health and safety and the lives, health and safety of their fellow workers, as well as the property of the owners. The inspector was to be subject to call and examination by the grand jury, which might recommend his removal from office if it saw fit. But so far as ventilation was concerned, the new bill was about as vague as the old, requiring merely that the operators provide "a proper system of pure air ventilation by satisfactory and effective modes, either natural or artificial."

Having expected a much stronger law, the miners did not hesitate to criticize both the statute and its sponsors. Dick was particularly touchy on the subject, insisting that he and Campbell had secured adequate legislation but that it had not been enforced.[32] However, he was anxious to make peace with his constituents, and wrote to the union members when they were in convention at the end of December 1899: "In the coming session of the Maryland General Assembly an effort will be made to amend the present mining law so that it will conform as much as possible to the wishes of the miners of the George's Creek region. . . . The present law is evaded to a very great extent, but I believe that the organization will be able to secure the active cooperation of the next grand jury toward a rigid enforcement of the entire law. This would demonstrate the merit of the law, notwithstanding the many adverse criticisms passed upon it. However, I would like your convention to carefully consider the whole proposition, and any suggestions you have to make in the way of amendments, etc., will be cheerfully accepted and I will do my utmost to have them incorporated in the law." [33]

According to the Baltimore *Sun* of February 16, 1900, the executive board of District 16 discussed amendments to the mining law, but instead of suggesting changes, adopted resolutions condemning criticism of Dick's bill of 1898.

When the new legislature met in 1900, Dick introduced an inspection bill designed to meet at least two of the objections to the existing law. To insure more frequent inspection of the mines, he proposed that the governor appoint two inspectors, each to be in charge of a separate district in the region.[34] Because he felt that in the past the mine inspectors had been "subservient to the operators at the expense of the miners," [35] Dick now proposed that the governor appoint the inspector from a list furnished by the local executive board of the United Mine Workers.[36] In order to deal with the practice of docking, which was still going on in spite of the law, he incorporated a provision which would limit the amount the weighmaster might deduct if foreign matter were present in the coal.[37] The bill was favorably reported and passed the Senate,[38] but was killed in the House.

It was characteristic of local politics that Dick's efforts on behalf of the miners should have been denigrated in the rough and tumble campaign of 1901. Hitting at ex-governor Lowndes, who was supporting the Republican candidate for the state Senate, the Lonaconing *Review* charged that Dick and the governor had concocted the 1898 mining law "to offset [the] ever increased unrest of the miners by additional, but not effective, legislation." [39] In this election year the miners did not vote on strictly party lines. Although their "usual affiliations [were] almost solidly Republican," [40] for the first time in many years they sent a Democrat to the Maryland House and elected a Democratic state senator.[41]

David J. Lewis, the new senator, was the son of a Welsh miner and had himself gone to work in the mines at the age of nine. When he was nineteen, he began studying law in his spare time, continuing to work in the mines until he became a practicing attorney in 1892. His election in 1901 was considered a victory for the miners because his opponent was "closely allied to the monied men of the county." [42]

At the time of his election delegate Francis J. Drum was secretary-treasurer of UMW District 16. Since two of the Republican delegates elected were also members of the United Mine Workers, the organization considered itself well represented in the forthcoming session. When the union miners met at Frostburg in November 1901 for the third annual convention of District 16, they discussed the mining laws and made recommendations for their amendment.[43] In his capacity as a district officer, delegate Drum was, of course, thoroughly informed of the wishes of the convention. Therefore the bill introduced in the Maryland House in February 1902 after a conference between Senator Lewis and the House members from Allegany and Garrett counties,[44] probably incorporated every significant change the miners asked for. The *United Mine Workers' Journal* of February 27 spoke of the proposed bill as repealing the Dick law of 1898 and wiping "all its fantastic and nonsensical provisions off the slate, to be replaced by some practical legislation . . . by far the best the State has yet seen."

Although the new law would make marked changes in the ventilation requirements, no opposition from the operators appears on the record. With almost unprecedented speed the bill passed the House. On March 4 Senator Lewis brought it before the Senate.[45] On March 6 the Cumberland *Evening Times* reported: "Senator Lewis scored a most remarkable achievement this afternoon by the passing of the mining bill without a speech and in two days. The bill was engrossed last night, the clerks working 15 hours without sleep to get it ready for today's session." The governor signed the bill, and it went into effect on March 24.

Since the mine inspection act of 1902 [46] was the first attempt at a definitive statute for Maryland, it deserves a more extensive description than we have given its three predecessors. Its most important provisions were technical in nature, making explicit the former general requirements for adequate ventilation. The new law called for the furnishing of at least one hundred cubic feet of air per minute for each man at work underground. It also set out specifications for openings, furnace shafts, passageways, overcasts, crosscuts, airways, trapdoors and bratticing. The mine inspector was to visit each mine at least once every two months, measure the amount of air in circulation, and post the readings in a conspicuous place for the men to see. Each mine foreman was also to measure the air current in his particular mine at least once a week with an anemometer supplied by the operator.

In former years the mine inspector's report had been printed at the whim of the legislature. The new law provided for the printing of a thousand copies of each report.

Incredible though it may seem, under the 1898 law the operators did not have to furnish maps of the mines. The law of 1902 not only required maps but also specified how detailed they should be.

Several subsections of the new law dealt with safety practices and the procedure to follow in case of accident. One of these subsections made it mandatory for operators to keep at each mine at least one properly constructed stretcher and woolen and waterproof blankets "for use in carrying away any person who may be injured."

Next to the improvements in ventilation, the sections of the statute which were of most immediate interest to the miners were those concerning the weighing of coal. Since 1878 the mine inspector had been designated inspector of weights and measures at the mines, but he had lacked the necessary scientific instruments for carrying out his duty. The new law required the state to furnish the appropriate weights and apparatus for testing the scales.

Previous legislation had provided that the miner be paid for all the coal loaded on a car, but nevertheless some weighmasters had continued the practice of docking in an arbitrary fashion.[47] The new law forbade any operator to "dock any miner in excess of five hundred pounds (cwts.) on any one car." The mining cars were to be stopped and uncoupled before weighing.

On the subject of checkweighmen it is necessary to elaborate a bit. Under Dick's law of 1898, the workers in any particular mine were to be allowed to employ a checkweighman at their own expense if a majority of them notified the operator or his agent in writing that this was their wish. Soon after the bill went into effect, Frank Stevens, a miner employed at the New York mine of Black, Sheridan and Wilson, circulated a petition for appointment of a checkweighman and was promptly discharged by the mine boss.[48] A committee of miners complained to the superintendent and succeeded in having Stevens reemployed, but nothing further was done about providing a checkweighman. Subsequently Senator Dick secured from Governor Lowndes, as a director of the company, a promise to instruct company officials to obey the law.[49]

However, as much as eighteen months later neither the miners of the Black, Sheridan and Wilson Company nor the miners of any other company in the region had employed a checkweighman. It is remotely possible that the need for putting the request in writing inhibited many who were dissatisfied with the weighing arrangements but who did not wish to be identified as trouble makers, but it seems much more likely that the George's Creek operators failed to comply with the law. Miners' committees repeatedly asked that the companies grant them their lawful privilege of placing checkweighmen at the scales. This issue was raised again and again in the communications between the miners and the operators preceding the strike of 1900.[50]

Senator Lewis' law of 1902 removed the requirement that the miners make their request for a checkweighman in writing. But whatever the reason, there were still no checkweighmen on the tipples in 1904, 1905, 1906, or 1907.[51] Yet in the smaller companies men continued to grumble about short weight, and on at least one occasion the exasperated mine inspector suggested that they either hire a checkweighman or stop complaining.[52]

In 1903 the Allegany County grand jury, upon instruction from the court, called for a report from the mine inspector and made further investigation of the efficacy of the mine inspection law.[53] The grand jurors reported that the statute was "a vast improvement on any of the previous enactments" but that a few company officials were disregarding some provisions of the law. They also found that the miners themselves were "generally ignoring the section of the law, respecting the burning of cheap and inferior oil, which hinders and nullifies the proper and healthful ventilation of mines put in operation by the coal companies." [54]

Mine inspector Thomas Murphy attempted to put an end to evasion of the law by several companies. He obtained warrants for the arrest of certain superintendents and foremen and later secured from the grand jury indictments against the operators, charging them with providing less than the amount of air required. When the cases against Hitchens Brothers Company, the Midland Mining Company, the American Coal Company, and the Piedmont-George's Creek Company came to trial, they were dismissed because the state's attorney had not drawn up the indictments properly.[55]

In 1904, amendments to the law were minor and did not involve ventilation.[56] During subsequent sessions of the legislature delegates from the coal region introduced amendments dealing with weighing and with the provision of additional stretchers, first aid supplies, ambulances, lavatories and drying rooms, but none of these measures was adopted. Thus the mining law of 1902 remained essentially unchanged until 1910, when further technical details as to cross-cuts for ventilation were added.[57]

During the period under discussion the miners tried unsuccessfully to obtain an eight-hour law. They also repeatedly introduced bills to prevent corporations from discriminating against members of labor unions. Introduction of this type of legislation ceased in 1908 when the judiciary committee reported unfavorably on a bill framed by D. E. Dick, "their conclusion being based on the decision of the United States Supreme Court that such a law is unconstitutional." [58] The Supreme Court's decision on the Erdman Act had just been published.*

In all, however, the legislative gains of the miners between 1888 and 1910 outweighed their setbacks. They obtained a long-sought semi-monthly pay bill in 1896,[59] and in 1900 secured a requirement that wages be paid in legal tender.[60]

An act of 1898 made it unlawful for mining and railroad companies doing business in Allegany County to operate company

* The Erdman Act passed by Congress in 1898 forbade railroads to require their employees to sign an agreement that they would not join labor unions. In *Adair* v. *United States* (208 U.S. 161) the Supreme Court ruled that this provision of the law was unconstitutional because it interfered with freedom of contract.

stores or for their officers or directors to have any interest in a general merchandise business in that county.[61] One of the principal merchants of Frostburg challenged the legality of the statute, and the case eventually went to the Maryland court of appeals.[62] Here the appellant unsuccessfully argued that "The people must be got at who are in a position of power over the men—to-wit; *the officers* of the companies. If these cannot be prohibited from selling to their men, then the law is in principle and has been found to be in practice useless." The court ruled that the act was "absolutely void" because it was in conflict with the Fourteenth Amendment to the Constitution of the United States. The Maryland legislature, said the judges, did not have the right to "single out the directors of one corporation, and solely because they are such directors, prohibit them from engaging in some other business open to the directors of all other corporations."

As soon as possible after the court decision, Senator Dick, who had been responsible for the passage of the law, prepared a new bill designed to "prevent continuance of alleged abuses of which miners and other workmen have complained."[63] A law which went into effect on April 7, 1900,[64] avoided discrimination by applying to "every corporation engaged in mining or manufacturing or operating a railroad in Allegany County." It did not specifically prohibit the operation of company stores, but stated: "nor shall such employees make any contract with [their] employers by which such employees shall be compelled to purchase their supplies, merchandise or goods from any private or company stores owned and operated by the said employers; nor shall said employers by means of such employment exercise any influence whatever on said hands by promises or threats to compel their employees to deal with any particular merchant or storekeepers." It also forbade employers to withhold wages for the payment of store accounts.

Among the laws which the miners secured after 1900, the statute which probably aroused most public interest was one which has been described as "the first act in the nature of a workmen's compensation act in this country."[65] In 1902 state senator David J. Lewis introduced a bill entitled "An Act to Create A Co-Operative Insurance Fund to be maintained by both employers and employees in certain perilous occupations, such as mining, quarrying, railroading, building and excavating."[66] After investigation and report by the legislative committee on insurance and loans,[67] the bill passed both houses without opposition and was signed by the governor on April 1.

The Maryland Bureau of Statistics and Information said of the insurance law: "This legislation was radical in its departure from the conservative lines heretofore followed in Maryland, and was probably the outcome of continued and strenuous appeals to the Legislature for an Employers' Liability Act, which would enable workingmen injured in the performance of their duties [to secure] compensation or partial compensation from their employer. The legislation is not only novel in Maryland, but so far as we know, it is the first attempt by a State to protect its citizens unable to protect themselves in the ordinary pursuits of a living at hazardous occupations. The aim of the act seems to be the mutual co-operation of the employer and employe, under the supervision of the Insurance Commissioner of the State, to protect the families of any persons who may be killed by accident, whether brought about by negligence of a co-employe or otherwise."[68]

The various mine inspection acts passed between 1876 and 1902 provided for suit against the operators if injury or death of a miner resulted from violation of or failure to comply with the law. However, as Senator Lewis pointed out in his argument for the compensation act: "The basic fact, upon which the proposed legislation rests, is that at least ninety per cent. of the accidents occurring in the occupations covered are caused by nobody's negligence, but by the risk naturally inherent to such perilous employments."[69] One of the novel features of the Maryland law was that it "recognized that the burden of the trade risk should not fall primarily on the workman."[70]

In brief, the cooperative insurance act of 1902 extended the liability of employers in certain hazardous occupations "to cases where employees were injured by the negligence of a fellow-servant and to cases where the injured employee negligently contributed to his own injury." [71] The law then proceeded to relieve the employers of their increased liability if they contributed to the insurance fund set up under the statute. Coal mine operators, for example, would pay in to the insurance fund $1.80 a year for each person employed by them within the state of Maryland. Half of this amount might be deducted from the employees' wages. From the fund thus created, the insurance commissioner was to pay $1,000 to the heirs of any person killed in the mines.

Enactment of the law covering the state as a whole was followed by the passage of a public local law designed specifically for the coal and clay mines of Allegany and Garrett counties.[72] The local law is perhaps unusual in that it has a preface which pleads the miners' case:

> Whereas, experience has shown that in certain perilous occupations followed in Allegany and Garrett Counties, in the state of Maryland by its workmen and artisans, a great number of fatal accidents annually occur, so that the frequency thereof may be estimated from year to year, in which perilous occupations it has been found by uniform experience that the unavoidable or trade risk is responsible for at least ninety-five per cent. of such fatal accidents; and
>
> Whereas, The persons who suffer such fatalities are usually poor, working from day to day to sustain themselves and their families, and unable to accumulate any estate for the sustenance of their widows and infant children in case of their untimely death; and
>
> Whereas, It is unjust and against public policy that in such perilous occupations the burdens of the trade risks should be entirely borne by the widows and orphans of such workmen; and
>
> Whereas, It is of grave necessity and importance that some method of partial indemnification, at least, should be available to the dependents of such persons as are inevitably killed in such perilous employment lest they come to undeserved suffering, and the State be required to undertake their support; therefore in consideration the following Act.

The coal companies were unmoved by this appeal to emotion, apparently confident that any person bringing suit against them was unlikely to win. The Bureau of Statistics and Information commented in its report for 1902 (p. 319): "It is unfortunate that we cannot report, nine months after the enactment of the legislation, that more corporations or employers have availed themselves of the opportunities offered by the Act. . . . It was the special aim of those who advocated the passage of the bill to reach the mining industry of western Maryland, but up to this time not one of those companies have registered their employes for the purpose of protecting them." At the annual convention of District 16, the members of the UMW deplored the failure of the companies to contribute to the insurance fund, pointing out that participation would cost all coal operators combined only $5,400 a year.[73] The law remained in operation for a little less than two years, during which "The coal companies, to a corporation and to a man, refused to avail themselves of the law, and never deposited a single dollar in the treasury of the insurance department."[74]

The success of the act, "and perhaps the success of such legislation in general in America, will depend very largely on the view the courts take of it," wrote George E. Barnett of Johns Hopkins University shortly after the law was passed.[75] The court of common pleas of Baltimore City clearly expressed its view in the case of *Andrew J. Franklin* v. *The United Railways and Electric Company of Baltimore*. In this suit brought by an employee of a street railway company the court held the compensation act to be unconstitutional.[76] No appeal was taken.

This pioneer legislation which originated in the mining region came to an end without in any way benefiting, as Lewis had hoped, "the laboring masses, from which he sprang."[77] The miners had to continue to rely on their mutual aid societies, to which the employers contributed nothing. The benefits financed under such arrangements were necessarily small. The mutual aid society of Ocean No. 7 mine, for example, could afford death benefits of only $100 per casualty.[78]

For several years after the adverse decision of the court, no attempt was made to amend the compensation act so that it would meet the test of constitutionality. Then in 1908 state senator John B. Shannon introduced an act proposing a tax of six mills per net ton on all coal mined in Allegany and Garrett counties, and a tax of forty cents a month to be paid by clay mine operators on each of their employees. The money would go to the county treasurers "to create a relief fund out of which the sum of two thousand dollars shall be paid for the use of the dependents of employees killed in coal or clay mining." [79] After extensive amendment, the bill passed the Senate but did not succeed in passing the House.

Senator Frederick N. Zihlman of Allegany County introduced substantially the same act in the Maryland Senate in 1910. The taxes proposed were five mills a ton on coal and 30 cents a month for each employee in the clay mines.[80] Former state senator David J. Lewis, who had been elected to Congress in 1908, returned from Washington to marshal the forces in favor of the act, in which he was acutely interested.[81] Immediately after its introduction, the bill was referred to the Senate committee on judicial proceedings. Lewis appeared before the committee and was undoubtedly responsible for the series of amendments which completely rewrote the proposed law.[82] Besides adding a preamble, or justification, the amendments provided payments for injuries as well as for deaths, thus going far beyond the compensation act of 1902. After favorable report from the committee on judicial proceedings, the bill passed through both houses without incident and the governor signed it on April 7. It had been "enthusiastically supported by the Federation of Labor and all miners' organizations," [83] and by the Allegany Trades Council.[84]

The preamble to the 1910 statute [85] opened with a strong statement for the times: "It is the duty of the Government to provide sustenance in the case of helpless indigence to those who are or may become paupers and charges upon the public and is the settled practice of Governments to do so." Thereafter it continued in much the same vein as the introduction to the law of 1902.

The statute itself is lengthy, its sponsors apparently having been so anxious to prevent any question of its legality that they spelled out its provisions in the minutest detail. An excellent brief summary is provided in the *Bulletin of the* [U.S.] *Bureau of Labor* for September 1910:

"A second law was passed by the legislature of [Maryland] in 1910 . . . establishing cooperative insurance funds for the coal and clay miners of Allegany and Garrett counties. Employers and employees are to make equal contributions to a fund to be collected and held by the treasurers of the counties. Administration devolves on the county commissioners. The maximum award of $1,500 is for the death of a person leaving dependents; while for maiming injuries a schedule of awards is provided, the maximum being $750,* though medical relief in the amount of $1 per working day for not more than 26 weeks may also be allowed. For injuries without maiming, $1 per working day may be allowed for not more than 52 weeks. Suit may be brought, but doing so bars compensation rights, and conversely the acceptance of benefits bars the right to sue." [86] In this last respect the act avoided the constitutional question raised concerning the law of 1902.

The new law gave neither operators nor employees any option about contributing to the fund. "Another constitutional question was avoided by calling the contributions of the employers and employees 'taxes,' thereby resting the compulsory power exercised by the State upon its taxing rather than upon its police power." [87] The employer contributed 27 cents a month for each of his employees. Each employee matched his employer's contribution with 27 cents of his own deducted from his wages. According to the treasurer of Allegany County, the tax levied on the employers would be sufficient to pay death claims; the tax on the employees would serve to pay claims for major injuries and for disability.[88]

When the amendments were being discussed the question

* Loss of both hands, $750; loss of both feet, $750; loss of one hand and one foot, $750; loss of either hand, $375; loss of either foot, $375; loss of entire sight of both eyes, $750; loss of entire sight of one eye, $375.

arose whether a widow might not be so inexperienced in the handling of money that she would spend a lump sum payment unwisely. To provide for this contingency, Section 6 of the Act allowed the county commissioners to determine "whether the age, habits and prudence" of the surviving dependents were "such as to render them fit persons to receive the principal of such relief money." Not only might the commissioners arrange to have the relief money paid in limited installments, but they might also, when the dependents were a mother and infant children, invest half the relief money in a home for them.

The cooperative insurance act of 1910 was patently an improvement over the act of 1902. The new law went into effect on May 1, 1910. During the first six operating months, the insurance fund received $48,616.95. In the same period, it paid out $41,266 in claims.[89] Administration of the law seems thus to have had a satisfactory beginning, although the mine inspector complained in his report for 1912–13 (p. 6): "The reports of non-fatal accidents have increased owing to the compensation paid through the Miners and Operators Co-Operative Relief fund to employes who lose time on account of injuries received while discharging their duties. Before the act there was little or no occasion to report minor accidents."

This workmen's compensation statute, providing for a relief fund, was the last of the major legislative reforms secured by the miners in the period under discussion.

Chapter 19

End of an Era

(1900-1910)

With the strike of 1900 the era of relatively civilized corporate opposition to the miners' unions in western Maryland was approaching its end. Within a few years the amalgamation of large coal interests in Fairmont, West Virginia, George's Creek, Maryland, and Somerset, Pennsylvania, brought typical bully-boy tactics to the region. But for a short period it seemed true, as purportedly remarked by an anonymous company official, that the corporations had no objection to their men belonging to as many unions as they liked, provided that union officials did not interfere between employers and employees.[1] Former superintendent Randolph, writing a historical sketch of the coal region for the Maryland geological survey in 1905, said: "Since [the strike of 1900] the United Mine Workers have maintained a few lodges with small and uncertain membership and have had paid representatives in the region for considerable periods, but they are not recognized in any way by the employers." [2]

During 1901 the national offices of the UMW were engaged in a special effort to organize the miners of West Virginia,[3] and therefore could spare little time for promotion in Maryland. Daniel Young, president of District 16, made a valiant effort to preserve the "remnant of organization left in the field," but found that the George's Creek mining population, intelligent and enlightened though it might be, was "unwilling and indifferent to stand by organization." [4] He felt that Maryland miners, in comparing their own situation with the intolerable conditions in West Virginia, lacked the desperation which might have led them to organize.[5]

Young held numerous meetings throughout the region in 1901, traveling long distances on foot for want of funds.[6] In May national headquarters sent Chris Evans, one of the UMW's "most prominent organizers," for a brief period to take "charge of the order's *propaganda* in this region." [7] In spite of these efforts, Maryland coal workers remained apathetic. Even the sons of men who had first promoted labor organization in George's Creek were not members of the UMW.[8]

In 1902 Maryland was left largely to its own devices while the UMW devoted all its resources and ingenuity to the great struggle with the anthracite producers. When the anthracite strike seemed to be nearing success, union organizers again appeared in western Maryland. Several hundred people met at Frostburg on September 18 to hear E. S. McCullough of Michigan and T. W. Morgan of Ohio deliver "earnest and strong" addresses.[9] This was the beginning of a series of meetings which led the editor of the *Frostburg Mining Journal* to report "considerable activity in the way of getting up a United Mine Workers' revival. . . ." [10]

The visiting UMW officials comported themselves in a manner calculated to reduce resistance to outsiders, with the result that the Cumberland *Evening Times* remarked on October 2: "Unlike the average organizer, they are men of intelligence, liberal minded men, who are not backward to acknowledge that the coal companies have a right to exist. . . . To bring men to a true understanding of fundamental principles of the order is their first object and organization is the second."

However, while the Maryland miners were attending lectures on such topics as the "Philosophy of Trade Unionism," a transaction of great moment to them was being consummated by the Consolidation Coal Company and the major coal interests of Fairmont, West Virginia, and Somerset County, Pennsylvania. This was the "interlinking" of the interests of Consolidation, the

Somerset Coal Company, and the Fairmont Coal Company "through the acquisition of control * in the three companies by practically the same interests."[11] Consolidation had capital stock amounting to $10,250,000; Fairmont, $17,500,000; and Somerset, $4,000,000. The combined companies, owning land enough "to make a good-sized State," expected to put out ten million tons a year. "The most striking change . . . was the election of Mr. C. W. Watson, representing the Fairmont and Somerset companies, as president of the Consolidation Company, in place of Mr. Charles K. Lord, who resigned."[12]

Watson was a member of a West Virginia family which had been engaged in the coal business since 1852. By 1901 the Watson family controlled the Fairmont region and had extensive interests in Somerset County, Pennsylvania. Through a son-in-law, Aretas Brooks Fleming, former governor of West Virginia, the Watsons also exercised political power in their home state. Watson himself was elected to the United States Senate in 1911.

Named as assistant to the president of the combined companies was Jere H. Wheelwright, a lawyer who had acted as secretary for two U.S. Senators from West Virginia—Stephen B. Elkins and John M. Camden. Through his association with Senator Camden, Wheelwright acquired coal properties and joined with C. W. Watson in forming the Fairmont Coal Mining Company. He became president of Consolidation when Watson was elected to the Senate.

It was no secret that the methods used by the Watson interests to combat the UMW were far more aggressive than any which had heretofore been employed along George's Creek. The Fairmont Coal Company, for example, had for several years employed "detectives" so effectively that its miners dared not join the union.[13]

Initially, however, apart from Watson's well-known aversion to the union, there seemed to be nothing ominous in the creation of the tri-state merger. On the contrary, wages went up almost

* Actual control went to Consolidation, according to its report to its stockholders for the year ending December 31, 1903 (p.7).

at once in Fairmont, and not only Consolidation, but all the operators acting as a group, promised an advance in Maryland. The Fairmont Coal Company gave its employees a 10 percent increase in January 1903 "as a reward for their fealty" in refusing to join the union.[14] At approximately the same time, the Cumberland companies announced that on April 1 they would raise the wages of all mine employees by an amount to be determined by the extent of their new contracts.[15] The year 1902 had been a most fortunate one for the Maryland producers, who received high prices for all the coal they could deliver during the anthracite stoppage. They also benefited from a temporary restriction on output at Clearfield.[16]

To appreciate the size of Maryland operators' profits at this time, we need only to consider the example of the George's Creek Coal and Iron Company, which on February 10, 1903, declared a 100 percent stock dividend "to distribute among the stockholders a large surplus of net earnings that had been accumulating for some time." [17] The company's president, J. J. Alexander, said that since 1887 the firm had paid a "steady 6% on the par value of its shares, which were "held very closely by the directors and a few others." It had also, out of undistributed profits, been able to purchase a large tract of West Virginia coal land near Fairmont. The amount of the stock dividend was $1,100,000.

Speculating on the possibility of a wage adjustment some months before it was made, a Lonaconing correspondent of the Cumberland *Evening Times* remarked: "The operators are not unlike mill owners who open up their hearts and grant a ten percent. advance after they have advanced the finished product 25 per cent." [18] In March the Maryland companies announced that they would pay 65 cents a ton for pick mining in veins more than four feet high.[19] This rate happened to coincide with the scale adopted by the delegates of UMW's District 16, meeting at Meyersdale, Pennsylvania, earlier in the month.[20] Meyersdale employers granted their workers a similar increase.[21] At the same time, the Fairmont Coal Company announced a second raise, amounting to about 5 percent, effective April 1.[22]

Viewing the higher wage as "bait to satisfy the miners for the time being and counteract the work of organization," [23] UMW's McCullough redoubled his efforts to stimulate interest and increase union membership in Maryland. At the beginning of 1903 District 16 (including George's Creek and the Meyersdale region of Pennsylvania) had only 514 paid-up members.[24] By the middle of April organizers McCullough and Morgan had established headquarters at Frostburg and Lonaconing.[25] Between January and May they proceeded without hindrance to work in the Maryland end of District 16, continuing their policy of "very conservative proselytizing." [26] On May 14 the *United Mine Workers' Journal* reported, "Close on to half of the miners of the entire district are now members in good standing of the United Mine Workers," but regrettably did not indicate the distribution of the membership between the George's Creek and Meyersdale areas.

The first overt opposition to organization in the George's Creek section came from the West Virginia-dominated Davis Coal and Coke Company mines on Franklin Hill, near the Piedmont end of the valley. The superintendent there allegedly discharged a number of men because they had joined the union.[27] However, Consolidation and the other large employers in Maryland showed no inclination to follow the Davis example, and organizing continued at a brisk pace throughout the rest of the summer with the aid of Bernard Rice of Pennsylvania, Robert Salmond of Indiana, and the UMW's vice-president, Thomas L. Lewis.[28]

In 1903 the Mine Workers were "especially anxious" to make sure of District 16.[29] Letters from McCullough and Salmond to the *United Mine Workers' Journal* indicate that their work in Maryland was to some extent hampered by a residual factionalism among the miners dating back to the strike of 1882. McCullough referred to the several occasions on which the men had struck and had been defeated, "one-half of the men going to work in one strike, and the other half going to work in another strike, thus causing a very bitter feeling to exist between the factions, which resulted in breaking up homes, churches and fraternal societies,* and prevented them getting together in a labor organization." [30]

When vice-president Lewis came to the region, he "devoted his efforts to the establishment of a good feeling between the miner and the operator," saying that they should work together in harmony for their common advantage. Insisting that the United Mine Workers of America were opposed to strikes in every form, he stated that "there need be no fear of a strike in the George's Creek field for some time to come," because miners and operators could adjust their difficulties by other means.[31] In reinforcement of his position, Lewis moved to stop the sporadic labor disputes which had been occurring in the Somerset, Pennsylvania, end of District 16 by issuing an "ultimatum" that henceforth no one would refuse to work unless the district president himself called the men out.[32]

The reward for the conservative and conciliatory program of the UMW was unprecedented success. The *United Mine Workers' Journal* of August 13 claimed as union members in good standing between 4,000 and 6,000 of the 7,500 mine employees in District 16. About half of these were from George's Creek.

In neighboring parts of West Virginia the mining companies exhibited a "persistency . . . in discharging workmen because they are members of the United Mine Workers of America." [33] So far as Maryland producers were concerned, officially the union simply did not exist. Thus, when the Mine Workers' president, John Mitchell, spoke to a great crowd at Midland on September 26, he referred to the closer relationship which his organization had brought about between miners and operators in eight of the greatest coal-producing states, but regretted "that in Maryland . . . the operators refuse to treat with the representatives of the union." [34] However, many of the George's Creek company officials obliquely recognized the union by arranging

* A writer who described himself as a firm believer in organization challenged McCullough's statement concerning home, church, and fraternal societies, but not that about labor organizations. *Frostburg Mining Journal*, June 13, 1903.

to have their mines working full time on "Mitchell Day," a Saturday. Although they could easily have granted a half-holiday, they did not, and instructed company railroad men on Friday "to make special effort to supply empty coal cars at various mines on Saturday." [35]

The behavior of the Cumberland operators was merely petty. For true vindictiveness one must turn to the Meyersdale, Pennsylvania, region, where a bitter strike broke out on December 15 following the announcement of a wage cut of ten cents a ton.[36]

Wages in both George's Creek and Meyersdale had risen to 65 cents a ton in April 1903. Although employers and employees had no written agreements, it had become the tradition that a rate once offered and accepted remained in force for a year, during which the operator made contracts for selling his coal at a price based on that rate.* Meyersdale miners felt that their employers had broken faith with them in not maintaining the 65-cent wage for the remainder of the year.[37] On the other hand, the sixteen Somerset County coal companies which reduced the mining rate alleged that they had to do so because the Consolidation Coal Company had cut the price of its George's Creek coal so radically that the Meyersdale operators could not compete.[38]

One of the principal Meyersdale operators was the Somerset Coal Company, in which Consolidation held a controlling interest. Any action taken by this company, with the implicit approval of Consolidation, was therefore of special concern to Maryland mine employees. Consolidation, the largest employer in the Cumberland field, was capable of influencing other operators there. J. H. Wheelwright, assistant to the president of the big Fairmont-George's Creek-Somerset combination, set the tone for the Meyersdale strike by saying on behalf of the Somerset Coal Company: "We will close about a half dozen mines and discharge the union men. About as many other mines we will keep open and only work non-union men. If there are any union

* This was one of the reasons that the 1894 strike in Maryland did not receive public approval. The miners had accepted the amount offered by the operators only a short while before the strike was called.

men in these mines they will be discharged. We have never recognized the union and won't do it now." [39]

Somerset immediately began to carry out its policy, and evicted strikers' families in midwinter to make room for nonunion men brought in from other Pennsylvania mines.[40] Many of these strikebreakers were recent Italian and Slavic immigrants. The Merchants' Coal Company, another large Meyersdale producer, was equally determined to rid itself of union men. Just after Christmas, T. T. Boswell, Merchants' president, was asked to comment on a news story "that the 500 striking miners had been ordered to vacate the company houses by today, that a severe blizzard was raging in that section, and that if the men were compelled to leave their homes much suffering would be entailed." [41] In the best Gowen or Baer manner Boswell replied: "The miners hold a 10 day lease on the company's houses. Ten days ago we notified them that they must vacate the houses in the stated time. I don't care how severe the weather is, if the men are not out of the houses by tomorrow we will put them out."

This was only the beginning. The Meyersdale strike lasted for sixteen months, with episodes of arson, riot, and bloodshed.[42] In the little town of Garrett both strikers and scabs paraded the streets armed with rifles and shotguns.[43] At Salisbury the general manager of the Merchants' Coal Company threatened to install "rapid fire guns" to protect company property.[44]

The miners of George's Creek were heartily tired of strikes. They had been induced to join the union only after months of patient work by organizers who emphasized harmony between employers and employees. Suddenly they found themselves allied by their UMW membership with men who had officially been called out on strike by district president Daniel Young, a Maryland miner. To many George's Creek miners it seemed only a matter of time before they would be dragged unwillingly into the Meyersdale struggle. Rather than risk this (and particularly mindful of Consolidation's identification with the antiunion manoeuvers of the Somerset Coal Company), the majority of Maryland miners gradually dropped out of the United Mine Workers.

If the union organizers had been able to remain in George's Creek, it is possible that the amount of backsliding might have been limited. The organizers could have pointed out that in recent years the UMW had found the sympathy strike less valuable than the relief money which came from regions still working. However, McCullough and Salmond went to Meyersdale, where they were joined by a new organizer, Barney J. Palmer, sent specifically to organize "all the foreigners in that place." [45] The thought that in the event of a strike they, too, might be replaced by aliens was enough to send a distinct shudder down the spines of the thoroughly Americanized workmen of the Cumberland region. Moreover, not even the organizers could have glossed over the fact that as the months went on the UMW was no nearer success in Meyersdale, despite the fact that it was supporting the strikers financially to the amount of some $5,000 a week.[46] All things considered, in 1904 the Maryland miners weighed the benefits of union membership and found them wanting. The paid-up membership of District 16 at the end of 1903 was 1,786.[47] As of November 1904, paid-up membership of District 16 was 29.[48] Since the Meyersdale miners on strike were excused from paying dues, we may assume that the 29 represented men from George's Creek. During the year the nine locals of the UMW in Maryland paid to national headquarters for dues, relief and other purposes a total of only $581.35.[49] One can see why the *United Mine Workers' Journal* of September 22, 1904, described George's Creek, Maryland, as "a place where the true spirit of trade unionism has gone to sleep."

At the risk of belaboring the point, we must recognize that in this instance again the Maryland miners were better off than their immediate neighbors. Their employers continued to pay them 65 cents a ton until April 1904, when they announced a cut, not of ten cents as in Meyersdale, but of five cents, which the men accepted without protest.[50] Fairmont operators at the same time announced a reduction of 7.5 cents a ton on pick work, and four cents a ton on machine work.[51] These cuts were based on the allegation that the demand for soft coal had generally decreased.* President Mitchell of the United Mine Workers had, in fact, recommended in March that all bituminous miners accept a wage cut rather than risk a great strike.[52] Central and western Pennsylvania, Maryland, West Virginia and Ohio miners reportedly voted "for the reduction and peace." [53]

However, the amount of the voluntary reduction suggested for the Maryland and Pennsylvania districts was only three cents a ton. The Meyersdale region continued on strike, since employers there had imposed a cut far exceeding this amount. Furthermore, according to organizer McCullough, it was important for the UMW to support the Meyersdale strike because that region was "being employed by the operators as the place for the insertion of the opening wedge that would be followed by a reduction throughout the entire country." [54]

The Meyersdale *Republican* criticised the George's Creek men for their self-interest in accepting less than the district scale, saying, "The men over in that end of the district cared but little for those over here, or they would have stood out for the price and the recognition of the union, neither of which they done." [55]

Although the extent of the Maryland defection from the union was not then fully known, it was so much a matter of concern that a Labor Day speaker at Frostburg devoted much of his address "to rapping and rapping good and hard the miners of the George's Creek region." J. H. Walker, a national organizer for the UMW, "claimed that thousands of miners were idle, thousands had accepted a reduction in wages throughout the country because the Creek miners had not taken a firmer stand for the union." [56] It was not by coincidence that Walker chose Fairmont, West Virginia, as a horrible example of an unorga-

* It is questionable whether the Maryland reduction was justified. Consolidation, for example, in 1904 earned nearly 10 percent on its capital stock and paid a regular dividend of 2 percent, plus an extra dividend also of 2 percent. Cumberland *Evening Times,* January 18, 1905.

nized region, "where he claimed a union miner's life was in danger and he could not secure relief through the law for attacks upon him." With Fairmont cowed, with Somerset about to be subdued, the George's Creek men could reasonably expect the heavy hand of the "triple alliance" * to fall next upon them.

On December 1, 1904, Consolidation's superintendent B. S. Randolph resigned, and John C. Brydon was named to fill his place.[57] Brydon, previously employed by the Somerset Coal Company, was a protégé of J. H. Wheelwright, vice-president of Consolidation, and an advocate of "West Virginia methods." The Cumberland press had publicized his utterances during the Somerset County strike, and two of them bear repeating: "We will not treat with the organization in any way, and I am the one who will say who can work and who can not," [58] and "We are prepared to fight the [union] question until h--l freezes over and then to renew the matter on the ice. Rather than attach ourselves to a recognition of union which would cause us only more trouble, we would close all our mines in this region." [59]

The mining population looked upon Brydon's advent "as a token of change all along the line," and Consolidation employees ran "from one place to another whispering of what was going to happen in this mine and in that office." [60] Before the new superintendent had been in charge of Consolidation operations for a full year, it was said, and apparently with good reason, that the company "has completely changed in point of ownership and management, since the West Virginia Coal Combine has taken charge, and the methods prevailing heretofore have been revolutionized as far as conducting the affairs of the combine is concerned. In West Virginia it means that labor unionism is a bar to a man earning a livelihood; and such is the policy which that combine is pursuing in our region. This method is causing great

* The Consolidation, Somerset, and Fairmont Coal Companies "form the triple alliance that control[s] extensive coal properties in the George's Creek region of Maryland, the Somerset region of Pennsylvania, and the Fairmont region of West Virginia." Cumberland *Evening Times*, January 18, 1904.

bitterness and animosity in our hitherto peaceful valley. The United Mine Workers organization, not wishing to see the methods prevailing in West Virginia inaugurated in our county, sent into our midst one of its trusted officials, Mr. Wm. Wardjon, to advocate the cause of unionism; the first move made by the combine was to place home spies among our miners in order to ascertain the progress made by the advocates of unionism." [61] The background for this statement follows.

Although the Meyersdale strike was still going on, early in 1905 the UMW felt it imperative to send organizers to George's Creek to try to renew interest in the union.[62] E. S. McCullough returned to Maryland, where he was joined by Thomas Haggerty of Pennsylvania, and the two planned a series of meetings at which they would speak on "the rights, privileges and duties of labor." [63] Thus it appears that at first national headquarters intended to pursue the conservative line which had resulted in considerable success before the Meyersdale difficulty. It is not possible to determine why an entirely different type of organizer was sent to Maryland in April, when William Wardjon, a member of the national executive board from Kansas, came to the mining region to spend a few weeks studying the situation with the intent of "more thoroughly organizing the men in this vicinity." [64] Wardjon was a fiery little Welshman who soon distinguished himself by a series of speeches which "rubbed official hair the wrong way." [65] He was assisted by James L. Wilson, also a national board member, from Meyersdale.

In reporting the meetings held by Wardjon and Wilson, the Cumberland *Evening Times* added cryptically: "It has been observed also that there are other strangers here, whence no man knows nor whereof their business, nor quo vadis." [66] The most plausible explanation of this veiled remark is that company spies and spotters had already been brought in from other regions. Wardjon's first report, dated May 15, said in part: "Even to be heard talking unionism means discharge [by Consolidation], or if a company spotter and tale-bearer (and they are legion) learns that a man attends a union meeting or has union tendencies and opinions, he pays the same penalty." [67]

By August the existence of company spies had proved to be no myth. Wardjon reported thirty men discharged from one mine on August 10 "because these spotters had reported them for talking to the organizers, and last week [the first week in September] several more were discharged because they had been heard to express themselves in favor of the organization." [68] At a large open-air meeting in Frostburg on August 23, Wardjon publicly accused superintendent Brydon of having a blacklist and named four men present in the crowd as the company spotters who had furnished the names for the list.[69] During the uproar which ensued, the four "coal company deputies" behaved in such a manner that they were arrested on charges of disorderly conduct and carrying concealed weapons, and were held for the October term of court.[70]

Up to this time the "deputies" had been acting purely as private police, detectives, spotters, or, as Wardjon said, "would-be thugs." [71] On August 30, J. H. Wheelwright moved to give these four employees status under the coal and iron police act of 1880,[72] which had last been invoked during the strike of 1882. Under this act, certain types of corporations might ask the governor to commission policemen "for the protection of the property of said corporation . . . , and for the preservation of peace and good order *on their respective premises*." (Italics supplied.) Wheelwright prepared a petition asking for the appointment of E. C. Sperry, R. J. McLain, William England, and William Ray "to act as policemen for the preservation and protection of the property of [the Consolidation Coal Company], and for the preservation of peace and good order on its premises." [73] Only England was a citizen of Maryland. Wheelwright, with his Fairmont-Somerset background, did not approach the governor directly, but asked Van Lear Black, of the Black Sheridan and Wilson Coal Company, to present the petition to the governor on his behalf.[74] Wheelwright must have known that all four of these men had been arrested on serious charges, but nevertheless he wrote to Black: "We believe that these men, who are in our employ, are very reliable and will not abuse the privilege, and think it of interest to all concerned." [75]

Although Wheelwright in his petition did not cite any circumstances which made special police necessary at this particular time, the governor issued the commissions on September 8.[76] He later explained, in answer to criticism: "My rule has been to grant all such requests. The fact that they are asked for is accepted by me as sufficient reason for making the appointments, and I have not deemed it necessary to go into the motives prompting the request for such appointments, but have taken it for granted that corporations would not ask for the appointment of special officers, who would entail extra expense, unless they were needed." [77]

The four guards looked on the acquisition of their new "State Police" badges as license to go "around the county, upon the streets and among the citizens of Allegany County carrying concealed weapons, and brandishing them, causing uneasiness, indignation and fear of bodily harm among the peaceful inhabitants of the community." [78] On the very day that the clerk of the circuit court recorded the commissions at Cumberland, one guard was arrested for instigating and abetting an assault on organizer Wardjon, and another for carrying concealed weapons on the streets of Frostburg.[79]

The reaction of nonmining elements of the community indicates to what an extent Consolidation's new management and policies had alienated public opinion. Almost immediately the governor began to receive complaints that he was helping the company "to introduce West Virginia methods of mining in the George's Creek region." [80] The Allegany Trades Council protested the commissioning of the deputies as unwarranted and pointed out that they were exceeding their authority.[81] At a mass meeting held in Frostburg on September 17 several thousand county residents resolved to ask the governor to make a thorough investigation and determine whether the coal company's application was justified.[82] The Democratic party organization in Allegany County appealed for revocation of the police commissions "as being unnecessary and tending to introduce a system of espionage and terrorism un-American and undemocratic, and as obnoxious and inimical to our free institutions." [83]

The Reverend A. H. Thompson, a Frostburg clergyman, expressed his belief that Consolidation should have "maintained a more conciliatory attitude thro' its representative Mr. Brydon." * He felt, moreover, that the "appointment of private detectives to shadow [the labor organizers] constantly, and the attempt to intimidate the men from attending meetings in the interest of organization . . . touches on the rights of man, and must provoke trouble among free, independent and liberty-loving people anywhere." [84]

Still another private citizen urged the governor to revoke the commissions because "there was not, and is not now, the slightest danger to the property of the Consolidation Coal Company from any source. These men whom you appointed are simply used by this company as spies upon the employees." [85]

While friends of civil liberty were doing their best to have the police commissions recalled, Wardjon himself furnished the coal company with apparent justification for its concern. The *United Mine Workers' Journal* of September 14 contained a letter in which Wardjon accused the "vile and inhuman monsters at the heads of the Fairmount [*sic*], the Somerset and the Consolidation Coal Companies . . . and their general managers in control of the mines" of "malicious and murderous attacks upon our organizers" carried out by the companies' armed guards. A few sentences in this same letter implied that the French revolutionaries had the right idea in dealing with their wealthy oppressors. Possibly the reference to the French Revolution caused Van Lear Black to see blood in the streets and hear the tumbrils rolling, for

* In addition to his campaign against the organizers, Brydon had hit at the union by refusing to let the Consolidation men take part in the Labor Day parade. Baltimore *Sun*, September 11, 1905; and *United Mine Workers' Journal*, September 14, 1905.

he sent a clipping from the newspaper * to Governor Warfield "as a sample of the speeches being made by labor organizers now in the coal regions of Maryland." [86]

Shortly after Wardjon's letter appeared, J. H. Wheelwright asked Black to inform the governor that the town council of Frostburg had called a special meeting to muzzle the organizer, prohibiting him from making speeches "similar to the ones he has made lately" on penalty of being forbidden to speak at all within the borough limits.[87] In actual fact the town officials had taken no such action, and one of the councilmen protested circulation of the false report by a notoriously anti-labor publication.[88]

Meanwhile Francis J. Drum, president of District 16 and a former member of the Maryland legislature, was preparing to challenge the appointment of the special police on legal as well as on moral grounds.[89] In a letter dated September 20, Drum asked the governor to annul the appointments of Ray, McLain, and Sperry because they were not citizens of Maryland. "Surely the time has not come," he wrote, "when a respected Governor of Maryland is obliged to go out of his own State into West Virginia and Pennsylvania to employ conservators of the peace and protectors of property." Pointing out further that the appointees were not law-abiding persons, but "men of violent, reckless and ungovernable dispositions," he enclosed for the governor's information copies of docket entries by local magistrates before whom the guards had appeared on charges of disturbing the peace and the unlawful use of firearms. Drum asserted that there had been no threat to Consolidation property and that the special officers had been "imported into this State from West Virginia and Pennsylvania as experienced spies and spotters, to dog the footsteps of such miners in Allegany county as desire to

* Black did not bother to send the governor as a sample the 90-minute, noninflammatory speech of UMW's vice-president on "Trades Unionism" reported in the *United Mine Workers' Journal* of September 28, 1905.

avail themselves of the privileges conferred by law of joining labor organizations to promote their common welfare." *

The governor responded to Drum's letter by asking him to refer specifically to the law which prohibited the appointment of nonresidents as special police,[90] and consulted William S. Bryan, the attorney general of Maryland, as to the propriety of the appointments. Bryan replied that he was aware of no requirement that special police be residents of the state. Furthermore, he wrote, "As these special policemen are paid by the corporation and are designated by it, I do not see any reason why the Governor (in the absence of any cause to suspect any abuse of power by such special policemen) should feel called on to determine whether there is any occasion for their appointment." [91]

Drum's next letter to Governor Warfield was in the form of a brief, citing legal precedents to sustain his argument for annulment of the commissions. The document again called attention to the injustice of appointing nonresidents as special police. Conceding the right of a corporation to employ private detectives to protect its property, Drum denied that this was the purpose of the Consolidation Coal Company, which wished only "to hunt down those desirous of joining the members of the United Mine Workers and their leaders, under the *badge* and in the *name* of the *sovereignty* of the State of Maryland." [92]

Governor Warfield forwarded this second letter to the attorney general, who rejected as unsatisfactory Drum's legal arguments and repeated his earlier statement that he knew of no requirement that the officers must be citizens of the state.[93]

* Ch. 267 of the Maryland *Laws* of 1884 provided for the incorporation of trades unions "for mutual assistance in securing the most favorable conditions for the labor of [their] numbers." As we have already seen, Judge Boyd of the Allegany County Circuit Court, in his opinion handed down in connection with the 1894 strike, upheld the right of the miners to join the union. In 1900, speaking of organizer Warner, Judge Boyd had said, "He has both the right to belong to the union and to organize local unions wherever he wishes to" provided he proceeds in a lawful manner.

Although he did not budge from his position, it is evident that the governor had some misgivings about how the power he had conferred was being used. In the letter which informed Drum of the attorney general's ruling, he said almost apologetically: "I did not intend to reflect in any way upon the action of labor organizations, or to restrain working men in the exercise of their right to organize for their mutual protection." Furthermore, he added, "I have had the men referred to instructed as to their duties, and I hope that there will be no further objection found by members of your organization as to their conduct." [94]

In its report for the October term of court in 1905, the Allegany County grand jury requested the governor in future to confine his appointments of special police to residents of the state.[95] He acknowledged the recommendation and said that he would follow the suggestion. In the present instance, he wrote, he had had no knowledge that three of the men named by Consolidation were not Maryland residents. However, having made the appointments, he saw no good reason for revoking them.[96] This was the end of the affair so far as the executive department was concerned.

There still remained the disposition of the various cases in the circuit court. Those involving the carrying of concealed weapons were assigned on the criminal docket for hearing on October 17.[97] The state's attorney dismissed the case against E. C. Sperry for lack of evidence. The second case, *State* v. *Robert J. McLain*, was heard before the court.

In delivering his charge to the grand jury on October 2, Judge Boyd had called particular attention "to the law referring to the carrying of concealed weapons, stating that no person other than a sworn public officer should be permitted to go armed. The court was especially emphatic in regard to the many violations of this law, and impressed upon the grand jury the importance of indicting all such violations." [98] In the light of this charge, the court's decision in the McLain case is amazing. When McLain was arrested, he had not yet been commissioned as a special policeman and certainly did not fall into the category of "a sworn public officer." Nevertheless, "after a full hearing the Court decided that the Consolidation Coal Co. had a right to employ men as private policemen and that among their other duties they were required to attend the meetings held under the auspices of William Wardjon; that from the testimony in the cause it was evident that the defendant had a reasonable apprehension of personal bodily harm and that as his duties required him to go to a place where he was liable to be in danger that he would under the statute be justified in protection and that the utterances made at the meeting would have a tendency to influence the minds of the speaker's hearers. Therefore there must be a verdict of not guilty." [99]

None of the remaining cases, either for carrying concealed weapons or for disorderly conduct, was heard; some were marked "non pros" and others simply "dismissed." [100] According to Drum, while court action was pending many more private guards came into the Maryland region.[101] The executive files do not contain requests for commissioning these men.

Let us now return to Wardjon, whose final burst of activity in the region may well have alienated those union members and prospective members who had not already been discouraged by company spies. Wardjon had been viciously attacked by the Lonaconing *Star*, a rabidly anti-union paper which was given to intemperate statements about the character and reliability of organization leaders. Republican in politics, the *Star* in 1905 supported the candidacy of David E. Dick for the Maryland Senate.

Dick had a long record as a champion of the miners. He had been arrested on charges of criminal conspiracy during the strike of 1894, when he had been identified as "a prominent local agitator." [102] Elected to the state Senate in 1897, he sponsored many labor bills and pushed through a revision of the mine inspection law, which had not been changed since 1878. In 1899 he helped William Warner organize the region under the UMW prior to the strike of 1900,[103] and during the strike he made at least three speeches in favor of the miners' cause.[104] Warner himself had described Dick as a man possessing "all that goes to make up a noble character, [who] even would give his life for his fellow workman."[105] Dick had, therefore, a large following in the mining towns.

In a letter refuting the *Star's* accusations, Wardjon referred to its editor as "bought and owned by the Consolidation Coal Company" and a "sworn enemy of organized labor," and warned the mining population against voting for Dick, presumably on the guilt by association principle.[106]

The Cumberland *Evening Times*, long identified as a Democratic newspaper, joyfully promoted the "contest between Senator Dick, backed by his friend of the Star, and Mr. Wardjon and the United Mine Workers."[107] During the campaign Wardjon made "some imprudent speeches" which conservative miners resented.[108] Many far-sighted friends of labor realized that the creation of factions at this particular time would be disastrous. Warning of this danger, "A Worker" wrote in a letter to the *Frostburg Mining Journal* of October 28: "There is a broad hint in circulation that the corporations are almost 'dying a-laughing in their sleeves' at the division caused in the ranks of union labor by the assaults made on Hon. David E. Dick by Mr. Wardjon, and *vice versa*." Dick's defeat would so antagonize his adherents that they would desert the union; Dick's victory would so discredit Wardjon as to have the same effect.

The extent of the division was demonstrated on election day. In this normally Republican area Dick's Democratic opponent defeated him by a plurality of only sixty-three votes.[109]

Wardjon left the region at the end of November to attend a United Mine Workers' conference at Indianapolis.[110] He did not return. At the end of November 1905, after seven months of intensive organizing in Maryland, the paid-up membership for all of District 16 was only 246.[111] It is not possible to determine whether the organizing drive failed primarily because Wardjon's interference in politics created "two bitter factions among the men themselves in the so-called organization move,"[112] or because "the Consolidation Coal Co. interests [had] put every possible barrier in Wardjon's way so as to prevent a close organization of the men being formed."[113]

When the United Mine Workers met in annual convention in January 1906, their president, John Mitchell, denounced the "aggregation of union wreckers, who have antagonized every effort we have made to organize the non-union fields," including Meyersdale and George's Creek.[114] The Meyersdale strike alone had cost the union $395,000, largely because of the opposition of the Consolidation-allied Somerset Coal Company.

Discussing the possibility of a national coal strike in the spring of 1906, the *Coal Trade Journal* reported on April 4: "At the offices of the Consolidation, Fairmont and Somerset Coal Cos., which combination represents the largest of the producers . . . , it was stated . . . that it was not anticipated that their men would walk out. The union is not recognized in the companies' mines, and it is understood that anything like unionism in solid array has been well wiped out in their mines in the Meyersdale and Somerset regions of Pennsylvania. The Maryland holdings of the Consolidation in Allegany, and the Fairmont region holdings (West Virginia) are practically without union workers, it is stated, and no trouble is anticipated in those directions."

Before 1905 few Maryland miners had been penalized merely for joining a labor organization. Almost all of those blacklisted at any time between 1870 and 1901 had been prominent in strikes or had otherwise made themselves obnoxious to some company superintendent. The regime of the triple alliance placed Maryland miners on a par with those in West Virginia, who dared not even be seen talking to an organizer lest they be discharged. By 1907 company spotters lurked in every George's Creek community, "gathering every bit of information they can concerning the movements of the union men. . . . They try to secure the names of the miners who join the union and . . . these men are given their walking papers by the companies."[115]

Although the UMW continued to claim locals in a few George's Creek locations and at Meyersdale, membership shrank almost to nothing. In June 1909, District 16 disappeared from the "Official Roster of U.M.W. of A." published in each issue of the *United Mine Workers' Journal.* In August 1909, W. B. Wilson listed Cumberland, Meyersdale, Virginia, and West Virginia as coal fields in which no organization existed.[116]

Money power, combined with suppressive methods, had succeeded in placing Maryland unmistakably among the unorganized fields. Consolidation, "the first large producer to make a determined fight on union domination," [117] in 1906 purchased all the stock holdings of the Baltimore and Ohio Railroad in the interlinked companies,[118] gaining control over the Pittsburgh and Fairmont Fuel Company and the Clarksburg Fuel Company, associated with the Fairmont Coal Company. In 1909 Consolidation completely absorbed these companies, along with the Fairmont Coal Company and the Somerset Coal Company, and became the largest coal mining concern in the United States.[119] In the same year, the corporation declared a cash dividend of $1,009,882.50 and a stock dividend of $6,150,000.[120]

Chapter 20

Sequel

The Union

For almost seventy years after the Civil War, the Maryland coal operators kept labor organizations from gaining more than a temporary foothold in the George's Creek region. Recognition of the union came only with the legislation of the New Deal, following a decade of decline and depression in the bituminous industry.

World War I brought a revival of unionism to UMW's District 16, which had been inactive since 1909. In the summer of 1916 the United Mine Workers began an organizing campaign in Maryland and the adjoining counties of West Virginia, and by the end of the year claimed an enrollment of 80 percent of the coal employees.[1] By late October of 1917 every mine in the district had a local union.[2]

The corporations bitterly opposed organization in 1916,[3] and even after the establishment of the wartime truce between management and labor, the Consolidation Coal Company was accused of anti-union activities in its West Virginia and Maryland divisions.[4] In a number of communications to Harry A. Garfield, the United States fuel administrator, District 16 brought specific charges against the company, which, it alleged, had not only refused to meet with the miners to discuss their grievances, but had also exerted pressure on the other operators of the region to decline a union invitation to a joint conference.[5] A resolution adopted by Local 2819, Frostburg, forwarded to the fuel administration, complained of the "arrogant and kaiser-like" attitude of Consolidation officials in refusing to adjust grievances relating to overcrowding, sanitation, and ventilation, and in discriminating against and blacklisting miners who protested against the working conditions.[6] The officers of District 16 asked the fuel administrator to arrange a meeting between officials of the Maryland division of Consolidation Coal and representatives of the United Mine Workers. Although an interoffice memorandum to the administrator reported a "nasty situation" in Maryland and recommended the calling of a joint conference,[7] nothing in the administration files indicates that action was taken at this time.

The miners of District 16 therefore held a special convention at Cumberland from February 28 through March 2, 1918, to prepare a list of grievances and forward them to Washington.[8] The stenographic transcript of the convention consists of 222 pages of detailed reports of complaints against every operator in the district.*

It was not until May, after a series of delays inspired by the operators, that meetings—not joint, but separate—took place between the fuel administration and the operators and between the fuel administration and the miners. J. H. Wheelwright, president of Consolidation, speaking for the operators before the meetings, reported that they "definitely declined to meet the representatives of the Mine Workers" in spite of Garfield's request that

* That complaints of short weight, at least, were not capricious, is borne out by the report of the U.S. Bureau of Standards on tests of scales made on a spot sample in Maryland mines in August 1917 at the request of the Allegany County state's attorney. The bureau found that at Consolidation's Hoffman mine the scale was out 620 pounds in favor of the company, an error of about 15 percent of the net weight of every car. The company scalemaster knew that this condition had existed for at least four years. At the mine of the Phoenix and George's Creek Mining Company near Westernport, the weights used by the company had been tampered with by the insertion of lead plugs disguised by a coat of black paint, resulting in an advantage to the company of as much as 5 percent of the weight of every car. Throughout the region cars were mismarked as to their weight, and there was a general practice of weighing while the cars were in motion.

they get together and try to "harmonize their differences." [9] The administration thereafter prepared an agreement which it submitted to operators and miners, asking for acceptance by May 11.[10] After slight adjustments in wording, both parties signed the agreement, which was in effect by the end of May.*

The contract provided for the handling of grievances by committees, the employment of checkweighmen, and the equitable distribution of cars. While they did not fully recognize the United Mine Workers and provided no arrangement for check-off, the Maryland operators acknowledged the right of their employees to join a union and agreed not to discriminate against union members. The representatives of District 16 agreed not to force recognition "during the life of the agreement," but reserved the right "to exercise our persuasive powers in order to induce the operators to treat with our organization." [11] The understanding was to remain in effect for the duration of the war, or in any event for at least two years. Thereafter it would be subject to revision on ninety days' notice by either of the parties.

Although as it later proved, the union in securing the agreement had at best established a temporary *modus vivendi*, the wartime experience was valuable in that it ended the ancient misunderstanding between the Frostburg and lower-creek miners. It also removed their distrust of outside organization and convinced them of the effectiveness of supporting the union even on nonlocal issues. When, on November 1, 1919, John L. Lewis, acting president of the UMWA, called a nation wide strike in

* Consolidation had planned to discourage union activity in its mines by drawing up its own agreement "before we have the intervention of the United Mine Workers." Forestalled by the union's complaint in Maryland, Consolidation intended to go ahead in West Virginia "while we have the right to deal with our men direct, to take advantage of this opportunity and put in a plan that is entirely satisfactory [to Consol, no doubt]." J. H. Wheelwright to H. A. Garfield, May 8, 1918 (Fuel Admin. Records, Executive Office General Correspondence, Box 16). A copy of the proposed agreement for the Maryland, West Virginia, Pennsylvania, and Kentucky divisions of Consolidation is contained in Fuel Admin. Records, Box 1359.

the bituminous fields, 4,000 George's Creek workers joined in support of union demands for a 60 percent wage increase and a six-hour day for miners in other states.[12] The Maryland men were still working under a contract not due to expire until May 20, and were said to be generally satisfied with their wages and working conditions.* Nevertheless, they followed the national leadership, picking up their tools again only after Lewis ordered all miners to return to work at the 14 percent wage increase suggested by fuel administrator Garfield pending the decision of the Bituminous Coal Commission created by the President of the United States.[13]

Similarly, Maryland miners joined in the soft coal strike which began April 1, 1922, when John L. Lewis called for a general suspension after operators in the central competitive field failed to negotiate a new contract with the union.[14] However, after the national strike ended on August 15, Maryland miners did not return to work, but carried on a fifteen-month struggle for union recognition. After pouring approximately three-quarters of a million dollars into this historically nonunion field, the international executive board of the UMW called off the strike in November 1923.[15] "West Virginia methods" characterized the operators' resistance, and for the first time since the development of the region, a coal strike in Maryland was accompanied by considerable violence.[16] The companies brought in, and allegedly armed, strikebreakers from Pittsburgh, Cleveland, and various West Virginia fields.[17] The Cumberland *Evening Times* reported on August 27 that the Consolidation Coal Company had "a large number of imported miners . . . housed in barracks at Eckhart." Two of Consolidation's armed guards were involved in the shooting of a miner on a public road where there was no question of trespass.[18] One of these guards had

* The sources differ on the attitude of Maryland men toward the strike. District 16's president charged the companies with violation of the contract in respect to short weight, overcrowding, bad air, unsanitary working conditions, and the failure to offer steady work. The companies at the same time were allegedly circulating rumors that the men were striking reluctantly.

previously been charged with assault with intent to kill after he threw a hand grenade into a picket line.[19]

District 16 voted to abide by the decision of the international board, and the men applied for reinstatement by their former employers. The superintendents placed them on waiting lists and told them they would be taken back when vacancies occurred if they were "acceptable." "Radicals and others who [had been] the source of trouble in the past" were not to be considered for reemployment.[20] Most mines resumed under open shop conditions.[21]

At the beginning of the 1922 strike, the union had claimed more than 5,000 members in the Cumberland coal region, not specifying whether this was a "paid" or a "total" membership.[22] At the end of the strike "paid" membership for all of District 16 was 371; the "total" membership, 1,887.[23] After this, the United Mine Workers in this area "all but went under." [24] In July 1925, Consolidation, which had eleven mines in its Maryland division, "announced publicly that it would operate non-union," [25] discontinued its operations in union properties, and resumed after the organization of company employee associations. By May 1927 the corporation had become entirely nonunion.[26] As in the past, the other operators followed Consolidation's lead, and the Maryland mines became almost entirely open shop.[27] However, in the Lonaconing area a few men in Local 2835 kept the cause of unionism alive.[28]

The loss of Maryland in 1923 was only one incident in the progressive decline of the UMWA, which, after a peak membership of 484,000 in 1922, could claim only 150,000 dues-paying members in 1932.[29] Even the Clearfield region, staunchly organized since the days of John Mitchell, was operating as nonunion.[30]

By 1932 the bituminous industry itself was in an unenviable position. Consequently, when the National Industrial Recovery Act passed in 1933, the northern Appalachian operators, at least, were willing to see what it might accomplish. Even before promulgation of the bituminous coal code, miners in all fields were flocking back to the union. During the last half of 1933 Maryland miners began to rebuild their locals, and at the close of the year the district was almost 100 percent organized.[31] Under the bituminous code, operators of the George's Creek and Upper Potomac Coal Control Association signed a working agreement with District 16 of the UMW on September 29, 1933.[32] Renewed in April 1934 in conformity with the Appalachian wage agreement, the contract between the operators and their employees applied to roughly 3,500 mine workers in the district and gave them a seven-hour day, five-day week, the right to checkweighmen, provision for handling grievances, semimonthly pay, and a wage for pick mining of 72.2 cents a short ton. It also forbade the employment inside the mine of any one under the age of seventeen.[33] The union shop and the check-off system in the region were generally in practice by 1939.[34]

Once organized and recognized, the Maryland miners remained firmly attached to the United Mine Workers and to the leadership of John L. Lewis. They supported the strike of 1939—the greatest bituminous shut-down in seventeen years—in which Lewis fought to have the UMWA recognized as the sole bargaining agent for the coal miners.[35] They joined Lewis' controversial wartime strike in 1943 for a contract which would include portal-to-portal pay.[36] In 1946 Maryland miners went along with the suspension called until operators agreed on a contract providing for contribution to a miners' welfare fund.[37] In 1948 they contributed to the "pension paralysis" which resulted from the operators' refusal to support a plan for pensioning miners aged sixty-five who had worked for twenty years or more.[38] And, in the last gasp of the region's significant economic life, its two thousand miners accepted the three-day week ordered by Lewis in the summer of 1949 for mines east of the Mississippi in order to stabilize employment in the bituminous fields.[39]

After six months of part-time work interspersed with periods of complete cessation in 1949–50, many of the Maryland mines did not reopen. Thereafter, in a moribund industry employing progressively fewer men each year, the question of unionism became less important than the question of mere survival.

The Industry

Maryland reached its peak production in 1907.[40] For the next eleven years, annual production settled down to a steady 4 to 4.5 million tons, rising to 4.75 million tons in 1917.[41] War prosperity brought about the reopening of properties idle for years and thought to be worked out, but the shortage of men throughout the region prevented working to full capacity.[42] In 1920, just before the onset of postwar depression, the George's Creek mines produced roughly 4 million tons and employed about 5,500 persons.[43] This was the last of the good years for the region.

The early 1920's marked the beginning of more than fifteen years of decline for the bituminous industry as a whole. The wartime demand for coal and the postwar rise in prices had led to the opening of many new mines. In 1915 there were 5,500 working coal mines in the United States; in 1919, almost 9,000.[44] Although production and prices began to fall after 1920, the number of mines increased until 1923. Coal profits disappeared in an era of unbounded competition. Between 1925 and 1940 bituminous operators reported an aggregate net loss of over $350 million.[45] Consolidation and one other company sustained almost a third of the loss.[46]

Between 1922 and 1932, Maryland production fell off to "the general level which prevailed between 1880–1885," that is, 1.5 to 2 million tons.[47] The number of operating companies dropped from 99 in 1923 to 72 in 1930, and of this number only a dozen were large enough to produce more than 50,000 tons a year apiece.[48] Of the pioneer companies in Maryland, Consolidation and the George's Creek Coal Company were still in operation. Consolidation was the principal employer, with a working force of more than 800 in 1930. Only nine other companies employed more than 100 men each, and 41 companies out of the 72 operating in 1930 employed fewer than 10 persons each. Twenty-one operators had only one or two men apiece.[49]

The mining force dropped from 5,548 in 1920 to 3,755 in 1922, and then declined only slightly until 1933. The actual decrease over the decade 1922 through 1932 averaged sixty-five men a year.[50] Because of the relatively small number of men involved and the gradualness of the decrease, the drop in mine employment caused no great distress. The region was not isolated, and transportation facilities made it possible for the unemployed to seek work elsewhere. In 1924 about 900 former miners worked at local manufacturing in Cumberland, and many others got jobs in the automobile and rubber industries in Michigan and Ohio.[51] Throughout 1925 and during much of 1926 miners were actually in short supply,[52] and when the great depression arrived, there was no existing large body of unemployed mining personnel in western Maryland.

In 1929, because of severe competition from West Virginia and Pennsylvania, few Maryland operators made profits.[53] By the early 1930's unemployment was widespread in the region's nonmining industries as well as in the mines.[54] Although they were selling coal at less than cost, most of the George's Creek companies kept going after a fashion, and by dint of reduced wages and part-time operation avoided wholesale dismissals. However, by the end of 1932 many mine families were in need of food and clothing. The coal companies in most cases did not collect rents on company houses, arranged for credit at grocery stores, divided the work among their employees, and established cooperative welfare projects.[55] On the eve of the New Deal a Maryland miner wrote, "Conditions here are deplorable, with no future prospects." He did not blame the employers, but explained that large operations in West Virginia were taking the market, while in his own region miners were reduced to two or three days' work a week at $2.50 a day.[56] Some of the smaller companies were operating only one day in every two weeks.[57]

Consolidation's employment policy during these difficult

years is to be commended. Although the company was in dire financial straits * and went into receivership in 1932,[58] it nevertheless continued to operate seven mines in its Maryland division with a minimum layoff of men.[59]

After passage of the National Industrial Recovery Act in June 1933, but before promulgation of the bituminous code, Maryland mines began working nearly every day, though profits were "unsatisfactory." Once the code had gone into effect, production began to rise slightly, continuing until 1948, but even during the war period the most the region turned out in any one year was a little over two million tons.[60] Employment, on the other hand, gradually decreased, the diminution being due in part to more efficient operation and in part to the attraction of war industries and other jobs which paid more than mining.[61]

Between 1937 and 1945 Pennsylvania bituminous production increased at the rate of seven million tons a year, and West Virginia production at the rate of eight million tons.[62] It was obvious that the Maryland industry was long past hope of recovering its former competitive position. The region's largest producer, Consolidation,† disposed of its Maryland properties by a lease-sale arrangement in 1944 and sold its Cumberland and Pennsylvania Railroad to the Western Maryland Railway Com-

* The magnitude of the financial problem is indicated by the 100 percent loss (about $25 million) sustained by the Rockefellers, who had a 38 percent interest in the company. Consolidation was reorganized in 1935 under Section 77-B of the National Bankruptcy Act. Love, *op. cit.*, p. 14; and "Coal: The 'Pitt Consol' Adventure," *Fortune*, July 1947, p. 104.

† On November 23, 1945, the merger of the Consolidation Coal Company with the Pittsburgh Coal Company under the name of Pittsburgh Consolidation Coal Company resulted in the formation of the largest commercial coal company in the world. "Coal: The 'Pitt Consol' Adventure," p. 104; and Consolidation Coal Company, *Consol News*, 100th Anniversary Edition, Vol. III, No. 1, First Quarter 1964, pp. 8–9. The "Pittsburgh" was subsequently dropped from the company name. In September 1966 the Continental Oil Company purchased the coal properties and related assets of the Consolidation Coal Company. Consol's mines produced 43 million tons in 1966, and revenues from its properties during the last quarter of the year were over $84 million. *Moody's Industrial Manual*, June 1967, p. 2,172.

pany in the same year.[63] Another chapter of Maryland coal history closed in 1952 when the George's Creek Coal Company, chartered in 1835 as the George's Creek Coal and Iron Company, went out of business.

After 1948 the Maryland companies found it increasingly difficult to compete with the larger, highly mechanized fields, and their annual production dropped well below the million-ton mark. Following the labor disruptions of 1949–50, as we have seen, a number of George's Creek and upper Potomac companies did not reopen, and sold or leased their properties.[64] Many of the concerns still operating in 1952 were small and inadequately financed.[65] Employment decreased from 2,049 in 1948 to 772 in 1952.[66] In 1954 the region produced little more coal than it had in 1854, and the working force had been reduced to about 600. Hardship fell most heavily on the older men, who at fifty were generally considered beyond the hiring age for mines and could not move to more prosperous regions.

In 1966 the Maryland coal industry employed a few more than 350 men. The center of mining has moved from Allegany to Garrett County. Strip mining has greatly expanded and annual production since 1963 has exceeded one million tons. Of the sixty-four companies operating in 1966, less than a dozen individually had ten or more workers. Many truck operations employ only one or two men, but even the larger companies are said to be working on a profit margin which precludes the payment of the union scale. There is, however, no hard-core group of unemployed mine workers; on the contrary, men with the appropriate skills are in demand.

The Land

Although strip mining began in the Maryland region in 1941, no state law governed this type of operation until 1955. By that time stripping accounted for almost one-half of total production. The 1955 law[67] required licensing and the posting of a bond of $100 per acre to insure compliance with reclamation provisions, which included filling, grading, and planting according to the specifications of the Maryland Bureau of Mines. In practice the statute in many cases merely resulted in forfeiture of the bond and abandonment of the stripped area as soon as it became unprofitable.[68] Raising the bond to $500 an acre in 1965 [69] had little more effect. Continued agitation by conservationists brought about the passage of a new strip-mining law in 1967,[70] but an examination of its provisions leads one to believe that it will be no more efficacious than the earlier statutes unless the operators show unprecedented cooperation. On the basis of hearings held in August 1967, the Maryland Bureau of Mines has published detailed rules and regulations governing strip mining operations and reclamation procedures, including provisions for replanting in accordance with the recommendations of the state forester. Since the regulations have been in effect only since October 1, 1967, it remains to be seen whether they will in fact bring about compliance with the letter and the spirit of the law.*

The western Maryland counties have by no means become a small-scale Kentucky.[71] But the fact remains that the rich scenery, the bold and picturesque views, the high mountain ridges with dense slopes and crests of forest trees, the stands of oak, chestnut, hickory, maple, and pine which delighted the eye of Yale's Benjamin Silliman in 1838[72]—all have been ravaged by indefensible mining methods which in the last quarter century have despoiled the George's Creek Valley and the high plateaus of Garrett County. In a hundred years of underground working the operators did not destroy the surface as it has been destroyed by strip mining in the past two decades. Mountain tops have been cut and terraced, leaving high bare walls which cannot be masked by planting. Tons of overburden have been pushed down the slopes in avalanches of sterile soil and boulders. Even when reclamation has been carried out, the land resembles a crazy quilt of improbable design and moth-eaten texture.

* In the spring of 1969, while this volume was in press, the Maryland legislature found it necessary to adopt still another law for the restoration of strip-mined land.

Appendixes

Appendix A

Coal or Coal and Iron Companies Incorporated in Maryland (1828-1849)

Legis. Session	Chapter	Company	Capital Stock	Authorized Acreage
1828	170	Maryland Mining Co.	$ 200,000 [a]	— [b]
1834	152	Savage Coal and Iron Co.[c]	10,000 to 50,000	5,000
1835	306	Town Hill Mining, Mfg. & Timber Co.[c]	200,000	—
1835	324	Boston and New York Coal Co.	500,000 [d]	5,000
1835	326	Allegany Mining Co.	100,000 plus 500,000 for every 1,000 acres added	4,000 plus land acquired in collecting debts
1835	328	George's Creek Mining Co.[e]	300,000 in addition to land	10,000 [f]
1835	334	Cumberland Coal Mining Co.	1,000,000	12,000 [g]
1835	365	Clifton Coal Company	500,000	5,000
1837	—	Maryland and New York Iron and Coal Co.	—	—
1838	356	Western Coal and Iron Co.[c]	20,000 to 500,000	5,000
1838	358	Western Coal and Iron Mining Co.[c]	20,000 to 500,000	5,000
1838	389	Mining and Mfg. Co. of Hibernia	15,000 to 500,000 [h]	5,000
1840	214	Cumberland Coal and Iron Co.	5,000,000	5,000
1840	229	Potomac Coal and Iron Co.	5,000,000	5,000
1843	272	Barrellville Mining Co.	500,000	5,000
1844	135	Frostburg Coal Co.	500,000	5,000
1844	220	New York Mining Co.	500,000	5,000
1845	137	John Brant's Iron and Coal Co.	500,000 plus 500,000 for each 1,000 acres added up to 4,000	4,000

1845	255	Washington Coal Co.	100,000 [i]	500 [i]
1846	318	Borden Mining Co.	200,000	10,000
1847	229	Valley Mining and Mfg. Co.	200,000	5,000
1847	289	Preston Coal Co.	250,000	2,000
1847	296	Buena Vista Coal Co. of Allegany County	500,000 to 1,000,000	2,000
1847	306	Withers Mining Company	1,000,000	5,000
1849	56	Parker Vein Coal Co.	500,000	3,000
1849	152	Westernport Coal Co.	200,000 to 1,000,000	10,000
1849	153	Swanton Coal and Iron Co.	500,000 to 1,000,000	5,000
1849	169	Astor Coal Mining Co.	500,000	2,000
1849	316	People's Mining Co.	200,000 to 1,000,000	10,000
1849	412	Llangollen Mining Co.[e]	150,000 to 250,000	—

[a] Increased to $1,000,000 in 1835.

[b] 5,000 acres by act of 1835.

[c] Not authorized to build railroad.

[d] Increased to $3,000,000 in 1841.

[e] Name changed to George's Creek Coal and Iron Co. same year.

[f] Value of land to be set by disinterested persons.

[g] No more than $600,000 to be invested in land, to be valued by disinterested persons.

[h] Increased to $1,000,000 in 1840.

[i] Stock increased to $350,000; land, to 2,000 acres in 1849.

Appendix B

Rules of the Lonaconing Residency

1. Every Department of the Works, whether Mining, Carpentry, Blacksmithing, Masonry, or Digging of any kind, is under the charge of a particular Manager, by whom all the hands in the respective Departments are immediately superintended and to whom every hand in his particular Department will be required to pay entire respect and obedience. These Managers are selected in the discretion of the Superintendant; and report to him the State & progress of the work, in their particular charge. They also report the time and wages of the hands in their Employ; by which Report the payments for wages are all governed.

2. Every person in the Employment of the Company will be required to be present at work on every day in the year excepting Sundays and Christmas Day: And the hours of Employment (excepting in special cases, which the Superintendant allows in his discretion) shall be from Sunrise to Sunset, with such intermission for meals as shall from time to time be appointed.

3. Signals are given, by tolling the great Bell of the Company, of the hours for beginning and leaving work; and every person, who is not in his place in the proper time, will be expected to account for his absence to his immediate Manager.

4. Absences from work will be punished by abatement from the wages at the discretion of the Manager; and where such absences are frequent, or without sufficient excuse, the individual who is thus guilty, will be discharged and reported by the Manager to the Superintendant.

5. No distilled spiritous Liquor shall be sold on the grounds of the Company, nor shall any distilled Liquor be used by any person, whilst he is actually engaged at work for the Company. Intoxica-

tion at all times and places is strictly prohibited. The Managers of the several Departments are required to enforce the strict observance of this Rule, by dismissing immediately from Service any person or persons under their respective charges, who may be guilty of a violation thereof. The Superintendant requires its observation also on the part of the Managers themselves; and all others in the Service of the Company are likewise in such case directly responsible to him.

6. All Brawling, Quarrelling, Fighting and Gaming are prohibited. The firing of guns, which has been more frequent of late than usual in the Valley of the works is also, as a dangerous and unnecessary amusement, forbidden in future. The Managers must aid in the enforcement of this Rule by reporting to the Superintendant all violations thereof, which may come under their knowledge.

7. The Company expects all persons in its Service to observe an orderly and decorous conduct towards each other; and if any controversy should arise between such persons, the Superintendant will, on application, use his good offices and authority to reconcile the parties on just and proper grounds.

8. Every person in the Service of the Company, whether employed by the day, month, or year, shall be liable to be dismissed at the discretion of the Superintendant and on his dismissal shall leave the grounds of the Company.

9. In like manner the Superintendant claims the power of determining at any moment any lease or renting of the Companys grounds or houses, whether the same be made by the day month or year: and it is expressly declared that no tenant of the Company shall board or lodge any person, who may be discharged from the service of the Company, nor shall he be permitted to engage in any manner in buying or selling distilled spiritous liquors on the Companys grounds. When work is contracted to be done by the Job, it is understood, that it is to be subject to the acceptance or rejection of the Contractor; and all persons entering into such contracts are to be subject to the operation of these Rules, and to be liable to have their contracts rescinded for a violation thereof.

10. Monthly settlements will be made with every individual in the regular employment of the Company. From the sum due him for labor will be deducted the amount of the accounts against him, at the Store, Mills and Post-Office, and his contribution for the Doctor and School fund as hereafter mentioned, and the balance

will be paid him in money or a Check on Baltimore or Cumberland, at the discretion of the Superintendant.

11. For promotion of the general health a Physician is settled at Lonaconing. Every able bodied man in the Service of the Company, who will contribute out of his wages a monthly sum of fifty cents for the support of the Physician, will be entitled to medical advice and assistance for himself and family without further charge.

12. The Physician will make a daily Report to the Superintendant of the general health of the Village, and the numbers of deaths and births within its limits; and he will also, from time to time specially report all nuisances, which may affect the health or comfort of any portion of the Inhabitants.

13. Every tenant of the Company is expected to preserve neatness and cleanliness about his premises; and more especially to pay attention to the condition of the Stye or other places, where he may keep his hogs and pigs. The Superintendant will insist on the observance of this Rule and will prohibit the keeping of hogs or pigs by any person who may disregard the same.

14. After the first day of January next [1839] no Dog shall be kept on the Company grounds, without special permission of the Superintendant, who may withdraw the same at any time in his discretion.

Appendix C

Constitution of Miners and Laborers' Protective and Benevolent Association of Allegany County, Maryland

Name

This Association shall be known by the name of "The Miners and Laborers' Protective and Benevolent Association of Allegany County, Maryland."

Objects

The objects of this Association shall be to promote the best interests of the Miners and Laborers socially, politically and morally, for the protection of their health and lives; to spread intelligence among them in general on all questions affecting their labor; to abolish as far as possible, all strikes; and to adopt, in their place, ARBITRATION.

Further to urge the necessity of all its members who are not citizens to become such that they may use their rights and privileges.

Membership

Any person working in or about the Mines of Allegany County, Maryland, sixteen years of age and upwards, may become a member of this Association by paying an entrance fee of fifty cents for the first three months, and signing this Constitution.

Organization and Officers

Section 1. In order of organization each pit, or mine, shall be deemed a Society; they shall have a President and one Vice-President, Secretary, Treasurer, and a Delegate, all of which shall constitute an Executive Committee. Each pit, or mine, shall have a Delegate whose duty it shall be to attend all Delegate Conventions. The officers of each pit, or mine, shall be elected every three months.

Section 2. The County Executive Board shall consist of a President and one Vice-President, Secretary and Treasurer, and a delegate from each pit, or mine. These officers shall constitute the Executive Committee, whose duty it shall be to have the oversight of all matters pertaining to the interests of the miners in the county, and to call special meetings. They shall be elected semi-annually, the first election to occur at the first meeting of the county convention.

Duty of County Officers

The County Secretary shall act as general Secretary for the County. He shall keep a directory of all the Secretaries of the various mines, and keep them posted from time to time in all things pertaining to their interests. The Society shall furnish him all newspapers necessary to keep him informed of the condition of coal matters in the States, and he shall receive such compensation for his labor as the Society may decide.

The County Treasurer shall hold such amounts of money as the County Executive Committee shall from time to time determine, to defray the expenses incurred in carrying out the objects of the Association. He shall give such security for the faithful performance of his duty as the Board may decide, and the Treasurer of each mine, or pit, shall give such security for the monies they may hold as the various mines, or pits, may determine.

Contributions

Each member of the Association shall pay the sum of fifty cents per month, which shall be used for no other purpose than to defray the expenses incurred in carrying on the Association. All monies shall be kept at the mine where collected, but the County Executive Committee may order a levy from each mine to be paid to the County Treasurer to be used for County purposes.

Benefits

1. Every member of this Association who has signed this Constitution in case of disability, or injuries received in his regular employ-

ment, shall receive the sum of five dollars per week for six months and two dollars and fifty cents per week thereafter until he is able to resume his employment.

2. Upon the death of a member entitled to the benefits of the Association, the sum of thirty dollars shall be appropriated for funeral expenses; and the President shall notify the different Lodges to attend his funeral.

3. That each Lodge shall manage their own affairs to suit themselves; pass such By-Laws as suit them; *provided*, such are not in conflict with this Constitution.

Appendix D

Work Regulations Published by District Assembly 25, Knights of Labor (1881-1882)*

The following resolutions have been adopted as a guide for the miners and laboring men of this coal region:

1. *Resolved,* That no miner in the employ of any Company in this region do any day's work underground for less than $2.50 per day, and any man that may be called to do any work may make his own contract, but in no case may he go below the standard of laborers' wages at any mine in which he is employed; and in no case whatever shall cars be made up; and in no case compulsion be used.

2. *Resolved,* That one half cent per ton for blacksmithing on and after February 1st, 1880, be the standing price paid.

3. *Resolved,* That we restrict the hours of labor to ten hours per day as a standing day's work, beginning at 7 a.m. and stopping at 5 p.m.

4. *Resolved,* That five cars be a standard day's work for two men.

5. *Resloved,* That boys at the age of 12 years are entitled to a half turn, but they shall come to the mine and claim said half turn; and that the oldest son of a widow shall be allowed a full turn at the discretion of mine boss, provided, the father died while in the employ of said company.

6. *Resolved,* That all timber shall be delivered at the place of working.

7. *Resolved,* That laying up roads for pillars and timbering old places be done by the day at $2.50 per day.

* Published as an advertisement in the *Frostburg Mining Journal* May 7, 1881, through September 2, 1882.

8. *Resolved*, That all heading work shall be 78 cents per ton, and ten cars for turning off rooms paid at the same price.

9. *Resolved*, That we, your employees, shall not maintain any driver or laborer in not performing his duty faithfully.

10. *Resolved*, That we shall not sustain any recklessness or carelessness to the destruction of life and property, for should any be guilty he shall throw himself liable to the discretion of the boss and must suffer the consequences.

11. *Resolved*, That we shall not maintain any laborer who shall cease before his time is up; and with respect to the drivers we shall not maintain any pulling up at half-past four if he should be required to make another trip if it should take him to quarter past five; or, in other words, give one day and take the other.

12. *Resolved*, That should any one be discharged it shall be the duty of any who are able to take his place until the case is investigated.

13. *Resolved*, That any man loading over two tons five upon a car shall be skipped one car for each offence.

14. *Resolved*, That 16 feet shall be the distance to shovel coal before getting a road laid up to the excavation.

15. *Resolved*, That the Companies shall furnish the tools usually furnished for digging this coal.

16. *Resolved*, That the distance for miners to run cars down their rooms be limited to 400 feet, and, for the first additional 100 feet, he shall receive 5 cents per ton extra. Over 500 feet, heading price to be paid.

Appendix E

List of "Victimized Members" Following Strike of 1882 (together with the "number in each family")*

L. A. 718, Frostburg

Rees D. Rees	6	William Burnel	8	Conrad Miller	7
William B. Baird	2	Edward Bradley	7	Thos. D. Morgan	6
Thos. M. Barrett	5	William Bradley	6	Wm. Matthews	5
John A. Shaw	6	William Morgan	4	George Stark	7
David Lewis	5	John Morgan	5	Wm. Struby	5
Solomon Rizey	9	John Bath	2	John Thomas	6
Michael Cassidy	3	Thos. Griffiths	8	William Dando	single
John Gane	2	David P. Thomas	4	George Zias	"
Edward Walsh	6	David Gelly	10	D. J. Williams	"
Thos. J. Williams	8	Michael Dolan	2	William Scally	"
John Footen	5	Jos. Crook	6	Thomas Powell	"
P. J. Thomas	3	James Davis	5	Ed. Freel	"
Samuel Thomas	15	William Jenkins	6	Mac. Lyons	"
Napthilia Philips	5	Samuel Biddington	3	Henry Struby	"
Owen England	6	John Donehue	7	G. Kreitzberg	"
Wm. G. Thomas	4	James Hanson	5	John M. Davis	"
James Berwes	5	John Henry	6	John Hart	"
F. A. Rowe	4			Morris Thomas	"

* As published in a circular "Appeal for Aid" issued by the national executive board of the Knights of Labor on behalf of the "uniformly blacklisted" miners of District Assembly 25, October 6, 1882.

L. A. 853, Pompey Smash

Jacob Legair	7	Patrick P. McGann	7	Francis Flanagan	single
Michael Duffy	7	James Cain	3	Sandy Plunck	"
William Brodrick	4	Peter F. Cain	2	Daniel Cain	"
John Creamer	8	Patrick Moody	7	John McGann, Jr.	"
Michael Nelson	7	Levi Hawthorne	4	Richard Cramer	"
Isaac Orton	3	James Baxter	2	Wm. Hawthorne	"
Robert Thompson	2	Daniel Bahan	single	Timothy Higgins	"
Thomas Cain	2	Richard Leake	"	William Flinn	"
James Askie	12			Morris Flinn	"

L. A. 1311, Eckhart

William Lewis	10	Edward Barrey	4	James Farley	3
Thomas Close	5	William Condon	9	John McGuire	3
Samuel Arnold	5	Frank Lemmart	7	John Rephonn	2
William Story	4	Archy Gracie	4	Willliam Horchler	2
William McMillan	10	John Willison	3	Michael Finn	3
George Horchler	6	John Byrnes	3	John Womsley	single
Harry Russell	8	Thomas Barry	3	William Lee	"
William Scriven	6			George Bohmer	"

L. A. 775, Ocean

George Robinson	8	James Edwards	6	Paul Burns	3
Peter Monahan	10	Edward Powell	3	James McKay	7
Samuel H. Vaughan	6	Garrett Burns	5	John Burns	single
		Llewellyn Williams	6		

L. A. 767, Barton

David Openman	8	Walter Bath	8	Daniel James	2
David Watkins	5	James Park	2	James Cooper	1
Joseph Watkins	3	David Penman	5	Arthur Stewart	9
		William Bacon	7		

Also "ten other families, whose names the delegate does not remember, that will average four to the family, —40."

L. A. 1205, Westernport

Henry Cook	9	Henry Parker	4	Samuel Vance	2
John McIntosh	7	John McGreavey	11	Henry Younker	1
Andrew Patrick	9	John Deck	10	Henry Deck	1
William Morrow	6	Isaac Clark	4	James Lamkin	4
Patrick Kenney	5	Thomas Flanigan	1	John C. Murphy	5
Henry Lee	10	Wm. How	7	Edward Lawlis	5
Alexander Rankin	9	Patrick McCarol	1	Matthew Calligan	7
John Flanigan	3	Thomas McCarol	2	John Beane	1

Joseph Logsdon	10	Peter Curent	3	John H. Jones	5
John White	11	Wm. Howard	5	Samuel Bradley	6
Patrick Casey	10	James Richie	1	Wm. Whitfield	9
George Lucas	6	James Casey	1	George Parker	1

L. A. 1513, Frostburg

Rocks Williams	5	Charles Thomas	3	Adam Lemmart	1
Thomas Hewitt	10	George Kear	2	John McKinley	1
		Widan Rees	1		

L. A. 848, Lonaconing

Isaac Arbogast	5	Edward Wilson	5	Daniel Nightengale	8
P. A. Crowe	4	Harry Devlin	4	Samuel Mills	9
Daniel Moran	7	Christopher Devlin	3	Wm. H. Nightengale	2
John Moran	3	John Thomas	1	John Kenlock	2
Harman Schaidt	3	Jos. Thomas	3	John Morgan	4
Joseph Brooks	6	Daniel McCray	6	James Bishop	5
Jacob Folsom	3	John Devlin	4	Terence Woods	2
Peter Joyce	5	Joe Thomson	6	Thomas Woods	4
Thos. Maloy	4	Samuel Brown	3	Robt. Summerville	5
Wm. Baird, Sr.	4	Robert Gardner	3	John W. Evans	3
James Wilson	8	George Thomas	3	Wm. Frazer	2
R. W. Price	3	Thomas McConnell	3	Henry Schaidt	single
James Hamilton	7	August Richer	7	Patrick Finigan	"
John Boals	8	Stewart McGuinty	7	Hugh Killbride	"
John E. Beverage	4	Martin Garry	8	Joe Wilson	"
		Joseph Wright	2		

L. A. 1319, Alleghany

Maxville Hamilton	3	Wm. Johns	3	Thomas Shea	1
Thomas Anderson	5	Wm. Grimes	5	Van Tharp	3
Robert Walsh	6	Robert Middleton	5	Isaac Shoemaker	10
John Stoddart	3	Francis Dean	9	James Powers	8
		James Plummer	6		

Notes, Sources, and Index

Notes

ABBREVIATIONS

Balt.	Baltimore
Cumb.	Cumberland
Frost.	Frostburg
GCC&I Co.	George's Creek Coal and Iron Co.
K. of L.	Knights of Labor
Md. Bur. Ind. Stat.	Maryland Bureau of Industrial Statistics and Information under its various names
MGA, House	Maryland General Assembly, House of Delegates
MGA, Senate	Maryland General Assembly, Senate
MHR	Maryland Hall of Records
MHS	Maryland Historical Society
NA	National Archives
UMW	United Mine Workers

Chapter 1. The Stable Bulwark

1. James W. Thomas and T. J. C. Williams, *History of Allegany County* (n.p., 1923), I, 81.
2. Charles E. Beachley, *History of the Consolidation Coal Co. 1864–1934* (N.Y., 1934), p. 2.
3. *Ibid.;* and William B. Clark, "Historical Sketch Embracing an Account of the Progress of Investigation Concerning the Physical Features and Natural Resources of Md.," *Md. Geological Survey* (Balt., 1897), I, 223.

4. Howard N. Eavenson, *First Century and a Quarter of American Coal Industry* (Pittsburgh, 1942), p. 229.

5. A. T. Shurick, *Coal Industry* (Boston, 1924), p. 11.

6. J. Thomas Scharf, *History of Western Md.* (Phila., 1882), II, 1433, 1475.

7. Beachley, p. 2; and Scharf, II, 1500.

8. Scharf, II, 1500.

9. Beverley S. Randolph, "History of the Md. Coal Region," *Md. Geological Survey* (Balt., 1905), V, 514–516.

10. *Ibid.;* and Beachley, pp. 4–5.

11. Scharf, II, 1435.

12. *Ibid.;* and Beachley, pp. 4–5.

13. Scharf, II, 1435; Beachley, pp. 4–5; and Randolph, pp. 514–516.

14. For description of the boating operations, see *ibid.*

15. Randolph, 514–516.

16. For a technical description of the coal deposits see *Md. Geological Survey*, V, 221–511.

17. Scharf, II, 1437–1438; Md. Bur. Ind. Stat., *Second Biennial Report, 1886–1887* (Annapolis, 1888), p. 198; and James Macfarlane, *Coal-Regions of America* (N.Y., 1873), pp. 239–243.

18. In the interests of uniformity I shall use this spelling throughout except where directly quoting. In the source material the apostrophe is not consistently used.

19. Beachley, pp. 5–6.

20. Joseph G. Blandi, *Md. Business Corporations 1783–1852* (Balt., 1934), p. 31.

21. Md. *Laws* 1828, Ch. 170.

22. Beachley, p. 15.

23. Md. *Laws* 1835, Ch. 251.

24. *Ibid.*, Ch. 326.

25. Act incorporating the Parker Vein Coal Company, Md. *Laws* 1849, Ch. 56.

26. See Appendix A.

27. Thomas and Williams, I, 225–226.

28. Clark, pp. 56–71.

29. *Ibid.*, pp. 57–63.

30. Macfarlane, p. 243.

31. Scharf, II, 1437–1438.

32. Md. Bur. Ind. Stat., *Fourth Biennial Report, 1890–91* (Annapolis, 1892), p. 200; and Randolph, p. 523.

33. Scharf, II, 1435.

34. *Ibid.*

35. Randolph, pp. 518–521; and Beachley, p. 20.

36. Shurick, pp. 11–13.

37. Thomas and Williams, I, 536.

38. Scharf, II, 1446–1447.

39. Thomas and Williams, I, 490–491.

40. *George's Creek Coal and Iron Co.* (n.p., 1836), pp. 3, 7, 25.

41. Scharf, II, 1500; and Cumb. *Civilian*, Dec. 29, 1848.

42. Beachley, p. 16.

43. Md. Bur. Ind. Stat., annual reports, 1900–1911.

44. Md. Bur. Ind. Stat., *Seventeenth Annual Report, 1908* (Balt., 1908), pp. 52, 120.

45. Irene D. Neu, *Erastus Corning, Merchant and Financier, 1794–1872* (Ithaca, 1960), p. 48; and Ele Bowen, *Rambles in the Path of the Steam-Horse* (Phila., 1855), pp. 254–255.

46. James M. Swank, *History of the Manufacture of Iron in All Ages* (Phila., 1892), p. 256.

47. Eavenson, p. 239.

48. Neu, p. 48.

49. Randolph, pp. 518–519.

50. "The Cumb. Coal Region," *Bankers' Magazine* [N.Y.], IV, 394–399.

51. J. G. Randall, *Civil War and Reconstruction* (Boston, 1953), p. 637.

52. Beachley, pp. 17–19.

53. *Ibid.*, p. 23.

54. *Ibid.*, p. 19.

55. *Ibid.*, p. 12; and William J. Nicolls, *Above Ground and Below in the George's Creek Coal Region* (Balt., 1898), p. 12.

56. Beachley, p. 20.

57. Arthur Lovell, *Borden Mining Company, Brief History* (Frost. 1938), pp. 7–8.

58. Md. *Laws* 1853, Ch. 192.

59. "The Cumb. Coal Region," pp. 394–399.

60. Md. Bur. Ind. Stat., *Seventh Annual Report, 1898* (Balt., 1899), pp. 237–238.

61. Beachley, pp. 12–13.
62. Scharf, II, 1430.
63. Beachley, p. 24.
64. *Ibid.*, pp. 51–52.
65. Shurick, pp. 162–163.
66. Scharf, II, 1316, quoting a Professor Johnson, the Navy's expert in the coal tests.
67. MGA, House, Document EE, *Report of the Select Committee Appointed to Prepare a Statement in Relation to the Resources of Md.* (Annapolis, 1865), p. 47.
68. Scharf, II, 1316–1317.
69. Eavenson, p. 499.
70. Donald L. Kemmerer and C. Clyde Jones, *American Economic History* (N.Y., 1959), pp. 169–170.
71. *Ibid.*, pp. 170–171.
72. Eavenson, Table 20, pp. 426–434.
73. Kemmerer and Jones, p. 164.
74. Shurick, p. 211.
75. Cumb. *Civilian and Telegraph*, June 9 and 23, 1859.
76. *Ibid.*, Oct. 24, 1872.
77. *Ibid.*, Nov. 27, 1873.
78. Md. Mine Insp., *Annual Report, 1898* (n.p., n.d.), p. 6.
79. U.S. Census Office, *Eleventh Census of the U.S.: 1890, XIV* (*Mineral Industries in the U.S.*) (Washington, 1892), p. 384.
80. Swank, pp. 369–370.
81. *Ibid.*, pp. 256, 369–370.
82. "The Cumb. Coal Region," pp. 394–399.
83. U.S. Census Office, *Seventh Census of the U.S.: 1850, Abstract of the Statistics of Manufactures* (n.p., n.d.), p. 37.
84. U.S. Bureau of the Census, *Thirteenth Census of the U.S.: 1910*, XI (*Mines and Quarries, 1909*) (Washington, 1913), p. 95.
85. Md. Bur. Ind. Stat., *Report, 1898*, pp. 237–238. Md. miners worked in terms of long tons, but the information is reported in the U.S. census, for example, in short tons.
86. *Ibid.*, *Nineteenth Annual Report, 1910* (Balt., 1911), p. 201.
87. U.S. Census Office, *Tenth Census of the U.S.: 1880*, XV (*Report on the Mining Industries of the U.S.*) (Washington, 1886), p. 685.
88. *Ibid.*, *Eleventh Census of the U.S.*, XIV, 348.

89. *Coal Trade Journal*, April 16, 1902.

90. Andrew Roy, *History of the Coal Miners of the U.S.* (3d ed.; Columbus, O., 1907), p. 464.

91. *Coal Trade Journal*, Nov. 6, 1907.

92. Scharf, II, 1312.

Chapter 2. Intelligent and Fearless

1. J. Thomas Scharf, *History of Western Md.*, II, 1343.

2. Dieter Cunz, *Md. Germans* (Princeton, 1948), p. 190.

3. Ms population census schedules, Allegany County, Md., 1840, NA.

4. See William M. Richards, "An Experiment in Industrial Feudalism at Lonaconing, Md., 1837–1860" (M.A. thesis, U. of Md., 1950).

5. Scharf, II, 1500.

6. GCC&I Co., "Memoranda and Register," Oct. 19, 1837. Hereafter this unpublished ms owned by Lloyd R. Stallings, Convent, N.J., will be referred to as GCC&I Co. Journal.

7. *Ibid.*, Dec. 12, 1838.

8. *Ibid., Sept. 24.*

9. Untitled ms report of directors, GCC&I Co., p. 4 (J. H. Alexander Papers, 1839–1851, MHS).

10. GCC&I Co. Journal, *passim.*

11. Simon Becker to J. H. Alexander, Jan. 29, 1839 (J. H. Alexander Papers, MHS).

12. Letter dated June 12, 1843, in James Lumsden, *American Memoranda: A Short Tour in the Summer of 1843* (Glasgow, 1844), p. 21.

13. Andrew Roy, *History of the Coal Miners of the U.S.*, p. 48.

14. James W. Thomas and T. J. C. Williams, *History of Allegany County*, I, 47; and W. L. Fairbanks and W. S. Hamill, *Coal-Mining Industry of Md.* (Balt., 1932), p. 180.

15. Henry Thomas Weld, *Report of the Md. and N.Y. Iron and Coal Co.'s Lands, in the County of Alleghany* [*sic*] *and State of Md.* (n.p., n.d.), p. 65. This company was in existence from 1838 to 1848.

16. Wayne G. Broehl, Jr., *Molly Maguires* (Cambridge, 1964), p. 73; and *Workingman's Advocate*, Nov. 7, 1868.

17. Cumb. *Civilian*, Sept. 20, 1850.

18. Walton H. Hamilton and Helen R. Wright, *Case of Bituminous Coal* (N.Y., 1925), p. 103.

19. Fairbanks and Hamill, p. 180.

20. Cumb. *Civilian and Telegraph*, May 29, 1873.

21. Robert P. Arnot, *History of the Scottish Miners from the Earliest Times* (London, 1955), p. 64; see also McDonald's testimony before the Royal Commission appointed to Inquire into the Rule of Trades Unions, *Workingman's Advocate*, July 11 and Oct. 31, 1868.

22. *Workingman's Advocate*, Nov. 30, 1867.

23. Cunz, pp. 384–385.

24. The portion of Allegany County which was taken to form Garrett County contained few miners.

25. Balt. *Sun*, June 1, 1882.

26. Md. Bur. Ind. Stat., *Report, 1890–91*, p. 213.

27. Balt. *Sun*, March 1, 1886.

28. *Frost. Mining Journal*, July 25, 1896, quoting Bel Air [Md.] *Democrat.*

29. *UMW Journal*, June 6, 1901.

30. Fairbanks and Hamill, p. 180.

31. *Frost. Mining Journal*, April 18, 1882; *Coal Trade Journal*, Oct. 25, 1893; *Coal Trade Journal*, Nov. 8, 1893; see also E. Howard Blackburn and William H. Welfley, *History of Bedford and Somerset Counties, Penn.* (N.Y., 1906), II, 527; and Roy, *Coal Miners*, p. 104.

32. See Richards for a description of a self-sufficient company-governed coal and iron "estate" in the 1830's.

33. Roy, *Coal Miners*, p. 70.

34. For details of the strike of 1882, see Ch. 14.

35. Cumb. *Telegraph*, Nov. 15, 1855.

36. *Ibid.*, March 6, 1856.

37. *Ibid.*, Oct. 30.

38. Roy, *Coal Miners*, pp. 64–65.

39. Cumb. *Civilian*, Nov. 17, 1848.

40. *Ibid.*, July 6, 1849.

41. Broehl, p. 83.

42. *Ibid.*, pp. 85, 86, 161, 181, 361–362.

43. *Md. Directory and State Gazeteer, 1887* (Balt., 1887), p. 9.

44. Cunz, p. 384.

45. *Ibid.*

46. Cumb. *Telegraph*, April 17, 1856.

47. *Ibid.*, Nov. 13.
48. *Ibid.*, Nov 20.
49. Cumb. *Civilian and Telegraph*, Sept. 27, 1860.
50. *Ibid.*, Sept. 11, 1862.
51. Roy, *Coal Miners*, p. 67.
52. Edward A. Wieck, *American Miners' Association* (N.Y., 1940), p. 70.
53. Roy, *Coal Miners*, pp. 48–49.
54. *Workingman's Advocate*, Oct. 31, 1868.
55. N.Y. *Evening Post*, Oct. 22, 1860.
56. *UMW Journal*, April 15, 1909.
57. Scharf, II, 1437.
58. See, for example, letter from a Frost. correspondent of the Cumb. *Union*, Oct. 7, 1865.
59. Md. Bur. Ind. Stat., *Report, 1890–91*, p. 214.
60. Penn. Dept. of Internal Affairs, *Annual Report, 1884*, Part III, *Industrial Statistics* (Harrisburg, 1885), p. 69.
61. Md. Bur. Ind. Stat., *First Biennial Report: 1884–1885* (Balt., 1886), pp. 77–78.
62. *Ibid., Report, 1890–91*, p. 214.
63. Balt. *Sun*, Feb. 17, 1891.
64. Md. Mine Insp., "First Annual Report of Peter Cain, Dec. 30, 1876" (unpublished ms, MHS), p. 13; the U.S. Dept. of the Interior library has a photostatic copy of this report.
65. Md. Bur. Ind. Stat., *Report, 1884–85*, p. 76.
66. *Ibid., Report, 1890–91*, p. 215.
67. *Ibid., Third Annual Report, 1895* (Balt., 1895), p. 182.
68. McDonald testimony, *Workingman's Advocate*, July 11, 1868.
69. *Frost. Mining Journal*, May 27, 1893.
70. Thomas and Williams, I, 525. Vol. II of this work presents at least 100 examples of the upward mobility of miners and their children.

Chapter 3. Sooty Demons of the Mine

1. N.Y. *Evening Post*, Oct. 22, 1860.
2. A. T. Shurick, *Coal Industry*, p. 57, describes the development

of methods in Md. to 1880; see also Md. Mine Insp. *Annual Report 1913–14* (Balt., n.d.), pp. 40–43.

3. Ele Bowen, *Rambles in the Path of the Steam-Horse*, pp. 294–296.

4. James Macfarlane, *Coal-Regions of America*, p. 247.

5. Andrew Roy, *History of the Coal Miners of the U.S.*, pp. 81–84; and Cumb. *Civilian and Telegraph*, Sept. 16 and 23, 1869.

6. *Coal Trade Journal*, Oct. 8, 1884.

7. Wm. J. Nicolls, *Above Ground and Below in the George's Creek Coal Region*, pp. 18–19.

8. *Coal Trade Journal*, March 5, 1902.

9. Howard N. Eavenson, *First Century and a Quarter of American Coal Industry*, pp. 239–241.

10. Md. Mine Insp., *Report, 1876*, pp. 9, 13–29; see also Md. Bur. Ind. Stat., *Report, 1890–91*, pp. 210–211.

11. Md. Mine Insp., *Report, 1876*, p. 9.

12. *Frost. Mining Journal*, Sept. 8, 1894.

13. Md. Mine Insp., *Report, 1876*, p. 12.

14. *Frost. Mining Journal*, July 7, 1883.

15. *Coal Trade Journal*, Sept. 14, 1898.

16. Md. Mine Insp., *Annual Report for 1899* (Cumb., n.d.), p. 3.

17. Balt. *Sun*, Aug. 27, 1900, quoting from a summary of the mine inspector's report.

18. Md. Mine Insp., *Annual Report 1910–11* (Cumb., n.d.), p. 7.

19. Md. Bur. Ind. Stat., *Report, 1884–85*, pp. 78–79.

20. N. Allen Stockton, "The Coal Mines of Md.," *Md. Geological Survey*, V, 549.

21. Charles E. Beachley, *History of the Consolidation Coal Co.*, p. 52; and *UMW Journal*, Dec. 12, 1907.

22. Roy, *Coal Miners*, p. 429.

23. Md. Mine Insp., *Report of Thomas Brown for Year Ending Dec. 31, 1881* (Annapolis, 1882), pp. 20–22.

24. Roy, *Coal Miners*, p. 81.

25. Md. Bur. Ind. Stat., *Report, 1884–85*, pp. 78–79; see also Macfarlane, p. 245.

26. Eavenson, pp. 239–241.

27. Md. Bur. Ind. Stat., *Report, 1890–91*, pp. 210–211.

28. Md. Commissioners of Mines, *Report to the Governor of Md.*, Senate Document I (Annapolis, 1880), p. 9.

29. Md. Bur. Ind. Stat., *Report, 1890–91*, p. 199.

30. Md. Mine Insp., *Report, 1881*, pp. 20–22.

31. *Ibid., Report, 1904–05*, pp. 5–6.

32. *Ibid., Annual Report 1909–10* (Cumb., 1910), p. 8.

33. Stockton, p. 549.

34. Bowen, p. 303.

35. Md. Bur. Ind. Stat., *Report, 1884–85*, pp. 78–79.

36. Md. *Laws* 1876, Ch. 173, Sec. 4.

37. Cumb. *Civilian and Telegraph*, May 29, 1873.

38. Md. Mine Insp., *Report, 1904–05*, p. 85. The last column has been added to the original table in order to show the averages over the 10-year period.

39. *Ibid., Report, 1898*, p. 7.

40. *Ibid., Annual Report 1908–09* (Cumb., 1909), p. 15; and *Annual Report 1906–07* (Lonaconing, Md., 1907), p. 38.

41. James W. Thomas and T. J. C. Williams, *History of Allegany County*, I, 422–423, 520–521; and Md. Mine Insp., *Report, 1913–14*, pp. 43–44.

42. *Coal Trade Journal*, Nov. 3, 1910; and *Frost. Mining Journal*, Dec. 10, 1910.

43. Md. *Laws* 1902, Ch. 139.

44. *Frost. Mining Journal*, Jan. 4, 1908.

45. Stockton, p. 549.

46. *Ibid.*, p. 546.

47. Md. Bur. Ind. Stat., *Report, 1884–85*, pp. 78–79.

48. *Ibid.*

49. *Ibid.*

50. *Coal Trade Journal*, Oct. 4, 1905.

51. Md. Mine Insp., Annual *Report 1903–04* (Balt., 1905), pp. 5–25.

52. *Ibid., Report, 1904–05*, pp. 11–62.

53. *Ibid., Annual Report 1905–[06]* (Lonaconing, 1907), pp. 27–68.

54. *Ibid., Report, 1908–09*, pp. 12–40.

55. *Ibid., Report, 1909–10*, pp. 21–57.

56. *Ibid., Report, 1910–11*, pp. 24–83.

57. *Frost. Mining Journal*, March 30, 1901; see also Md. Mine Insp., *Report, 1904–05*, pp. 25–26.

58. Md. Mine Insp., *Report, 1904–05*, pp. 25–26.

59. William Bullock Clark and Edward B. Mathews, "Md. Mineral Industries 1896–1907," *Md. Geological Survey* (Balt., 1909), VIII, 106, 164, 177; data as of 1907.

60. Roy, *Coal Miners*, p. 450.

61. Md. *Laws* 1876, Ch. 173.

62. Md. Mine Insp., *Second Annual Report of Owen Riordan for Year Ending Dec. 31, 1879* (Annapolis, 1880), pp. 10–11.

63. Roy, *Coal Miners*, pp. 81, 140.

64. Md. Mine Insp., *Report, 1876*, p. 8.

65. *Ibid.*, *Second Annual Report of Peter Cain, Dec. 31, 1877*, House of Delegates Document H (Annapolis, 1878), pp. 6–16.

66. Roy, *Coal Miners*, pp. 445, 448.

67. Cumb. *Civilian*, May 23, 1851.

68. Md. Commissioners of Mines, p. 37 of appendix; and Md. Mine Insp., *Report, 1881*, pp. 5–19.

69. Md. Mine Insp., *Report, 1876*, p. 9.

70. *Ibid.*, *Report, 1881*, p. 22.

71. *Ibid.*, pp. 22–23.

72. *Coal Trade Journal*, July 8, 1885.

73. Md. Bur. Ind. Stat., *Report, 1890–91*, p. 210.

74. *Ibid.*, *Report, 1898*, p. 212; and Md. Mine Insp., *Report, 1903–04*, pp. 5–25; see also Md. Bur. Ind. Stat., *Report, 1890–91*, pp. 214–215; and *UMW Journal*, April 18, 1901.

75. Md. *Laws* 1902, Ch. 124.

76. Md. Mine Insp., *Report, 1903–04*, pp. 3–4.

77. *Ibid.*, *Report, 1905–[06]*, p. 16.

78. *Ibid.*, *Annual Report 1912–13* (Cumb., n.d.), p. 8.

79. *Ibid.*, *Report, 1905–[06]*, p. 16.

80. *Ibid.*

81. Md. Bur. Ind. Stat., *Report, 1908*, p. 48.

Chapter 4. Half Turns

1. James Macfarlane, *Coal-Regions of America*, p. 244.

2. Md. Bur. Ind. Stat., *Report, 1884–85*, p. 78.

3. K. of L. regulations were published weekly in the *Frost. Mining Journal* from May 7, 1881, to Sept. 2, 1882; they had first appeared in the *Journal of United Labor* on Dec. 15, 1880.

4. *Frost. Mining Journal,* April 8, 1876.

5. Md. Bur. Ind. Stat., *Third Biennial Report, 1888–89* (Annapolis, 1890), p. 180.

6. *Frost. Mining Journal,* May 27, 1893.

7. Md. Bur. Ind. Stat., *Report, 1884–85,* pp. 219–220.

8. *Ibid., Report, 1890–91,* p. 215.

9. *Ibid., First Annual Report, 1892* (Balt., 1893), p. 212.

10. *Frost. Mining Journal,* May 25, 1907.

11. Md. Bur. Ind. Stat., *Report, 1898,* p. 225.

12. Owen R. Lovejoy, "Child Labor in the Soft Coal Mines," in National Child Labor Committee, *Proceedings of the Third Annual Meeting* (N.Y., 1907), p. 5.

13. U.S. Census Office, *Ninth Census of the U.S.: 1870,* III (*The Statistics of the Wealth and Industry of the U.S.*) (Washington, 1872), pp. 383–384.

14. *Ibid.,* p. 762.

15. *Ibid.,* I, xxviii–xxix.

16. *Ibid., Tenth Census of the U.S.,* XV, 681.

17. *Coal Trade Journal,* April 14, 1886.

18. Md. Bur. Ind. Stat., *Report, 1884–85,* p. 164. The yearly figure is given in the report as $200.50, but this is obviously an error in computation.

19. *Ibid., Report, 1888–89,* p. 180.

20. U.S. Census Office, *Eleventh Census of the U.S.,* XIV, 350.

21. *Ibid.*

22. *Ibid.,* p. 349.

23. *Ibid., Special Reports, Mines and Quarries, 1902,* p. 712.

24. Md. Bur. Ind. Stat., *Sixteenth Annual Report, 1907* (Balt., 1908), p. 39.

25. *Ibid.,* p. 58.

26. *Ibid., Report, 1908,* p. 240.

27. David J. McDonald and Edward A. Lynch, *Coal and Unionism, History of the American Coal Miners' Unions* (Silver Spring, Md., 1939), p. 74.

28. U.S. Bureau of the Census, *Thirteenth Census of the U.S.,* XI, 95.

29. Penn. Dept. of Internal Affairs, *Annual Report, 1886,* Part III, *Industrial Statistics* (Harrisburg, 1887), p. 43.

30. Md. *Laws* 1874, Ch. 3.

31. *Ibid.*, 1876, Ch. 125.

32. U.S. Industrial Commission, *Report on Labor Legislation* (Washington, 1900), V, 51, 221.

33. Md. *Laws* 1902, Ch. 124.

34. See Md. Mine Insp., *Report, 1877*, p. 5; *Report, 1905–[06]*, pp. 24, 71; and *Report, 1899*, pp. 6–7.

35. Md. *Laws* 1912, Ch. 731, Sec. 8.

Chapter 5. Tonnage, Turns, and Off-Takes

1. GCC&I Co. Journal, July 12, Aug. 29, and Sept. 10, 1839.

2. Howard N. Eavenson, *First Century and a Quarter of Amercan Coal Industry*, p. 12.

3. Md. *Laws* 1878, Ch. 157, Sec. 12.

4. Andrew Roy, *History of the Coal Miners of the U.S.*, pp. 64–65.

5. *Ibid.*, p. 227.

6. Md. Mine Insp., *Report, 1879*, p. 12.

7. Md. Bur. Ind. Stat., *Report, 1884–85*, p. 81.

8. *Ibid.*, *Report, 1888–89*, p. 180.

9. *Ibid.*, *Report, 1890–91*, p. 243.

10. As published—in more comprehensive form than here indicated—this table contained three minor arithmetical errors. When corrected, the totals balanced properly; therefore, the corrected figures, where pertinent, have been inserted in the abbreviated table. In the last column the figure $200.00 originally read $121.00 and $5.13 read $105.13; all other figures are given correctly as in the original table.

11. J. Thomas Scharf, *History of Western Md.*, II, 1503.

12. Eavenson, pp. 239–241.

13. N. Allen Stockton, "The Coal Mines of Md.," p. 541.

14. Advertisement run in *Frost. Mining Journal*, May 7, 1881, to Sept. 2, 1882.

15. Balt. *Sun*, June 17, 1882.

16. Letter to the Balt. *Sun*, June 14, 1882.

17. Arthur E. Suffern, *Coal Miners' Struggle for Industrial Status* (N.Y., 1926), pp. 15–17; see also Md. Mine Insp., *Report, 1908–09*, p. 6.

18. Md. Bur. Ind. Stat., *Report, 1884–85*, p. 81.

19. *Ibid.*, p. 164.

20. Robert P. Arnot, *History of the Scottish Miners*, p. 17.

21. *Coal Trade Journal*, Oct. 17, 1906.

22. *Ibid.*, May 10, 1905.

23. *Ibid.*, April 29, 1908.

24. Md. Bur. Ind. Stat., *Report, 1898*, pp. 237–238.

25. *Ibid.*

26. See, for example, Cumb. *Telegraph*, May 7, 1857; Cumb. *Civilian and Telegraph*, Sept. 18, 1862, and July 2, 1863; and Md. Bur. Ind. Stat., *Report, 1890–91*, pp. 207–208.

27. See Consolidation Coal Co., *Report to the Stockholders, Dec. 1871* (N.Y., 1872), p. 14; James Macfarlane, *Coal-Regions of America*, p. 258; and Md. Mine Insp., *Report of Thomas Brown for the Year Ending Dec. 31, 1883* (Annapolis, 1884), p. 5.

28. Macfarlane, p. 258.

29. Md. Mine Insp., *Report, 1883*, p. 5.

30. *Coal Trade Journal*, March 14, 1883.

31. *Frost. Mining Journal*, June 2, 1883.

32. Md. Bur. Ind. Stat., *Report, 1884–85*, pp. 155–157.

33. *Ibid.*, pp. 80–81.

34. *Ibid.*, *Report, 1898*, pp. 221–222.

35. Md. Mine Insp., *Report, 1904–05*, p. 38.

36. Md. Bur. Ind. Stat., *Report, 1884–85*, p. 80; and Wm. J. Nicolls, *Above Ground and Below in the George's Creek Coal Region*, pp. 20–21.

37. Md. *Laws*, 1878, Ch. 157.

38. Md. Bur. Ind. Stat. *Report, 1884–85*, p. 80.

39. *Ibid.*, *Report, 1890–91*, pp. 215–216.

40. *Frost. Mining Journal*, April 26, May 3, May 10, and Sept. 27, 1873; and Cumb. *Civilian and Telegraph*, Sept. 25, 1873.

41. Md. Mine Insp., *Report, 1876*, pp. 29–31.

42. Md. *Laws* 1878, Ch. 157, Sec. 13.

43. *Ibid.*, Sec. 12.

44. Letter from Thomas Brown, mine insp., to *Frost. Mining Journal*, Feb. 19, 1881. Very soon after this letter was published, the mine owners ordered new scales.

45. Md. Bur. Ind. Stat., *Report, 1884–85*, p. 220.

46. *Frost. Mining Journal*, Aug. 11, 1877.

47. Md. *Laws* 1898, Ch. 34.

48. *Frost. Mining Journal,* Dec. 14, 1901.

49. Eavenson, pp. 239–241.

50. Md. *Laws* 1880, Ch. 273.

51. *Ibid.,* 1900, Ch. 453.

52. Md. Bur. Ind. Stat., *Report, 1892,* p. 219.

53. Stockton, p. 560.

54. Regulations of D. A. 25, K. of L., as advertised in *Frost. Mining Journal,* May 7, 1881, to Sept. 2, 1882.

55. See, for example, Bur. Ind. Stat. reports for 1884–85, p. 83; 1886–87, p. 136; 1890–91, p. 209; and Stockton, p. 560.

56. Md. *Laws* 1868, Ch. 471, Sec. 217.

57. Md. Bur. Ind. Stat., *Report, 1890–91,* p. 217.

58. See *Md. Laws* 1898, Ch. 493; and Md. *Laws* 1900, Ch. 453.

59. *Ibid., 1880,* Ch. 273, Sec. 1.

60. *Ibid.,* Sec. 3.

61. Md. Bur. Ind. Stat., *Report, 1886–87,* p. 217.

62. Md. *Laws* 1892, Ch. 445.

63. *Ibid.,* 1900, Ch. 433.

64. K. of L. regulations published in *Frost. Mining Journal,* May 7, 1881, to Sept. 2, 1882.

65. Md. Mine Insp., *Report, 1883,* p. 44.

66. *Ibid., Report, 1912–13,* p. 5.

Chapter 6. One Continuous Street and Town

1. George W. Hughes, *Extracts from Reports of an Examination of the Coal Measures Belonging to the Md. Mining Co. in Allegany County; and of a Survey for a Railroad from the Mines to the Chesapeake and Ohio Canal at Cumb.* (Washington, 1837), pp. 16–17.

2. Benjamin Silliman, *Extracts from a Report Made to the Md. Mining Co.* (N.Y., 1838), pp. 4–5.

3. N.Y. *Evening Post,* Oct. 22, 1860.

4. J. Thomas Scharf, *History of Western Md.,* II, 41.

5. *Coal Trade Journal,* Oct. 6, 1897.

6. W. L. Fairbanks and W. S. Hamill, *Coal-Mining Industry of Md.,* p. 253.

7. James W. Thomas and T. J. C. Williams, *History of Allegany County*, I, 535.

8. Scharf, II, 1439.

9. William M. Richards, "An Experiment in Industrial Feudalism at Lonaconing, Md.," p. 10.

10. Untitled ms report of directors, GCC&I Co., 1839, p. 10 (J. H. Alexander Papers, 1839–51, MHS).

11. Balt. *Sun*, Nov. 24, 1842.

12. From advertisement of trustees' sale of the Md. Mining Co. in *Cumb. Miners' Journal*, July 2, 1852.

13. *Ibid.*, March 4, 1853.

14. *Ibid.*

15. Richards describes the "industrial feudalism" of this company.

16. Scharf, II, 1466–1467.

17. Thomas and Williams, I, 539.

18. *Cumb. Miners' Journal*, Sept. 24, 1852.

19. Richards, pp. 12–13.

20. James Renwick, report to GCC&I Co., 1836, contained in MHS folder labeled "Georges Creek Coal and Iron Co., 1836."

21. Scharf, II, 1500.

22. GCC&I Co., report of directors, 1839, pp. 2–3 (J. H. Alexander Papers, MHS).

23. Referred to in the notes as GCC&I Co. Journal.

24. GCC&I Journal, Dec. 22, 24 and 26, 1838.

25. Cumb. *Civilian*, Aug. 20, 1847.

26. Md. Mining Co., *Report to the Stockholders, 1847* (N.Y., n.d.), p. 2.

27. *Ibid.*, *Report for June, 1848* (N.Y., June 6, 1848), p. 4.

28. Cumb. *Civilian*, May 25 and June 1, 1849; *Cumb. Miners' Journal*, Jan. 14, April 29 and Aug. 19, 1853; and Franklin Coal Co., *First Annual Report of the President and Directors* (Balt., 1855), p. 5.

29. Hampshire and Balt. Coal Co. of Va., *Report to the Directors* (n.p., 1864), p. 6.

30. Consolidation Coal Co., *Report, 1871*, p. 13.

31. *Ibid.*, *Report to the Stockholders, for the year 1873* (N.Y., 1874), p. 10.

32. *Ibid.*, p. 7.

33. Cumb. *Civilian and Telegraph*, April 23, 1874.

34. *Coal Trade Journal,* Oct. 6, 1897, and Feb. 2, 1898.

35. Md. Coal Co., *Annual Report, Dec. 13, 1870* (N.Y., 1870), p. 7.

36. *Coal Trade Journal,* Sept. 25, 1889.

37. Penn. Bureau of Statistics of Labor and Agriculture, *First Annual Report, 1872–73* (Harrisburg, 1874), p. 489.

38. Scharf, II, 1466.

39. Andrew Roy, *History of the Coal Miners of the U.S.,* p. 48.

40. Midland Coal and Iron Co., *Report on the Valuable Coal Properties of The Midland Coal and Iron Co. and The Manchester Mining and Manufacturing Co.* (N.Y., 1856), p. 10.

41. Hampshire and Balt. Coal Co., *Report, 1864,* p. 6.

42. Scharf, II, 1503.

43. *Ibid.*

44. *Frost. Mining Journal,* Sept. 30, 1899.

45. *Ibid.,* April 15, 1882; see also Balt. *American,* March 15, 1882.

46. Md. Bur. Ind. Stat., *Report, 1884–85,* pp. 76–77.

47. Penn. Dept. of Internal Affairs, *Annual Report, 1902,* Part III, *Industrial Statistics* (n.p., 1903), p. 5.

48. See also Md. Bur. Ind. Stat., *Report, 1892,* p. 219; and *Frost. Mining Journal,* Feb. 28, 1891.

49. Penn. Dept. of Internal Affairs, *Annual Report, 1886,* Part III, *Industrial Statistics,* pp. 27–28. For other evidence on living conditions in Penn. and W. Va. see various reports of Penn. Dept. of Internal Affairs; *UMW Journal; Workingman's Advocate;* U.S. Industrial Commission Reports, Vol. XIII, *Report on the Relations and Conditions of Capital and Labor Employed in the Mining Industry* (Washington, 1901); Peter Roberts, *Anthracite Coal Communities* (N.Y., 1904); and U.S. Coal Commission Report, Part III, *Report on the Bituminous Mine Workers and Their Homes* (Washington, 1925), pp. 1399–1646.

50. GCC&I Co. Journal, entries for Sept. 18, 1837, Jan. 18, 1838, and March 4, 1839.

51. Thomas Weld to Robert Graham, July 20, 1845 (GCC&I Co. Letterbook, MHS).

52. W. B. Hazlehurst to Robert Graham, July 15, 1845 (GCC&I Co. Letterbook, MHS).

53. Midland Coal and Iron Co., *Report, 1856,* p. 11.

54. Ele Bowen, *Rambles in the Path of the Steam-Horse*, pp. 254–255.

55. N.Y. *Evening Post*, Oct. 22, 1860.

56. Md. Bur. Ind. Stat., *Report, 1884–85*, p. 76.

57. Balt. *Sun*, April 5, 1882.

58. Balt. *American*, March 16, 1882.

59. Balt. *Sun*, Feb. 17, 1891.

60. Richard Lowitt, ed., "Frost. 1882: German Strikers vs German Strikebreakers," in Society for the History of the Germans in Md., *Twenty-eighth Report* (Balt., 1953), p. 75.

61. Md. State Board of Health, *Annual Report 1904* (Balt., 1906), p. 101.

62. Scharf, II, 1505.

63. *UMW Journal*, Feb. 11, 1909.

64. U.S. Census Office, *Tenth Census of the U.S.*, I, 207.

65. *Md. Directory* (Balt., 1880), p. 15.

66. *Ibid.;* see also Cumb. *Civilian*, July 29, 1877; and Scharf, II, 1495.

67. *Md. Directory, 1880*, p. 3.

68. Scharf, II, 1439.

69. Cumb. *Civilian and Telegraph*, May 20, 1873.

70. Letter from Mayor Peter Lammert in Balt. *Sun*, March 20, 1903, quoted by *Frost. Mining Journal*, March 21, 1903.

71. Cumb. *Civilian and Telegraph*, Nov. 30, 1871.

72. Md. Bur. Ind. Stat., *Report, 1898*, p. 8. There is apparently no existing file of either newspaper.

73. Thomas and Williams, II, 809–810; see also *Portrait and Biographical Record of the Sixth Congressional District, Md.* (N.Y., 1898), pp. 140–141.

74. Thomas and Williams, I, 566–568, and II, 903–904.

75. *Ibid.*, II, *passim.*

76. Md. *Laws* 1870, Ch. 77.

77. GCC&I Co. Journal, July 30, 1839; and Balt. *Sun*, Sept. 1, 8, 15 and 22, 1854.

78. Cumb. *Civilian*, March 25, 1877.

79. Md. State Board of Health, *Report*, 1904, p. 99.

80. Md. Bur. Ind. Stat., *Report, 1884–85*, pp. 76–77.

81. *Md. Directory and State Gazeteer, 1887*, p. 25.

82. *Frost. Mining Journal,* June 4, 1892.

83. Md. State Board of Health, *Biennial Report, 1888* (Annapolis, 1888), pp. 146–147; see also Md. State Board of Health, *Twelfth Biennial Report for the Two Years Ending Dec. 31, 1897* (Balt., 1898), pp. 6–7.

84. *Ibid., Report, 1888,* p. 180.

85. *Ibid.,* p. 342.

86. *Frost. Mining Journal,* April 10, 1880.

87. Thomas and Williams, II, 1248–1249.

88. Balt. *Sun,* March 21, 1903.

89. Roy, *Coal Miners,* p. 50.

Chapter 7. Some Keep a Cow

1. Balt. *Sun,* Feb. 17, 1891.

2. Cumb. *Civilian and Telegraph,* Dec. 23, 1875.

3. Md. Mining Co., *Report for June 1848,* p. 7.

4. Midland Coal and Iron Co., *Report, 1856,* p. 11.

5. Quoted in Howard N. Eavenson, *First Century and a Quarter of American Coal Industry,* pp. 239–241.

6. J. Thomas Scharf, *History of Western Md.,* II, 1503.

7. *Frost. Mining Journal,* March 31, 1877.

8. Md. Bur. Ind. Stat., *Report, 1884–85,* pp. 76–77.

9. Penn. Dept. Internal Affairs, *Annual Report, 1884,* Part III, p. 6.

10. *Coal Trade Journal,* Oct. 6, 1897.

11. Midland Coal and Iron Co., *Report, 1856,* p. 10.

12. *Md. Directory, 1880,* pp. 4, 15, 21, 24, 25, 26.

13. *Md. Directory and State Gazeteer, 1887,* pp. 11, 29, 30.

14. *Coal Trade Journal,* Oct. 6, 1897.

15. *Ibid.,* Feb. 2, 1898.

16. GCC&I Co. Journal, *passim.*

17. Eavenson, pp. 239–241.

18. *UMW Journal,* May 18, 1893.

19. Md. Bur. Ind. Stat., *Report, 1893,* p. 12.

20. Penn. Bureau of Statistics of Labor and Agriculture, *Report, 1872–73,* pp. 446, 457.

21. Eavenson, pp. 239–241.

22. Letter to Wm. Alexander, April 26, 1840, pp. 72–73 in GCC&I Co. Letterbook.

23. Md. Bur. Ind. Stat., *Report, 1884–85*, pp. 189–197.

24. Penn. Dept. of Internal Affairs, *Report, 1884*, Part III, pp. 40–42.

25. GCC&I Co. Report from J. H. Alexander and P. T. Tyson, dated Balt., Oct. 1836 (J. H. Alexander Papers, MHS).

26. Benjamin Silliman, *Extracts from a Report Made to the Md. Mining Co.*, pp. 4–5.

27. N.Y. *Evening Post*, Oct. 22, 1860.

28. Md. Commissioners of Mines, *Report, 1880*, p. 9.

29. Balt. *American*, April 6, 1882.

30. Md. Bur. Ind. Stat., *Report, 1884–85*, pp. 76–77.

31. Balt. *American*, April 6, 1882.

32. *Md. Directory and State Gazeteer, 1887*, pp. 25–27.

33. Md. Mine Insp., *Report, 1883*, p. 18.

34. *Coal Trade Journal*, Oct. 6, 1897.

35. Cumb. *Civilian*, Jan. 31, 1851.

36. Eavenson, pp. 239–241.

37. *Frost. Mining Journal*, Oct. 3, 1874.

38. *Ibid.*, Jan 22, 1876.

39. Md. Bur. Ind. Stat., *Report, 1884–85*, p. 191.

40. *Ibid.*, *Report, 1890–91*, p. 215.

41. Penn. Dept. of Internal Affairs, *Report, 1884*, Part III, p. 58.

42. Letter to Richard Wilson in GCC&I Co. Letterbook, p. 10.

43. Scharf, II, 1503.

44. Letter from Lonaconing "Miner" in *Workingman's Advocate*, Nov. 30, 1867.

45. *Ibid.*, Dec. 14.

46. Md. *Laws* 1868, Ch. 471, Sec. 16.

47. *Luman* v. *Hitchens*, 90 Md. 14.

48. Richard Lowitt, ed., "Frost. 1882: German Strikers vs. German Strikebreakers," p. 75.

49. Md. Bur. Ind. Stat., *Report, 1884–85*, pp. 84–85.

50. *Ibid.*, *Report, 1890–91*, p. 217.

51. Md. *Laws* 1898, Ch. 493.

52. *Frost. Mining Journal*, Dec. 2, 1899.

Chapter 8. Polite Culture and General Refinement

1. W. M. Wardjon in *UMW Journal*, May 15, 1905.
2. Md. Bur. Ind. Stat., *Report, 1898*, pp. 223–224.
3. *Ibid., Report, 1892*, p. 213.
4. Balt. *Sun*, Feb. 17, 1891.
5. Ms population census schedules, Allegany County, Md., 1860, NA.
6. Md. Bur. Ind. Stat., *Report, 1890–91*, p. 214.
7. Md. Mine Insp., *Annual Report for the Years Ending Dec. 31, 1892–93* (Balt., 1894), p. 4 (1892); see also Md. Bur. Ind. Stat., *Report, 1898*, p. 223.
8. Balt. *Sun*, Feb. 17, 1891.
9. Md. Bur. Ind. Stat., *Report, 1890–91*, p. 214.
10. Dieter Cunz, *Md. Germans*, p. 386.
11. Andrew Roy, *Recollections of a Prisoner of War* (Columbus, O., 1909), pp. 148–149.
12. Cumb. *Civilian and Telegraph*, Nov. 18, 1875.
13. Md. Bur. Ind. Stat., *Report, 1892*, pp. 212–213.
14. *Ibid., Report, 1890–91*, pp. 213–214.
15. Md. Mine Insp., *Report, 1892–93*, p. 4 (1892).
16. GCC&I Co. Journal, Feb. 6, 1839.
17. *Ibid.*, Feb. 11.
18. *Ibid.*, Rule No. 10, beginning of Vol. II.
19. James W. Thomas and T. J. C. Williams, *History of Allegany County*, I, 537.
20. Cumb. *Civilian and Telegraph*, Jan. 19, 1860.
21. Ms population census schedules, Allegany County, Md., 1860, NA.
22. Cumb. *Civilian and Telegraph*, Jan. 12, 1860.
23. Cumb. *Democratic Alleganian*, March 17, 1860; Md. *Laws* 1860, Ch. 335.
24. Cumb. *Democratic Alleganian*, May 5, 1860.
25. Cumb. *Civilian and Telegraph*, Nov. 21, 1861.
26. Thomas and Williams, I, 411.
27. J. Thomas Scharf, *History of Western Md.*, II, 1355.

28. Journal of Proceedings of the Board of School Commissioners of Allegany County, Aug. 15, 1865; hereafter this document will be referred to as School Comm. Journal.

29. *Ibid.*

30. *Ibid.*, Aug. 30.

31. *Ibid.*

32. Cumb. *Civilian and Telegraph*, Oct. 12, 1865.

33. School Comm. Journal, Oct. 24, 1865.

34. Howard N. Eavenson, *First Century and a Quarter of American Coal Industry*, pp. 239–241.

35. Cumb. *Union*, Jan. 27, 1866.

36. School Comm. Journal, April 18, 1866.

37. Cumb. *Civilian and Telegraph*, March 29, 1866.

38. *Ibid.*, Aug. 8, 1867.

39. *Ibid.*, Jan. 14, 1869.

40. *Ibid.*

41. Cumb. *Civilian and Telegraph*, March 7, 1872.

42. *Ibid.*, April 24 and Oct. 2, 1873.

43. *Ibid.*, July 8, 1875.

44. Cumb. *Civilian*, Oct. 1, 1876.

45. *Ibid.*, Feb. 18, 1877.

46. Scharf, II, 1355–1356.

47. Md. Commissioners of Mines, *Report, 1880*, p. 9; see also Md. Bur. Ind. Stat., *Report, 1884–85*, p. 78.

48. Md. State Board of Education, *Annual Report, 1881* (Annapolis, 1882), p. 55.

49. Md. Bur. Ind. Stat., *Report, 1890–91*, p. 214.

50. U.S. Census Office, *Eleventh Census of the U.S.*, XIV, 350.

51. See Table 4, Ch. 4.

52. Md. Bur. Ind. Stat., *Report, 1892*, pp. 212–213.

53. *Frost. Mining Journal*, May 27, 1893.

54. Md. *Laws* 1884, Ch. 39.

55. School Comm. Journal, Oct. 17, 1884.

56. *Ibid.*, April 1886, p. 484.

57. *Frost. Mining Journal*, April 10, 1886; see also School Comm. Journal, Nov. 20, 1886.

58. Md. *Laws* 1904, Ch. 90.

59. *Frost. Mining Journal*, Feb. 27, 1892.

60. See William M. Richards, "An Experiment in Industrial

Feudalism at Lonaconing, Md.," pp. 24–27; and Thomas and Williams, I, 546–548.

61. GCC&I Co. Journal, Feb. 17, March 24 and April 14, 1839.

62. *Ibid.*, Jan. 5.

63. *Ibid.*

64. *Ibid.*, June 2 and Aug 4.

65. *Ibid.*, June 2.

66. *Ibid.*, Feb. 27.

67. *Ibid.*, Nov. 3.

68. *Ibid.*, Nov. 16.

69. *Ibid.*, Nov. 20.

70. *Ibid.*, Jan. 5.

71. Richards, p. 26.

72. Scharf, II, 1506.

73. *Ibid.*, 1503.

74. *Ibid.*, 1506–1507.

75. *Ibid.*, 1497.

76. *Ibid.*

77. See, for example, Cumb. *Telegraph*, Aug. 30, 1855, and Aug. 28, 1856; and Cumb. *Democratic Alleganian*, Jan. 9, 1858.

78. Cumb. *Civilian and Telegraph*, Nov. 28, 1867.

79. *Ibid.*, May 23, 1872.

80. Scharf, II, 1479–1486.

81. Sidney Webb, *Story of the Durham Miners* (London, 1921), pp. 21–24; and Andrew Roy, *History of the Coal Miners of the U.S.*, p. 15.

82. Cumb. *Civilian and Telegraph*, July 1, 1869.

83. Md. Mine Insp., *Report for 1878* (n.p., n.d.), p. 28.

84. Balt. *Sun*, Feb. 17, 1891.

85. Md. Bur. Ind. Stat., *Report, 1884–85*, pp. 78–79.

86. *Md. Directory and State Gazeteer, 1887*, p. 9.

87. Balt. *American*, April 6, 1882.

88. See Scharf, II, 1469, 1492; and *Md. Directory and State Gazeteer, 1887*, pp. 11, 23, 29.

89. Scharf, II, 1469–1471, 1490–1493, 1507–1508.

90. *Md. Directory and State Gazeteer, 1887*, p. 9.

91. Cumb. *Telegraph*, Feb. 28, 1856.

92. Cumb. *Civilian*, Feb. 6, 1868.

93. Cumb. *Civilian and Telegraph*, Jan. 2, 1873.

94. Scharf, II, 1471.

95. *Ibid.*, 1492.

96. *Md. Directory, 1880*, p. 21.

97. Letter to David Owen Gay, Nov. 8, 1839 in GCC&I Co. Letterbook, pp. 29–30.

98. *Md. Directory and State Gazeteer, 1887*, p. 25.

99. Cumb. *Alleganian*, Jan. 26, 1876.

100. GCC&I Co. Journal, March 1, 1839.

101. Cumb. *Civilian*, Nov. 23, 1871.

102. Quoted in Cumb. *Civilian and Telegraph*, May 20, 1873.

103. Balt. *Sun*, Feb. 17, 1891.

104. *Frost. Mining Journal*, Feb. 21, 1891.

105. Cumb. *Civilian*, Jan. 30, 1876.

106. Scharf, II, 1471.

107. *Ibid.*, 1509.

108. *Ibid.*, 1494.

109. *Frost. Mining Journal*, May 24, 1884.

110. George Korson, *Coal Dust on the Fiddle* (Phila., 1943), pp. 17–19.

111. Balt. *Sun*, April 12, 1900.

112. Cumb. *Civilian*, Jan. 26, 1849.

113. *Ibid.*, July 6, 1849, and July 5, 1850.

114. *Ibid.*, July 12, 1850.

115. *Ibid.*, Oct. 11.

116. Cumb. *Democratic Alleganian*, July 12, 1857.

117. Cumb. *Civilian and Telegraph*, Sept. 15, 1870.

118. See, for example, Cumb. *Civilian*, July 5, 1850; Cumb. *Telegraph*, June 28, 1855; Cumb. *Democratic Alleganian*, July 12, 1857, and July 10, 1858.

119. *UMW Journal*, April 15, 1909.

120. Cumb. *Civilian and Telegraph*, June 7, 1866.

121. *Md. Directory and State Gazetter, 1887*, pp. 11, 23, 25–27, 29–30, 31, 33.

122. *Frost. Mining Journal*, March 21, 1891.

123. Cumb. *Civilian and Telegraph*, Aug. 12, 1869.

124. *Ibid.*, July 13, 1871.

125. *Ibid.*, July 6.

126. Cumb. *Democratic Alleganian*, Jan. 29, 1859.

127. Cumb. *Civilian and Telegraph*, Aug. 27, 1874.

128. *Frost. Mining Journal,* May 11, 1872.

129. *Ibid.,* Jan. 29, 1876.

130. Webb, p. 18; see also the English coalfield ballad, "The Cock Fight," in A. L. Lloyd, *Come All Ye Bold Miners* (London, 1952), p. 36.

131. *Frost. Mining Journal,* Feb. 24, 1894.

132. Cumb. *Civilian and Telegraph,* Sept. 12 and Oct. 10, 1867.

133. *Ibid.,* Oct. 9, 1873.

134. *Ibid.,* Oct. 23.

135. *Ibid.,* Feb. 11 and March 4, 1875.

136. GCC&I Co. Journal, July 12, 1839.

137. *Ibid.,* July 27.

138. Cumb. *Civilian and Telegraph,* July 21, 1879.

139. Cumb. *Civilian,* April 23, 1876.

140. Cumb. *Civilian and Telegraph,* June 24, 1869.

141. *Coal Trade Journal,* Aug. 24, 1881.

Chapter 9. The Antebellum Years (1835–1860)

1. GCC&I Co., Report of Directors, 1839, p. 9 (J. H. Alexander Papers, MHS).

2. *Ibid.*

3. GCC&I Co. Journal, Jan. 19, 1839.

4. *Ibid.,* Feb. 6.

5. *Ibid.,* Feb. 18.

6. *Ibid.,* Oct. 2.

7. *Ibid.,* "Rules of the Lonaconing Residency," beginning of Vol. II.

8. GCC&I Co., Report of Directors, 1839, p. 10A (J. H. Alexander Papers, MHS).

9. GCC&I Co. Journal, Aug. 3, 1839.

10. GCC&I Co., *George's Creek Coal and Iron Co.* (n.p. [1849]), undated report of C. E. Detmold, lessee of the iron facilities, MHS. Internal evidence dates the report in 1849.

11. Henry Thomas Weld, *Report of the Md. and N.Y. Iron and Coal Co.'s Lands,* pp. 64–65; and Capt. John Pickell, U.S. Engineer, to Matthew St. Clair Clark, pres. of Md. Mining Co., May 26, 1835,

in George W. Hughes, *Extracts from Reports of an Examination of the Coal Measures Belonging to the Md. Mining Co.*, p. 6.

12. GCC&I Co. Journal, Dec. 1, 1837.
13. *Ibid.*, Jan. 2, Feb. 19, and Feb. 27, 1838.
14. *Ibid.*, March 1.
15. *Ibid.*, March 6 and 14.
16. *Ibid.*, April 21.
17. *Ibid.*, May 7.
18. *Ibid.*, June 11.
19. *Ibid.*, April 13 and July 9.
20. *Ibid.*, July 12 and 19, 1839.
21. *Ibid.*
22. *Ibid.*, July 26.
23. *Ibid.*, Aug. 5.
24. *Ibid.*, Aug. 3.
25. *Ibid.*, Aug. 29.
26. *Ibid.*, Sept. 10.
27. Cumb. *Civilian*, Dec. 11, 1846.
28. *Ibid.*, May 7, 1847; and Balt. *Sun*, May 8, 1847.
29. Cumb. *Civilian*, May 11, 1847.
30. Balt. *Sun*, April 5, 1882.
31. Cumb. *Civilian*, June 29, 1849; and Balt. *Sun*, June 23, 1849.
32. Cumb. *Civilian*, June 22, 1849.
33. *Ibid.*, June 29.
34. *Ibid.*, Feb. 15, 1850.
35. *Ibid.*, Jan. 31, 1851; see also Beverley S. Randolph, "History of the Md. Coal Region," p. 517; and Andrew Roy, *History of the Coal Miners of the U.S.*, p. 51.
36. Cumb. *Civilian*, Jan. 24, 1851.
37. Roy, *Coal Miners*, pp. 51–52.
38. *Ibid.*, p. 51; see also Edward A. Wieck, *American Miners Association*, p. 49.
39. J. Thomas Scharf, *History of Western Md.*, II, 1503.
40. *Cumb. Miners' Journal*, Oct. 10, 1851.
41. From Assignee's Report to Stockholders of the Md. Mining Co., published in *Cumb. Miners' Journal*, Dec. 5, 1851.
42. Md. *Laws* 1847, Ch. 228.
43. *Cumb. Miners' Journal*, July 2, 1852.
44. Md. *Laws* 1852, Ch. 229.

45. *Cumb. Miners' Journal,* Dec. 17, 1852.

46. *Ibid.,* May 27, 1853.

47. *Ibid.,* June 10.

48. *Ibid.,* June 17.

49. *Ibid.,* June 10.

50. *Ibid.,* June 17.

51. *Ibid.,* June 24.

52. *Ibid.*

53. Letter to the editor of the Balt. *Sun,* Jan. 20, 1854.

54. This strike is fully described in the Balt. *Sun,* Jan. 9–April 21, 1854.

55. Balt. *Sun,* Jan. 20, 1854.

56. Letter signed "Truth," dated Jan. 17, Balt. *Sun,* Jan. 20, 1854.

57. Unsigned letter dated Frost., Md., Jan. 7, 1854, Balt. *Sun,* Jan. 10, 1854.

58. *Ibid.*

59. *Ibid.*

60. Letter signed "Truth," Balt. *Sun,* Jan. 20, 1854.

61. Balt. *Sun,* Jan. 23, 1854; for subsequent letters on this strike see Balt. *Sun,* Feb. 22 and March 1, 1854.

62. *Ibid.,* Jan. 27.

63. *Ibid.,* Feb. 10 and 13, quoting various Cumb. papers.

64. *Ibid.,* Feb. 13 and 24, quoting various Cumb. papers.

65. *Ibid.,* Feb. 25, quoting *Cumb. Miners' Journal;* and March 3, quoting Cumb. *Telegraph.*

66. Balt. *Sun,* April 1, 1854, quoting *Cumb. Miners' Journal.*

67. *Ibid.,* March 24, quoting *Cumb. Miners' Journal.*

68. *Ibid.,* March 3, quoting Cumb. *Telegraph.*

69. *Ibid.,* March 24, quoting Cumb. *Telegraph.*

70. *Ibid.,* March 17, quoting Cumb. *Telegraph.*

71. Roy, *Coal Miners,* pp. 52–53.

72. Balt. *Sun,* March 31, 1854, quoting Cumb. *Telegraph.*

73. *Ibid.,* April 1, quoting *Cumb. Miners' Journal.*

74. *Ibid.,* April 5 and 7.

75. Thos. L. Patterson to W. S. Ringgold, April 6, 1854 (records of the National Park Service, Record Group 79, Records of the Chesapeake and Ohio Canal Co., Box 205, NA, hereafter referred to as C&O Canal Records).

76. Balt. *Sun,* May 5, 1854.

77. Ele Bowen, *Rambles in the Path of the Steam-Horse*, pp. 296–297.

78. Balt. *Sun*, June 23, 1854.

79. *Ibid.*, July 28 and Aug. 7.

80. *Ibid.*, Aug. 25, quoting Cumb. *Telegraph.*

81. *Ibid.*, April 6, 1855, quoting Cumb. *Telegraph.*

82. *Ibid.*

83. *Ibid.*, May 11, quoting Cumb. *Telegraph.*

84. Cumb. *Telegraph*, June 21, 1855.

85. *Ibid.*

86. *Ibid.*

87. *Ibid.*, July 12.

88. *Ibid.*, Oct. 18, 1855, and Jan. 3, 1856; and Franklin Coal Co., *First Annual Report*, p. 7.

89. Cumb. *Telegraph*, Feb. 14 and May 22, 1856.

90. *Ibid.*, June 5.

91. According to Saward's *Coal Trade Annual* for 1878, p. 23, the rate remained at 30 cents per ton from Aug. 1855 to June 1862.

92. Statistics of the Cumb. Coal Trade, Cumb. *Civilian and Telegraph*, Jan. 20, 1870.

93. Cumb. *Telegraph*, July 29, 1858.

Chapter 10. The War and Readjustment (1860–1868)

1. Cumb. *Civilian and Telegraph*, Sept. 6, 1860.

2. Andrew Roy, *Recollections of a Prisoner of War*, p. 149.

3. Cumb. *Civilian and Telegraph*, Oct. 4, 1860.

4. N.Y. *Evening Post*, Oct. 22, 1860.

5. Cumb. *Civilian and Telegraph*, Nov. 15, 1860.

6. *Ibid.*

7. Allan Nevins, *War for the Union: The Improvised War, 1861–1862* (N.Y., 1959), pp. 80–84.

8. Pottsville (Penn.) *Miners' Journal*, April 27, 1861.

9. J. Thomas Scharf, *History of Western Md.*, II, 1503–1504.

10. Cumb. *Civilian and Telegraph*, Aug. 29, 1861.

11. Scharf, II, 1504.,

12. Roy, *Recollections*, pp. 148–149.

13. Harold R. Manakee, *Md. in the Civil War* (Balt., 1961), p. 121.

14. Pottsville *Miners' Journal*, Jan. 11, 1862; and Statistics of the Cumb. Coal Trade published in Cumb. *Civilian and Telegraph*, Jan. 20, 1870. The production figures given in the 1860 census of manufactures are for eight establishments only; according to the census bureau's *Special Reports, Mines and Quarries: 1902*, pp. 1–6 the "statistics of mines continued to be taken . . . only in a scattering way" until the census of 1870, and the degree of completeness of the statistics of mining was uncertain.

15. Cumb. *Civilian and Telegraph*, Jan. 20, 1870.

16. *Ibid.*

17. Alfred Spates to W. S. Ringgold, May 6, 1861 (C&O Canal Records, Box 215).

18. Spates to Ringgold, May 13, 1861 (C&O Canal Records, Box 215).

19. Pottsville *Miners' Journal*, Aug. 24, 1861.

20. *Ibid.*, Jan. 11, 1862; it is not clear whether the word "returns" refers to statistical returns or returned empty cars.

21. A. C. Greene to Ringgold, April 7, 1862 (C&O Canal Records, Box 216).

22. *Ibid.*, April 11.

23. *Ibid.*, April 18.

24. *Ibid.*, April 29.

25. Frederick E. Saward, *Coal Trade Annual*, V (N.Y., 1878), 23.

26. C. B. Conant, "Coal Fever: The Price and Prospects of Anthracite Coal," *Merchants' Magazine and Commercial Review*, May 1865, p. 359.

27. Greene to Ringgold, June 12, 1862. (C&O Canal Records, Box 216).

28. *Ibid.*, June 21.

29. *Ibid.*, Aug. 12.

30. Saward, *Coal Trade Annual*, V (1878), 23.

31. Greene to Ringgold, Nov. 13, 1862 (C&O Canal Records, Box 216).

32. *Ibid.*, Dec. 2.

33. Spates to Ringgold, Dec. 5, 1862 (C&O Canal Records, Box 216). Spates was in charge of the Cumb. division of the canal.

34. Pottsville *Miners' Journal*, Jan. 10, 1863.

35. Cumb. *Civilian and Telegraph*, Jan. 20, 1870.

36. Saward, *Coal Trade Annual*, V (1878), 23.

37. Greene to Ringgold, Jan. 13, 1863 (C&O Canal Records, Box 217).

38. Pottsville *Miners' Journal*, Jan. 23, 1864.

39. Cumb. *Civilian and Telegraph*, Jan. 20, 1870.

40. *Ibid.*, July 2 and Aug. 13, 1863.

41. Greene to Ringgold, Sept. 2, 1863 (C&O Records, Box 217).

42. Canal Board to Lloyd Lowe (C&O Canal Records, Letter Book M, p. 122).

43. Lowe to Ringgold, March 14, 1864 (C&O Canal Records, Box 218).

44. Washington *National Republican*, March 17, 1864.

45. Greene to Ringgold, March 18, 1864 (C&O Canal Records, Box 218).

46. *Ibid.*, March 23.

47. Edward A. Wieck, *American Miners Association*, p. 142.

48. Wayne G. Broehl, Jr., *Molly Maguires*, p. 90. Broehl's entire Chapter IV on the violence in this region during the war is interesting; see also Conant, p. 360.

49. Pottsville *Miners' Journal*, Feb. 13, 1864.

50. *Ibid.*, Feb. 20.

51. Lowe to Ringgold, March 24, 1864 (C&O Canal Records, Box 218).

52. Conant, p. 359.

53. Md. Coal Co., *Report, 1870*, p. 8.

54. Hampshire and Balt. Coal Co., *Report, 1864*, p. 11.

55. Cumb. *Alleganian*, Sept. 21, 1864.

56. Greene to Ringgold, Sept. 29, 1864 (C&O Canal Records, Box 218); see also Cumb. *Alleganian*, Oct. 5, 1864.

57. Cumb. *Civilian and Telegraph*, Jan. 20, 1870.

58. Spates to Ringgold, Oct. 11, 1864 (C&O Canal Records, Box 218).

59. Cumb. *Alleganian*, Dec. 14, 1864.

60. This letter is in the possession of the MHS.

61. Spates to Ringgold, April 26, 1865 (C&O Canal Records, Box 219).

62. Cumb. *Alleganian*, May 10, 1865; and Cumb. *Civilian and Telegraph*, May 11, 1865.

63. Cumb. *Alleganian,* May 24, 1865; and Cumb. *Civilian and Telegraph,* May 25, 1865.

64. Cumb. *Alleganian,* May 31, 1865.

65. *Ibid.*

66. Cumb. *Civilian and Telegraph,* May 31, 1865.

67. *Ibid.,* June 8.

68. Cumb. *Alleganian,* June 7 and 14, 1865.

69. Central Coal Co., American Coal Co., Consolidation Coal Co., Borden Mining Co., Hampshire Coal Co., and Cumb. Coal and Iron Co., to President and Directors of the C&O Canal Co., June 12, 1865, asking for a reduction in tolls (C&O Canal Records, Box 219).

70. Cumb. *Alleganian,* Jan. 31, 1866; and Cumb. coal statistics, *Civilian and Telegraph,* Jan. 20, 1870.

71. MGA, House, *Report of Select Committee, 1865,* p. 45.

72. Cumb. *Alleganian,* April 4, 1866; and Cumb. *Civilian and Telegraph,* April 5, 1866.

73. Cumb. *Civilian and Telegraph,* April 12, 1866.

74. Cumb. *Alleganian,* April 11 and 18, 1866; Cumb. *Union,* April 28, 1866; Cumb. *Civilian and Telegraph,* May 3, 1866. The *Alleganian* of May 2, 1866, reported settlement at 60 cents a ton, but this figure does not agree with that given in Saward's *Coal Trade Annual,* 1878, or with the two newspapers cited above.

75. Cumb. *Civilian and Telegraph,* July 4, 1866.

76. *Ibid.,* March 14, 1861.

77. *Ibid.,* July 4, 1866.

78. *Ibid.*

79. *Ibid.,* July 12.

80. *Ibid.,* Jan. 20, 1870.

81. Md. Coal Co., *Report, 1870,* p. 8.

82. *Ibid.*

83. Cumb. *Alleganian,* Dec. 9, 1868; and Cumb. *Civilian,* Dec. 10, 1868.

84. William M. Richards, "An Experiment in Industrial Feudalism at Lonaconing, Md.," p. 42.

85. See Cumb. *Alleganian,* Dec. 30, 1868, and Jan. 6, Jan. 27, and Feb. 17, 1869; Cumb. *Civilian,* Dec. 17, 1868, and Feb. 18, 1869; and Balt. *Sun,* Feb. 19, 1869.

86. Greene to Hon. Jas C. Clark, Sept. 14, 1870 (C&O Canal Records, Box 229).

Chapter 11. Big Business: Small Union (1869–1875)

1. Charles E. Beachley, *History of the Consolidation Coal Co.*, pp. 17–19 gives the details of Consolidation's initial acquisition of assets.

2. *Ibid.*, p. 21.

3. *Ibid.*, pp. 20–21.

4. *Ibid.*

5. *Ibid.*, p. 23; and Beverley S. Randolph, "History of the Md. Coal Region," p. 521.

6. Frederick E. Saward, *Saward's Coal Freight Circular*, Vol. I (April 1869–April 1870), No. 12.

7. *Ibid.;* and Cumb. *Alleganian*, Feb. 23, 1870.

8. Cumb. *Alleganian*, Feb. 23, 1870.

9. Md. Mine Insp., *Report, 1883*, p. 5.

10. Frank Julian Warne, *Coal Mine Workers, A Study in Labor Organization* (N.Y., 1905), pp. 187–188.

11. Chris Evans, *History of UMW of America from the Year 1860 to 1890* (n.p., n.d.), I, 13.

12. Andrew Roy, *History of the Coal Miners of the U.S.*, p. 65.

13. E. Welbourne, *Miners' Unions of Northumberland and Durham* (Cambridge, England, 1923), p. 105; and Alexander McDonald, testimony before Royal Commission, *Workingman's Advocate*, July 11, Aug. 1, Aug. 15, and Aug. 29, 1868.

14. Robert Page Arnot, *The Miners: A History of the Miners' Federation of Great Britain* (London, 1949), pp. 44–45.

15. *Workingman's Advocate*, Nov. 30, 1867.

16. *Ibid.*, Dec. 2, 1869.

17. Arthur E. Suffern, *The Coal Miners' Struggle for Industrial Status*, p. 25; and Evans, I, 22.

18. Beachley, p. 24.

19. C. Slack to J. C. Clark, July 11, 1870 (C&O Records, Box 229); A. C. Greene to Clark, Sept. 14, 1870 (C&O Canal Records, Box 229); and Md. Coal Co., *Report, 1870*, p. 8.

20. Slack to Clark, July 11, 1870 (C&O Canal Records, Box 229).

21. James Macfarlane, *Coal-Regions of America*, p. 258; see also Consolidation Coal Co., *Annual Report, 1871*, p. 14; and Howard N. Eavenson, *First Century and a Quarter of American Coal Industry*, p. 239.

22. *Frost. Mining Journal, April 26,* 1873; and Cumb. *Alleganian,* April 30, 1873.

23. Cumb. *Civilian and Telegraph,* May 1, 1873.

24. Cumb. *Alleganian,* April 30, 1873; Cumb. *Civilian and Telegraph,* May 1, 1873; and *Frost. Mining Journal,* May 3, 1873.

25. Balt. *Sun,* May 2, 1873.

26. Cumb. *Civilian and Telegraph,* May 1, 1873.

27. *Ibid.,* quoting Cumb. *Daily Times.*

28. *Frost. Mining Journal,* May 3, 1873.

29. Cumb. *Daily Times,* May 6, 1873; Balt. *Sun,* May 9, 1873; and *Frost. Mining Journal,* May 10, 1873.

30. Balt. *Sun,* May 9, 1873; and Cumb. *Alleganian,* May 7, 1873.

31. Cumb. *Alleganian,* May 7, 1873; see also Balt. *Sun,* May 9, 1873; and Cumb. *Daily Times,* May 6, 1873.

32. Cumb. *Alleganian,* May 21, 1873.

33. Roy, *Coal Miners,* p. 78.

34. Cumb. *Civilian and Telegraph,* May 15, 1873; and Cumb. *Alleganian,* May 21, 1873.

35. Cumb. *Civilian and Telegraph,* May 29, 1873; Balt. *Sun,* May 30, 1873; Cumb. *Alleganian,* May 28, 1873; and *Frost. Mining Journal,* May 31, 1873.

36. Cumb. *Civilian and Telegraph,* May 29, 1873.

37. Penn. Bureau of Statistics of Labor and Agriculture, *Report, 1872–73,* pp. 315–380. It is evident from miners' correspondence that they subscribed to labor newspapers such as the *Workingman's Advocate* and later the *Miners' National Record;* the *National Labor Tribune* of May 2, 1874, claimed a thousand readers in the Md. mining region.

38. Cumb. *Civilian and Telegraph,* May 29, 1873.

39. Balt. *Sun,* May 9, 1873.

40. Cumb. *Civilian and Telegraph,* May 29, 1873.

41. *UMW Journal,* Aug. 20, 1908.

42. Cumb. *Daily Times,* May 26, 1873.

43. See Ch. 9.

44. From Brown obituary, Cumb. *Evening Times,* April 15, 1905.

45. Cumb. *Daily Times,* May 26, 1873.

46. Cumb. *Civilian and Telegraph,* May 29, 1873.

47. Cumb. *Alleganian,* June 4, 1873. The total delegation consisted of: George Hoskin and William Bane, Borden Mines; William

Evans and James Murphy, New Hope; James Gatehouse and George Goodwin, Consolidation; E. M. Davis, Midlothian and Johnson; John Mayer and Benjamin Jenkins, Furnace Mines; Thomas Brown and Peter Cain, Pompey Smash; John Ray, Miller and Midland; William McVay and Peter Monaghan, Ocean; Richard Rolands, Pekin; John Mansfield, Caledonia; Joseph Ware and A. M. Sample, Hampshire; Samuel McGowan and John Costello, Virginia.

48. *Frost. Mining Journal*, June 21, 1873; and Cumb. *Civilian and Telegraph*, June 26, 1873.

49. *Frost. Mining Journal*, Aug. 9, 1873.

50. Cumb. *Civilian and Telegraph*, Aug. 14, 1873.

51. *Frost. Mining Journal*, Sept. 6, 1873.

52. Cumb. *Civilian and Telegraph*, Sept. 25, 1873; and *Frost. Mining Journal*, Sept. 27, 1873.

53. Cumb. *Alleganian*, Sept. 24, 1873; see also Cumb. *Civilian and Telegraph*, Sept. 25, 1873.

54. *Frost. Mining Journal*, Oct. 11, 1873.

55. *Ibid.*, Jan. 10, 1874.

56. *Ibid.*, Dec. 13, 1873.

57. *Ibid.*, Jan. 17, 1874.

58. Md. Coal Co., *Annual Report of the Md. Coal Co. of Allegany County, Md., for the Year 1873* (N.Y., 1874), pp. 6–7.

59. Cumb. *Civilian and Telegraph*, Feb. 5, 1874.

60. *Frost. Mining Journal*, May 16, 1874.

61. *Workingman's Advocate*, June 6, 1874.

62. Joseph G. Rayback, *History of American Labor* (N.Y., 1959), pp. 117–118.

63. Roy, *Coal Miners*, pp. 150–153.

64. *Frost. Mining Journal*, Feb. 14, 1874.

65. Cumb. *Alleganian*, Oct. 15, 1873.

66. *Workingman's Advocate*, Nov. 1, 1873.

67. Cumb. *Alleganian*, Nov. 26, 1873.

68. *Frost. Mining Journal*, May 23, 1874; Cumb. *Civilian and Telegraph*, May 28, 1874; and Cumb. *Alleganian*, May 27, 1874.

69. *Workingman's Advocate*, May 30, 1874.

70. *Frost. Mining Journal*, May 30, 1874; and *Workingman's Advocate*, June 6, 1874.

71. *Frost. Mining Journal*, May 30, 1874.

72. *Ibid.*

73. *Ibid.*, Aug. 8.

74. *National Labor Tribune*, July 4, 1874.

75. *Frost. Mining Journal*, Oct. 10, 1874.

76. *Ibid.*, Jan. 10.

77. *Ibid.*, Nov. 14; *Workingman's Advocate*, Nov. 7, 1874; and *Miners' National Record*, Nov. 1874.

78. Evans, I, 70.

79. Suffern, pp. 25–31.

80. *Ibid.*, and *Workingman's Advocate*, Oct. 30, 1875.

81. Md. Bur. Ind. Stat., *Report, 1898*, pp. 237–238.

82. Roy, *Coal Miners*, p. 155.

83. Cumb. *Civilian and Telegraph*, March 26, 1874.

84. *Frost. Mining Journal*, June 20, 1874; and Cumb. *Civilian and Telegraph*, June 25, 1874.

85. *Frost. Mining Journal*, Oct. 3, 1874; see also *Workingman's Advocate*, Oct. 24, 1874.

86. Letter from J. James dated Dec. 7, 1874, in *Miner's National Record*, Jan. 1875.

87. Pottsville *Miners' Journal*, April 23, 1875.

88. *Ibid.*, Oct. 1; and *Miners' National Record*, Dec. 1875.

89. *Miners' National Record*, Dec. 1875.

90. *Ibid.*, Oct. 1875.

91. Pottsville *Miners' Journal*, May 14, May 21, Oct. 1 and Oct. 22, 1875; *Workingman's Advocate*, July 17 and Oct. 9, 1875; *Miners' National Record*, June, Sept., and Oct., 1875; and *National Labor Tribune*, Oct. 9, 1875.

92. *Miners' National Record*, June 1875.

93. *Ibid.*, July.

94. *Frost. Mining Journal*, July 24, 1875.

95. *Miners' National Record*, Sept. 1875.

96. *Ibid.*, Nov. 1875.

97. Cumb. *Daily Times*, March 11, 1874.

98. *Ibid.*

99. *Ibid.*, March 19.

100. *Ibid.*, March 23.

101. Welbourne, p. 113.

102. Cumb. *Daily Times*, March 19, 1874.

103. Cumb. *Alleganian*, April 1, 1874.

104. *National Labor Tribune*, Sept. 25, 1875.

105. *Frost. Mining Journal*, Nov. 14, 1874; and Cumb. *Alleganian*, Nov. 18, 1874.

106. Cumb. *Daily Times*, Feb. 4, 1881.

Chapter 12. Depression: Emergence of the Knights of Labor (1876–1880)

1. Md. Coal Co., *Annual Report of the Md. Coal Co. of Allegany County, Md., for the Year 1876* (N.Y., 1877), p. 4.

2. Charles E. Beachley, *History of the Consolidation Coal Co.*, p. 24.

3. *Ibid.*, and Balt. *Sun*, Feb. 26, 1877.

4. Md. Coal Co., *Report, 1876*, p. 4.

5. Andrew Roy, *Coal Mines* (Cleveland, 1876), p. 254.

6. Md. Coal Co., *Report, 1876*, p. 5.

7. Md. *Laws* 1876, Chs. 64 and 80.

8. Cumb. *Alleganian*, April 7, 1876; *Frost. Mining Journal*, April 8, 1876; and Cumb. *Civilian*, April 9, 1876.

9. Cumb. *Civilian*, April 9, 1876.

10. *Ibid.*, and Cumb. *Alleganian*, April 10, 1876.

11. Cumb. *Civilian*, April 9, 1876.

12. Cumb. *Alleganian*, April 11, 1876; *Frost. Mining Journal*, April 15, 1876; Balt. *Sun*, April 12, 1876; and Cumb. *Civilian*, April 16, 1876.

13. Balt. *Sun*, April 13, 1876.

14. *Frost. Mining Journal*, April 15, 1876; and Cumb. *Civilian*, April 16, 1876.

15. Cumb. *Alleganian*, Feb. 28, 1876.

16. *Ibid.*, April 11.

17. *Ibid.*, April 18.

18. *Ibid.*, April 25.

19. Cumb. *Alleganian*, April 21, 1876.

20. Cumb. *Civilian*, April 16, 1876.

21. Cumb. *Alleganian*, April 22, 1876.

22. *Ibid.*, April 25.

23. *Ibid.*, May 9; and Cumb. *Alleganian and Times*, Aug. 24, 1876.

24. Cumb. *Alleganian*, March 23, 1876.

25. Cumb. *Alleganian and Times*, May 29, 1876.

26. Cumb. *Civilian*, June 4, 1876.
27. Cumb. *Alleganian and Times*, June 22, 1876.
28. *Ibid.*, Aug. 7.
29. *Ibid.*, Oct. 31.
30. *Ibid.*, Dec. 21.
31. *Ibid.*, Dec. 23.
32. *Ibid.*, Dec. 30.
33. Joseph G. Rayback, *History of American Labor*, p. 129.
34. Cumb. *Alleganian and Times*, Dec. 30, 1876.
35. Balt. *Sun*, Jan. 20, 1877; and Cumb. *Alleganian and Times*, Jan. 20, 1877.
36. Cumb. *Alleganian and Times*, Jan. 15, 1877; *Frost. Mining Journal*, Jan. 20, 1877; Balt. *Sun*, Jan. 20, 1877; and Cumb. *Civilian*, Jan. 14, 1877.
37. Cumb. *Alleganian and Times*, Jan. 15, 1877.
38. *Ibid.*, Jan. 17.
39. *Ibid.*
40. *Ibid.*, Jan. 20; and Balt. *Sun*, Jan. 20, 1877.
41. *Ibid.*
42. *Frost. Mining Journal*, Jan. 20, 1877.
43. Cumb. *Alleganian and Times*, Jan. 25, 1877.
44. *Ibid.*, Jan. 20.
45. *Ibid.*, Jan. 27.
46. *Ibid.*
47. *Ibid.*
48. *Frost. Mining Journal*, Feb. 3, 1877.
49. Cumb. *Alleganian and Times*, Feb. 12, 1877; and Balt. *Sun*, Feb. 23, 1877.
50. Cumb. *Alleganian and Times*, Feb. 13, 1877.
51. Balt. *Sun*, Feb. 23, 1877.
52. Cumb. *Alleganian and Times*, March 5, 1877.
53. Cumb. *Civilian*, March 11, 1877.
54. *Frost. Mining Journal*, March 17, 1877.
55. Cumb. *Civilian*, March 18, 1877.
56. *Ibid.*
57. *Ibid.*
58. Cumb. *Alleganian and Times*, March 23, 1877.
59. *Ibid.*, Jan. 27; and *Frost. Mining Journal*, Jan. 27, 1877.

60. Consolidation Coal Co., *Report to the Stockholders for the Year Ending Dec. 31, 1877* (Balt., 1878), p. 5.

61. *Frost. Mining Journal*, March 31, 1877.

62. Cumb. *Alleganian and Times*, March 19, 1876.

63. Cumb. *Civilian*, May 22, 1877.

64. Balt. *Sun*, June 18 and 19 and July 2, 1877.

65. *Ibid.*, July 26.

66. Cumb. *Alleganian and Times*, July 21, 1877.

67. Cumb. *Civilian*, July 29, 1877; and *Frost. Mining Journal*, Aug. 4, 1877.

68. Balt. *Sun*, July 23 and 24, 1877.

69. *Ibid.*, July 31.

70. *Ibid.*, July 30.

71. *Frost. Mining Journal*, Aug. 4, 1877.

72. Balt. *Sun*, Aug. 1, 1877.

73. *Ibid.*, July 30.

74. *Ibid.*, Aug. 1.

75. Cumb. *Alleganian and Times*, Aug. 1, 1877; and Balt. *Sun*, July 31, 1877.

76. Cumb. *Alleganian and Times*, Aug. 3, 1877.

77. Balt. *Sun*, Aug. 3, 1877.

78. *Ibid.*

79. Cumb. *Alleganian and Times*, Aug. 3, 1877.

80. Balt. *Sun*, Aug. 18, 1877.

81. *Ibid.*, Aug. 6, 9, 10 and 11.

82. Cumb. *Alleganian and Times*, Sept. 7, 1877.

83. *Ibid.*, Jan. 22, 1878.

84. *Ibid.*, Jan. 28.

85. *Ibid.*

86. *Ibid.*

87. *Ibid.*

88. *Ibid.*

89. *Frost. Mining Journal*, Feb. 2, 1878.

90. Cumb. *Alleganian and Times*, Feb. 1, 1878.

91. *Ibid.*, Feb. 5.

92. *Ibid.*, March 19.

93. *Ibid.*, March 15.

94. *Ibid.*, March 21.

95. *Ibid.*, Jan. 20 and 25, 1877.

96. *Ibid.*, Aug. 14 and 21.

97. Consolidation Coal Co., *Report to the Stockholders for the Year Ending Dec. 31, 1878* (Balt., n.d.), p. 5.

98. John McBride, "Coal Miners," in George E. McNeill, ed., *Labor Movement: The Problem of Today* (Boston, 1877), pp. 251–252; and Richard Lowitt, "Frost. 1882: German Strikers vs German Strikebreakers," p. 75.

99. In fact, it is evident from correspondence between T. V. Powderly and other officers that they distrusted trades unions.

100. Norman J. Ware, *Labor Movement in the United States, 1860–1895* (N.Y., 1929), p. 211.

101. K. of L., *Record of the Proceedings of the Second Regular Session of the General Assembly, Held at St. Louis, Mo., Jan. 14–17, 1879* (n.p., n.d.), p. 65.

102. T. V. Powderly, *Thirty Years of Labor, 1859 to 1889* (Columbus, O., 1889), p. 628.

103. K. of L., *Record of the Proceedings of the Third Regular Session of the General Assembly, Held at Chicago, Ill., Sept. 2–6, 1879* (n.p., n.d.), p. 115.

104. *Frost. Mining Journal*, Aug. 16, 1879.

105. Joe Brooke to T. V. Powderly, Nov. 29, 1880 (T. V. Powderly Papers, Archives of The Catholic University of America, Washington, D.C., Box 3).

106. Md. Mine Insp., *Report, 1883*, p. 8.

107. *Ibid.*

108. *Frost. Mining Journal*, Aug. 2, 1879; and Cumb. *Daily Times*, Aug. 28, 1879.

109. *Frost. Mining Journal*, Aug. 23, 1879; and Cumb. *Daily Times*, Aug. 26, 1879.

110. See Balt. *Sun*, Sept. 8–Oct. 28, 1879, for coverage of the strike.

111. K. of L., *Proceedings, 1879*, p. 115.

112. U.S. Census Office, *Tenth Census of the U.S.*, IV, 681.

113. Cumb. *Daily Times*, Sept. 4, 1879; Balt. *Sun*, Sept. 8, 1879; and *Frost Mining Journal*, Sept. 13, 1879.

114. E. B. Jones and James W. Bishop to Powderly, Sept. 29, 1879 (Powderly Papers, Box 2).

115. Powderly to Jones, Oct. 6, 1879 (Powderly Papers, Letter Book [Powderly and Stephens], 4 July–7 Oct. 1880, p. 164).

116. Balt. *Sun*, Oct. 29, 1879.

117. *Ibid.*, Oct. 23–Nov. 3; and Cumb. *Daily Times*, Oct. 20 and 21, 1879.

118. Cumb. *Daily Times*, Oct. 20, 1879.

119. *Ibid.*, Dec. 30.

120. A. Willison to Governor William T. Hamilton, Feb. 7, 1880 (Executive Papers, Misc. 1880, MHR).

121. Md. Commissioners of Mines. *Report, 1880*, pp. 5–6.

122. Cumb. *Daily Times*, Feb. 3, 1880.

123. Regulations of K. of L., as advertised in *Frost. Mining Journal*, May 7, 1881, to Sept. 2, 1882.

124. *Frost. Mining Journal*, Feb. 23, 1880.

125. Cumb. *Daily Times*, Feb. 23, 1880; Md. Mine Insp., *Report, 1883*, p. 6; and Penn. Dept. of Internal Affairs, *Annual Report, 1881–82*, Part III, *Industrial Statistics* (Harrisburg, 1883), p. 147.

Chapter 13. Ventilation and Inspection Legislation (1876–1880)

1. Cumb. *Daily Times*, Sept. 15, 1873.

2. Cumb. *Civilian and Telegraph*, Sept. 25, 1873.

3. *Ibid.;* see also Cumb. *Alleganian*, Nov. 12, 1873; and *Frost. Mining Journal*, Nov. 6, 1873.

4. MGA, House, *Journal of Proceedings, Jan. Session, 1874* (Annapolis, 1874), pp. 1102, 1233–1234, 1277–1278.

5. MGA, Senate, *Journal of Proceedings, Jan. Session, 1874* (Annapolis, 1874), pp. 987, 1028.

6. Cumb. *Alleganian*, Jan. 27, 1876.

7. *Ibid.*

8. *Ibid.*, Feb. 1 and 2; and *Frost. Mining Journal*, Feb. 5, 1876.

9. Andrew Roy, *Coal Mines*, p. 279.

10. Cumb. *Alleganian*, March 1, 1876.

11. *Ibid.*

12. *Ibid.*, March 6.

13. *Ibid.*, March 7.

14. *Ibid.*, March 11.

15. Md. *Laws* 1876, Ch. 173.

16. Md. Mine Insp., *Report, 1877*, pp. 6–15.

17. *Ibid., Report, 1876,* p. 5.

18. A. C. Greene to Wm. H. Smith, March 24, 1876 (Executive Papers, Applications 1876, MHR).

19. Md *Laws* 1876, Ch. 173, Sec. 2.

20. *Ibid.*, Sec. 5.

21. Thomas Rooney to Governor John Lee Carroll, April 1, 1876 (Executive Papers, Applications 1876, MHR).

22. Greene to Thomas G. McCulloh, April 17, 1876 (Executive Papers, Applications 1876, MHR).

23. Noted on affidavit of the clerk, dated May 23, 1876 (Executive Papers, Misc. Papers, MHR).

24. William R. Barnard to Governor Carroll, April 19, 1876 (Executive Papers, Applications 1876, MHR).

25. McCulloh to Carroll, April 21, 1876 (Executive Papers, Applications 1876, MHR).

26. Cumb. *Alleganian and Times,* Feb. 12, 1878.

27. Md. Mine Insp., *Report, 1877,* p. 5.

28. *Ibid., Report, 1876,* pp. 31–32.

29. Cumb. *Civilian,* Jan. 21, 1877.

30. Md. Commissioners of Mines, *Report, 1880,* p. 6.

31. Md. Mine Insp., *Report, 1879,* p. 13.

32. *Frost. Mining Journal,* June 10, 1876.

33. MGA, House, *Message of John Lee Carroll, Governor of Md., to the General Assembly, at the Regular Session, Jan. 1878* (Annapolis, 1878), Document A, p. 31.

34. Cumb. *Alleganian and Times,* Jan. 23, 1878.

35. *Frost. Mining Journal,* June 14, 1879.

36. Cumb. *Alleganian and Times,* Oct. 10, 1877.

37. *Frost. Mining Journal,* April 13, 1878.

38. *Ibid.*

39. Md. Mine Insp., *Report, 1876,* pp. 29–31.

40. Cumb. *Alleganian,* Jan. 19 and 27, 1877.

41. *Ibid.*, Aug. 1 and 3.

42. *Frost. Mining Journal,* Aug. 4, 1877.

43. *Ibid.*, Aug. 11.

44. Balt. *Sun,* Aug. 22, 1877; *Frost. Mining Journal,* Aug 25, 1877; and Cumb. *Civilian,* Aug. 26, 1877.

45. Cumb. *Alleganian and Times,* Jan. 25, 1878; and *Frost. Mining Journal,* Jan. 26, 1878.

46. *Frost. Mining Journal*, Feb. 9, 1878.
47. *Ibid.*
48. Cumb. *Alleganian and Times*, Feb. 15, 1878.
49. *Ibid.*
50. *Ibid.*
51. *Frost. Mining Journal*, March 9, 1878.
52. Cumb. *Alleganian and Times*, March 11, 1878.
53. *Ibid.*, March 12 and 14.
54. *Frost. Mining Journal*, July 20, 1901.
55. Cumb. *Alleganian and Times*, March 15 and 19, 1878.
56. *Frost. Mining Journal*, Feb. 1, 1879.
57. *Ibid.*, June 14.
58. Md. *Laws* 1878, Ch. 371.
59. Cumb. *Civilian*, June 17, 1877.
60. Cumb. *Alleganian and Times*, June 10, 1878.
61. *Ibid.*
62. *Frost. Mining Journal*, Oct. 12, 1878.
63. See *Frost. Mining Journal*, Nov. 22 and 29, and Dec. 6, 1879.
64. *Ibid.*, Nov. 22.
65. *Ibid.*, Nov. 29.
66. J. B. Oder to Governor Carroll, undated (late 1878 or early 1879) (Executive Papers, Applications 1879, MHR).
67. Md. Commissioners of Mines, *Report, 1880*, p. 27.
68. *Ibid.*, p. 50.
69. *Ibid.*
70. *Ibid.*, p. 4.
71. *Ibid.*, p. 5.
72. *Frost. Mining Journal*, Feb. 7, 1880.
73. Md. Commissioners of Mines, *Report, 1880*, pp. 46–47.
74. *Frost. Mining Journal*, Feb. 7, 1880.
75. Andrew Roy, *History of the Coal Miners of the U.S.*, pp. 52–53.
76. *Frost. Mining Journal*, Aug. 9, 1879.
77. *Ibid.*, Feb. 14, 1880.
78. Md. Mine Insp., *Report, 1879*, p. 12.
79. Letter from "Miner," *Frost. Mining Journal*, Feb. 7, 1880.
80. Cumb. *Daily Times*, Aug. 4, 1879.
81. *Ibid.*, Sept. 26 and Oct. 8.
82. *Ibid.*, Nov. 7.

83. MGA, House, *Special Message from His Excellency Wm. T. Hamilton to the General Assembly of Md., Feb. 12, 1880* (Annapolis, 1880), Document L, p. 5.

84. MGA, Senate, *Journal of Proceedings, Jan. Session, 1880* (Annapolis, 1880), p. 353.

85. *Ibid.*, pp. 386, 460, 503, 559–560.

86. MGA, House, *Journal of Proceedings, Jan. Session, 1880* (Annapolis, 1880), pp. 825, 1106.

87. *Frost. Mining Journal*, Feb. 28, 1880; and MGA, Senate, *Journal, 1880*, p. 1158.

Chapter 14. The Knights of Labor and the Great Strike of 1882

1. An "Appeal for Aid," issued Oct. 6, 1882 (Powderly Papers, Box 5) lists 234 names of "blacklisted" members. The names of the secretaries of the local assemblies were printed in the *Journal of United Labor* throughout 1880, 1881, and 1882; the names of various delegates who consulted with the operators during the strike appeared in local newspapers.

2. K. of L., *Record of the Proceedings of the Fourth Regular Session of the General Assembly, Held at Pittsburgh, Pa., Sept. 7–11, 1880* (n.p., d.d.), pp. 208–209.

3. Cumb. *Daily Times*, July 23, 1880.

4. Charles E. Beachley, *History of the Consolidation Coal Co.*, pp. 26–27.

5. Cumb. *Daily Times*, Feb. 23 and May 28, 1880; Md. Mine Insp., *Report, 1883*, p. 6; and Penn. Dept. of Internal Affairs, *Report, 1881–82*, Part III, p. 147.

6. Cumb. *Daily Times*, Feb. 23, April 20, May 28 and May 31, 1880.

7. K. of L., *Proceedings, 1880*, pp. 211–213.

8. John Herron to Powderly, Dec. 22, 1879 (Powderly Papers, Box 2).

9. Md. Mine Insp., *Report, 1883*, p. 7.

10. *Frost. Mining Journal*, July 10, 1880.

11. Cumb. *Daily Times*, July 26, 1880.

12. *Ibid.*, July 5; and *Frost. Mining Journal*, July 10, 1880.

13. Cumb. *Daily Times*, July 15, 1880; and *Frost. Mining Journal*, July 17, 1880.

14. Cumb. *Daily Times*, July 21, 1880.

15. *Ibid.*, Dec. 31, and Feb. 2, 1881.

16. *Frost. Mining Journal*, Feb. 4, 1882; and Cumb. *Daily Times*, Feb. 6, 1882.

17. Cumb. *Daily Times*, Feb. 25, 1882.

18. *Ibid.*, Feb. 23.

19. *Ibid.*, March 7; and *Frost. Mining Journal*, March 11, 1882.

20. *Ibid.*

21. Balt. *Sun*, March 11, 1882.

22. Md. Mine Insp., *Report, 1883*, p. 13.

23. Balt. *Sun*, April 5, 1882.

24. Cumb. *Daily Times*, March 21, 1882; and *Frost. Mining Journal*, March 25, 1882.

25. Balt. *Sun*, March 21, 1882; see also Md. Mine Insp., *Report, 1883*, p. 15.

26. Cumb. *Daily Times*, Feb. 25, 1882.

27. Balt. *Sun*, March 30, 1882; see also John G. Vaughn to Powderly, March 17, 1882 (Powderly Papers, Box 4).

28. Balt. *Sun*, April 8 and June 22, 1882.

29. *Ibid.*, June 22.

30. Md. *Laws* 1880, Ch. 460.

31. *Ibid.*, Ch. 22.

32. Norman J. Ware, *Labor Movement in the U.S. 1860–1895*, p. 120.

33. *Ibid.*, pp. 120–126.

34. Powderly to Cornelius Cotter, May 25, 1882 (Powderly Papers, Box 5).

35. Ware, p. 130

36. John H. Jones to Powderly, March 30, 1882 (Powderly Papers, Box 4); Balt. *Sun*, April 1, 1882; and Cumb. *Daily Times*, April 4, 1882.

37. "Appeal for Aid" (Powderly Papers, Box 5).

38. Balt. *American*, April 7, 1882; see also Balt. *Sun*, March 30, 1882; Balt. *American*, April 6, 1882; and Cumb. *Daily Times*, May 3, 1882.

39. Balt. *Sun*, March 31, 1882.

40. Cumb. *Daily Times*, and Balt. *Sun*, April 1, 1882.

41. Balt. *Sun,* April 7, 1882; and Cumb. *Daily Times,* April 10, 1882.

42. Balt. *Sun,* April 3, 1882.

43. Penn. Dept. of Internal Affairs, *Report 1881–82,* Part III, pp. 148–149.

44. Cotter to Powderly, April 3, 1882 (Powderly Papers, Box 4).

45. Cumb. *Daily Times,* April 17, 1882.

46. Robert Layton to Powderly, May 13, 1882 (Powderly Papers, Box 5).

47. *Ibid.,* April 18 (Powderly Papers, Box 4).

48. Cumb. *Daily Times,* April 20, 1882.

49. Layton to Powderly, April 28, 1882 (Powderly Papers, Box 4).

50. Gilbert Rockwood to Powderly, May 2, 1882 (Powderly Papers, Box 5).

51. Cumb. *Daily Times,* May 5, 1882.

52. Layton to Powderly, April 29, 1882 (Powderly Papers, Box 4).

53. *Ibid.,* May 15 (Powderly Papers, Box 5).

54. Penn. Dept. of Internal Affairs, *Report, 1881–82,* Part III, p. 150.

55. Layton to Powderly, May 13, 1882 (Powderly Papers, Box 5).

56. Powderly to Layton, May 17, 1882 (Powderly Papers, Letter Book 1, 17 April 1882–26, July 1882).

57. Layton to Powderly, May 15, 1882 (Powderly Papers, Box 5).

58. Penn. Dept. of Internal Affairs, *Report, 1881–82,* Part III, pp. 151–154.

59. Powderly to Cotter, May 25, 1882 (Powderly Papers, Letter Book 1).

60. Balt. *Sun,* April 21, 1882.

61. Cumb. *Daily Times,* May 23, 1882.

62. Balt. *Sun,* May 30, 1882.

63. *Ibid.,* May 29, 30, and 31.

64. *Ibid.,* June 6.

65. *Ibid.,* May 31.

66. *Ibid.*

67. Balt. *Sun,* and Cumb. *Daily Times,* June 1, 1882.

68. Balt. *Sun,* June 1, 1882.

69. Balt. *Sun*, and Cumb. *Daily Times*, June 8, 1882.

70. Thomas Brown to Governor Hamilton, June 9, 1882 (Executive Papers, Opinions 1882, MHR).

71. Balt. *Sun*, June 2, 1882.

72. *Ibid.*, June 3 and 6; and *Journal of United Labor*, June 1882.

73. Balt. *Sun*, June 5 and 13, 1882.

74. Cumb. *Daily Times*, June 7 and 9, 1882.

75. Balt. *Sun*, July 17, 1882.

76. *Ibid.*, June 9, 13, 19, 26, July 8 and 15; and Cumb. *Daily Times*, June 10, 14, 15, 19, July 11 and 14, 1882.

77. Balt. *American*, April 9, 1882.

78. *Ibid.*, May 31.

79. *Ibid.*

80. Balt. *Sun*, May 31, 1882.

81. *Ibid.*

82. *Ibid.*, June 6.

83. *Ibid.*, June 12 and 13.

84. *Ibid.* June 2.

85. *Ibid.*, June 13, 16, and 17.

86. *Ibid.*, June 13.

87. *Ibid.*

88. *Ibid.*, June 8 and 16.

89. *Ibid.*, June 15.

90. *Ibid.*, June 17.

91. Henry Joseph Browne, *Catholic Church and the Knights of Labor*, pp. 80–81.

92. *Ibid.*

93. *Frank Leslie's Illustrated Newspaper*, June 17, 1882.

94. Balt. *Sun*, June 16, 1882; and Cumb. *Daily Times*, June 17, 1882.

95. Balt. *Sun*, and Cumb. *Daily Times*, June 17, 1882.

96. Cumb. *Daily Times*, June 17, 1882.

97. Powderly to John H. Jones, June 17, 1882 (Powderly Papers, Letter Book 1).

98. Cumb. *Daily Times*, June 20, 1882.

99. Penn. Dept. of Internal Affairs, *Report, 1881–82*, Part III, pp. 156–157.

100. M. McPadden to Powderly, June 24, 1882 (Powderly Papers, Box 5).

101. *Journal of United Labor*, Dec. 1882. Up to this time only

one other assessment had ever been levied by national headquarters. The assessment for Maryland was of 10 cents a week on every member in good standing, and covered the period July 5 through Aug. 25, 1882.

102. Circular issued by A. C. Rankin, master workman, D. A. 3, March 17, 1883 (Powderly Papers, Box 7).

103. Printed circular from secretary of D. A. 9, July 13, 1882 (Powderly Papers, Box 5).

104. Balt. *Sun*, June 30, 1882.

105. Rees D. Rees to Powderly, July 1, 1882 (Powderly Papers, Box 5).

106. Penn. Dept. of Internal Affairs, *Report, 1881–82*, Part III, p. 161.

107. Md. Mine Insp., *Report, 1883*, p. 23.

108. *Ibid.*, p. 24.

109. Balt. *Sun*, July 8, 1882.

110. *Ibid.*, July 13.

111. *Ibid.*

112. Cumb. *Daily Times*, July 15, 1882.

113. Balt. *Sun*, July 8, 1882.

114. *Ibid.*, July 11.

115. Cumb. *Daily Times*, July 11, 1882.

116. *Ibid.*, July 14, 1882.

117. James A. Briscoe to Governor Hamilton, July 14, 1882 (Executive Papers, 1882 Letters and Applications, MHR).

118. Balt. *Sun*, July 15, 1882.

119. *Ibid.*, July 24, 25, and 27.

120. Md. Mine Insp., *Report, 1883*, p. 25.

121. Balt. *Sun*, July 28, 1882.

122. Thomas Brown to Hamilton, July 28, 1882 (Executive Papers, 1882 Letters and Applications, MHR).

123. McPadden to Powderly, July 20, 1882 (Powderly Papers, Box 5).

124. *Ibid.*, July 21.

125. Penn. Dept. of Internal Affairs, *Report, 1881–82*, Part III, p. 160.

126. Powderly to McPadden, July 25, 1882 (Powderly Papers, Letter Book 1).

127. Cotter to Powderly, July 29, 1882 (Powderly Papers, Box 5).

128. Rockwood to Powderly, July 21, 1882 (Powderly Papers, Box 5).

129. Md. Mine Insp., *Report, 1883*, p. 24.

130. Balt. *Sun*, Aug. 1, 1882.

131. *Ibid.*, Aug. 3.

132. *Ibid.*, Aug. 4.

133. *Ibid.*

134. Atty. Gen. C. J. M. Gwinn to Hamilton, Aug. 5, 1882; and Gwinn to Hamilton, Aug. 8, 1882 (Executive Papers, Opinions 1882, MHR).

135. Balt. *Sun*, Aug. 4, 5, and 7, 1882.

136. *Ibid.*, Aug. 7.

137. *Ibid.*, Aug. 9.

138. McPadden to Powderly, Aug. 10, 1882 (Powderly Papers, Box 5).

139. Balt. *Sun*, Aug. 16, 1882.

140. *Ibid.*, Aug. 11.

141. *Ibid.*, Aug. 5.

142. *Ibid.*, Aug. 10.

143. *Ibid.*, Aug. 15.

144. *Ibid.*

145. *Ibid.*, Aug. 18; and Cumb. *Daily Times*, Aug. 18, 1882.

146. Balt. *Sun*, Aug. 19, 1882.

147. *Ibid.*, Aug. 21.

148. Cumb. *Daily Times*, Aug. 21, 1882.

149. Balt. *Sun*, Aug. 22, 1882.

150. *Ibid.*, Aug. 23.

151. *Ibid.*

152. Richard Lowitt, "Frost. 1882: German Strikers vs German Strikebreakers," p. 78.

153. Balt. *Sun*, Aug. 29, 1882.

154. Cumb. *Daily Times*, Aug. 24, 1882; and Balt. *Sun*, Aug. 31, 1882.

155. "Appeal for Aid" (Powderly Papers, Box 5).

156. William B. Baird to Powderly, Oct. 11, 1882 (Powderly Papers, Box 5).

157. Penn. Dept. of Internal Affairs, *Report, 1881–82*, Part III, pp. 162–163.

158. *Frost. Mining Journal*, April 15, 1882.

159. Balt. *Sun*, July 3, 13, 18, 20, 28, 31, Aug. 1, 5, 7, 10, 12, and 17, 1882.

160. *Ibid.*, Aug. 26.

161. Cumb. *Daily Times*, Aug. 3, 1882.

162. Appeal on behalf of Walter Edwards (Powderly Papers, Box 7); for Combs' contribution see letters from John W. Evans, *National Labor Tribune*, July 29 and Aug. 5, 1882.

163. McPadden to Powderly, Aug. 10, 1882 (Powderly Papers, Box 5).

Chapter 15. Decade of Uncertainty (1882–1892)

1. *Coal Trade Journal*, XXI, No. 35, 553.

2. Balt. *American*, Aug. 30, 1882; however D. A. 25 appealed to its members after the strike to observe the 5-car limit as far as possible. *National Labor Tribune*, Oct. 28, 1882.

3. See Table 13.

4. D. A. 40 reported no members in the *Proceedings* of the General Assembly for 1884.

5. Cumb. *Daily Times*, Sept. 22, 1882.

6. George H. Arnold to Powderly, Oct. 10, 1882 (Powderly Papers, Box 5); Powderly to Arnold, Oct. 14, 1882 (Powderly Papers, Letter Book for period 29 July 1882–10 April 1883); and Powderly to Robert Layton, Oct. 14, 1882 (Powderly Papers, Letter Book, 29 July 1882–10 April 1883).

7. William B. Baird to Powderly, Oct. 11, 1882 (Powderly Papers, Box 5).

8. Layton to Powderly, Jan. 16, 1883 (Powderly Papers, Box 6). According to the *National Labor Tribune* of May 26, 1883, Layton was present at the organizing convention of the Amalgamated Association of Miners of the U.S.

9. *National Labor Tribune*, Jan. 13 and April 7, 1883; see also David J. McDonald and Edward A. Lynch, *Coal and Unionism, A History of the American Coal Miners' Unions*, p. 20.

10. *National Labor Tribune*, May 19, May 26, and Aug. 25, 1883; see also *Frost. Mining Journal*, Feb. 9, 1884.

11. Norman J. Ware, *Labor Movement in the U.S. 1860–1895*, p. 212.

12. K. of L., *Proceedings of the Seventh Regular Session of the General Assembly, Held at Cincinnati, O., Sept. 4–11, 1883* (n.p., n.d.), p. 438 (Document 57).

13. M. F. Flannigan to Powderly, Oct. 10, 1883 (Powderly Papers, Box 7).

14. Rees D. Rees to Powderly, Jan. 31, 1884 (Powderly Papers, Box 8).

15. Circular Appeal (Powderly Papers, Box 7).

16. Md. Mine Insp., *Report, 1883,* pp. 32–33.

17. *Coal Trade Journal,* XXI, No. 39, 616.

18. *Frost. Mining Journal,* June 2, 1883; and *National Labor Tribune,* July 14, 1883.

19. Md. Mine Insp., *Report, 1883,* p. 30.

20. *Ibid.*

21. *Frost. Mining Journal,* March 1 and 8, 1884.

22. Md. *Laws,* 1884, Ch. 427.

23. *Coal Trade Journal,* Oct. 8, 1884; and *Frost. Mining Journal,* Nov. 22, 1884.

24. Md. Bur. Ind. Stat., *Report, 1884–85,* pp. 236–237 and footnote.

25. *Ibid.*

26. *Ibid.,* pp. 76–77

27. *Ibid.,* p. 81.

28. Md. Bur. Ind. Stat., *Report, 1898,* pp. 237–238.

29. K. of L., *Record of the Proceedings of the Eighth Regular Session of the General Assembly, Held at Phila., Pa., Sept. 1–10, 1884* (n.p., n.d.), pp. 590–591.

30. *Ibid.*

31. Md. Bur. Ind. Stat., *Report, 1884–85,* p. 81; derived from figures for annual earnings and number of days worked.

32. Penn. Dept. of Internal Affairs, *Report, 1884,* Part III, pp. 47–48.

33. *Frost. Mining Journal,* April 5, 1884.

34. Md. Bur. Ind. Stat., *Report, 1884–85,* p. 81.

35. Penn. Dept. of Internal Affairs, *Report, 1884,* Part III, p. 4.

36. *Frost. Mining Journal,* June 7, 1884.

37. *Ibid.,* Nov. 22; and *Coal Trade Journal,* Nov. 19, 1884.

38. *Frost. Mining Journal,* Nov. 22, 1884.

39. *Ibid.,* Nov. 29.

40. See Table 13.

41. *Frost. Mining Journal*, Nov. 29, 1884.

42. *Ibid.*, Dec. 6.

43. Quoted in Cumb. *Daily Times*, Dec. 2, 1884.

44. *Coal Trade Journal*, April 21, 1885.

45. Cumb. *Daily Times*, April 17, 1885.

46. *Frost. Mining Journal*, April 25, 1885.

47. *Ibid.*, April 25.

48. Cumb. *Daily Times*, April 29, 1885; and *Frost. Mining Journal*, May 2, 1885.

49. *Coal Trade Journal*, May 13, 1885.

50. Cumb. *Daily Times*, May 15, 1885.

51. *Ibid.*

52. *Ibid.*, May 16 and 18.

53. *Ibid.*, May 16.

54. Md. Bur Ind. Stat., *Report, 1884–85*, footnote, p. 15.

55. Md. *Laws* 1884, Ch. 427

56. *Ibid.*, Ch. 39.

57. *Frost. Mining Journal*, Feb. 16, 1884.

58. Cumb. *Daily Times*, Feb. 23, 1885; and *Frost. Mining Journal*, Jan. 23, 1886.

59. *Frost. Mining Journal*, May 2, 1885.

60. Md. Bur. Ind. Stat., *Report, 1884–85*, Table p. 164.

61. *Ibid.*, p. 83.

62. W. B. Baird to Powderly, Feb. 13, 1886 (Powderly Papers, Box 16).

63. Penn. Dept. of Internal Affairs, *Annual Report, 1885*, Part III, *Industrial Statistics* (Harrisburg, 1886), pp. 169–170.

64. Md. Bur. Ind. Stat., *Report, 1884–85*, footnote p. 15.

65. Chris Evans, *History of UMW of America*, I, 137; and Andrew Roy, *History of the Coal Miners of the U.S.*, pp. 227–228.

66. Liguori Logsdon to Powderly, Jan. 30, 1886 (Powderly Papers, Box 16).

67. Ware, p. 213.

68. From copy of printed circular (Powderly Papers, Box 17); see also Cumb. *Daily Times*, Feb. 19, 1886; *Frost. Mining Journal*, Feb. 20 and 27, 1886; and *Coal Trade Journal*, Feb. 24, 1886.

69. Powderly Papers, Box 17.

70. *Frost. Mining Journal*, Feb. 20, 1886.

71. Roy, *Coal Miners*, pp. 237–239; Evans, I, 175–185; and Arthur

E. Suffern, *Coal Miners' Struggle for Industrial Status*, pp. 35–49.

72. *Frost. Mining Journal*, Feb. 27, 1886.

73. Powderly Papers, Box 17.

74. Balt. *Sun*, March 4, 1886; and Cumb. *Daily Times*, March 3, 1886.

75. Md. Bur. Ind. Stat., *Report, 1886–87*, pp. 71–72.

76. Balt. *Sun*, April 9, 1886.

77. *Ibid.*

78. Cumb. *Daily Times*, April 13, 1886.

79. K. of L., *Record of Proceedings of the General Assembly of the K. of L. of America, Vol. IV. Commencing with Session at Richmond, Va., Oct. 4 to 20, 1886* (published by the General Assembly, 1886), pp. 203, 282.

80. Beverley S. Randolph, "History of the Md. Coal Region," pp. 524–525; and Charles E. Beachley, *History of the Consolidation Coal Co.*, p. 27.

81. Cumb. *Daily Times*, June 27, 1882.

82. *Ibid.*, Feb. 19, 1886; and *Coal Trade Journal*, Feb. 24, 1886.

83. Md. Bur. Ind. Stat., *Report, 1886–87*, pp. 71–72.

84. Balt. *Sun*, March 9, 1886.

85. *Frost. Mining Journal*, March 20, 1886.

86. Cumb. *Daily Times*, March 23, 1886.

87. *Ibid.*, April 5.

88. *Ibid.*, April 14.

89. *Frost. Mining Journal*, April 17, 1886.

90. Cumb. *Daily Times*, April 16, 1886.

91. Balt. *Sun*, April 27, 1886.

92. *Ibid.*, April 30.

93. *Ibid.*, March 8.

94. *Frost. Mining Journal*, May 8, 1886; Cumb. *Daily Times*, May 8, 1886; and Balt. *Sun*, May 10, 1886.

95. Balt. *Sun*, May 17, 1886.

96. Cumb. *Daily Times*, May 15, 1886.

97. Balt. *Sun*, May 18 and 19, 1886; and *Coal Trade Journal*, May 19, 1886.

98. Cumb. *Daily Times*, May 18, 1886.

99. *Ibid.*, May 20.

100. *Ibid.*, May 15.

101. *Ibid.*, May 25.

102. *Ibid.*, May 22.
103. *Ibid.*, May 24.
104. *Ibid.*, May 26.
105. *Ibid.*, May 25.
106. *Coal Trade Journal*, June 2, 1886.
107. Md. Bur. Ind. Stat., *Report, 1886–87*, pp. 136–137.
108. *Ibid.*, p. 136.
109. Md. Bur. Ind. Stat., *Report, 1898*, pp. 237–238.
110. *Frost. Mining Journal*, Jan. 22, 1887.
111. *Ibid.*, Feb. 26.
112. Cumb. *Daily Times*, April 1, 1887.
113. *Frost. Mining Journal*, March 5, 1887; and *Coal Trade Journal*, March 9, 1887.
114. *Frost. Mining Journal*, March 12, 1887.
115. Md. Mine Insp., *Report, 1887*, p. 4.
116. *Ibid.*, p. 5.
117. *Frost. Mining Journal*, June 11, 1887.
118. *Ibid.*, July 9.
119. *Ibid.*, Sept. 1, 1888.
120. *Ibid.*, March 17.
121. Md. Bur. Ind. Stat., *Report, 1898*, pp. 237–238.
122. Md. Bur. Ind. Stat., *Report, 1886–87*, p. 137.
123. Md. Mine Insp., *Report, 1888*, p. 2.
124. *Ibid.*
125. *Ibid.*
126. *Ibid.*, p. 7.
127. *Frost. Mining Journal*, March 3 and 10, 1888; and *Coal Trade Journal*, March 7, 1888.
128. *Frost. Mining Journal*, March 17, 1888.
129. U.S. Census Office, *Eleventh Census of the U.S.*, XIV, 350. It should be borne in mind that the census figures for Penn., Va., and W.Va., are for their total bituminous areas, not merely those in direct competition with the Md. field.
130. Cumb. *Daily Times*, Jan. 7, 1891.
131. Md. Bur. Ind. Stat., *Report, 1898*, pp. 237–238.
132. Md. Bur. Ind. Stat., *Report, 1890–91*, p. 223.
133. *Ibid.*, p. 243.
134. *Ibid.*, pp. 212–213.
135. Md. Mine Insp., *Report, 1892*, p. 5.

136. *Ibid.*

137. Md. Bur. Ind. Stat., *Report, 1892*, p. 219.

138. K. of L., *Record of the Proceedings of the Ninth Regular Session of the General Assembly, Held at Hamilton, Ont., Oct. 5–13, 1885* (n.p., n.d.), pp. 126–127.

139. K. of L., *Proceedings, 1886* (Richmond), p. 326.

140. *Missouri Republican*, May 20, 21, and 23, 1886.

141. *Journal of United Labor*, July 10, 1886.

142. Cumb. *Daily Times*, Jan. 10, 1887; *Coal Trade Journal*, Jan. 12, 1887; and *National Labor Tribune*, Jan. 22, 1887.

143. *Journal of United Labor*, Jan. 27, 1889.

144. Ware, pp. 215–217; see also *Frost. Mining Journal*, Jan. 5 and Oct. 5, 1889, for references to continued K. of L. activity in this region.

145. Ware, p. 218.

146. Evans, II, 60; and Roy, *Coal Miners*, pp. 253–257.

147. Evans, II, 60.

148. Ware, p. 218.

149. Cumb. *Daily Times*, March 18, 1891.

150. *Ibid.*

151. *Frost. Mining Journal*, March 21, 1891.

152. *Ibid.*

153. *Ibid.*, and Cumb. *Daily Times*, March 18, 1891.

154. *Coal Trade Journal*, March 25, 1891.

155. Cumb. *Daily Times*, March 19, 1891.

156. *Ibid.*

Chapter 16. An Episode of Sympathy (1894)

1. Md. Bur. Ind. Stat., *Report, 1898*, pp. 237–238.

2. *Coal Trade Journal*, Jan. 17, 1894.

3. *Ibid.*, Jan. 24.

4. *Ibid.*, Feb. 14.

5. Md. Bur. Ind. Stat., *Report, 1895*, p. 172.

6. *Ibid.*, p. 175.

7. *Ibid.*, p. 174; and Balt. *Sun*, March 23, 1894.

8. Balt. *Sun*, March 23, 1894; and *Frost. Mining Journal*, March 24, 1894.

9. Balt. *Sun*, March 23, 1894.

10. See also Md. Bur. Ind. Stat., *Report, 1884–85*, p. 83, and *Report, 1886–87*, p. 136.

11. Md. Bur. Ind. Stat., *Report, 1895*, p. 178.

12. Quoted in *Frost. Mining Journal*, March 10, 1894.

13. Balt. *Sun*, March 24, 1894.

14. Andrew Roy, *History of the Coal Miners of the U.S.*, p. 303.

15. Frank Julian Warne, *Coal-Mine Workers, A Study in Labor Organization*, pp. 213–214; and Beverley S. Randolph, "History of the Md. Coal Region," p. 525.

16. Balt. *Sun*, April 12 and 21, 1894.

17. Flannigan's testimony during contempt trials, Cumb. *Daily Times*, May 26, 1894.

18. *Frost. Mining Journal*, April 21, 1894.

19. *Ibid.*, May 26.

20. *Ibid.*, April 21.

21. N.Y. *Times*, July 6, 1913.

22. Balt. *Sun*, April 21, 1894.

23. Roy, *Coal Miners*, p. 305.

24. Balt. *Sun*, May 8, 1894.

25. *Ibid.*, May 15; see also *UMW Journal*, May 10, 1894.

26. Balt. *Sun*, May 9, 1894.

27. *Ibid.*, Sept. 4, 1882, and May 15, 1894.

28. Md. Bur. Ind. Stat., *Report, 1895*, p. 178.

29. Randolph, p. 525.

30. Md. Bur. Ind. Stat., *Report, 1895*, p. 177.

31. *Journal of the K. of L.*, May 17, 1894.

32. Balt. *Sun*, May 15, 1894.

33. *Ibid.*, May 9, 10, and 11.

34. *Ibid.*, May 12.

35. *Ibid.*, May 14.

36. *Ibid.*

37. *Frost. Mining Journal*, May 19, 1894.

38. Balt. *Sun*, May 14, 1894.

39. *Ibid.*

40. *Ibid.*

41. *Ibid.*, May 18.

42. *Ibid.*, May 19.

43. *Ibid.*, May 17.

44. *Ibid.*, May 21.
45. *Ibid.*
46. *Ibid.*
47. *Ibid.*, May 22; and Md. Bur. Ind. Stat., *Report 1895*, pp.182–183.
48. Balt. *Sun*, May 24, 1894.
49. Md. Bur. Ind. Stat., *Report, 1895*, p. 182.
50. Balt. *Sun*, May 28, 1894.
51. Md. Bur. Ind. Stat., *Report, 1895*, p. 185.
52. Balt. *Sun*, May 29, 1894.
53. *Ibid.*, May 30.
54. *Ibid.*, June 1; and Md. Bur. Ind. Stat., *Report, 1895*, pp. 182–193.
55. Md. Bur. Ind. Stat., *Report, 1895*, p. 185.
56. *Ibid.*, pp. 186, 188–189.
57. Balt. *Sun*, May 28, 1894.
58. *Ibid.*
59. *Ibid.*, May 30 and 31.
60. *Ibid.*, June 2.
61. *Ibid.*, June 4.
62. *Ibid.*, June 6.
63. C. F. Mayer to A. B. Fleming, June 4, 1894 (Aretus Brooks Fleming Collection, W.Va. University Library, Morgantown, W.Va.).
64. Baltimore *Sun*, June 6, 1894.
65. *Ibid.*
66. *Ibid.*
67. *Ibid.*, June 7.
68. Cumb. *Daily Times*, June 2, 1894; and Balt. *Sun*, June 8, 1894.
69. Balt. *Sun*, June 4 and 11, 1894.
70. *Ibid.*, June 7.
71. Cumb. *Daily Times*, May 24, 1894; and Balt. *Sun*, May 25, 1894.
72. Balt. *Sun*, June 11, 1894.
73. *Ibid.*
74. *Ibid.*, June 7.
75. *Ibid.*, June 11.
76. *Coal Trade Journal*, June 6, 1894.
77. Balt. *Sun*, June 12, 1894.

78. *Ibid.*
79. *Ibid.*
80. *Ibid.*, June 14.
81. *Ibid.*
82. *Ibid.*, June 16.
83. *Ibid.*, June 18.
84. *Ibid.*, June 14, 15, and 18.
85. *Ibid.*, June 16.
86. Cumb. *Daily Times*, June 16, 1894.
87. Balt. *Sun*, June 20, 1894.
88. *Ibid.*, June 19 and 21.
89. *Ibid.*, June 21 and 22; and Cumb. *Daily Times*, June 21, 1894.
90. Balt. *Sun*, June 27, 1894.
91. *Ibid.*, June 28.
92. *Ibid.*, June 29.
93. *Ibid.*, July 2.
94. *Ibid.*, June 29.
95. *Ibid.*, July 6 and 13.
96. *Ibid.*, July 13.
97. C. F. Mayer to A. B. Fleming, July 7, 1894 (Aretus Brooks Fleming Collection).
98. *Frost. Mining Journal*, Oct. 27, 1894.
99. Balt. *Sun*, July 6, 1894.
100. Letter signed Jas. P. O'Brien in *UMW Journal*, Aug. 9, 1894.
101. *Frost. Mining Journal*, Aug. 18, 1894; and *Coal Trade Journal*, Aug. 15, 1894.
102. Md. Bur. Ind. Stat., *Report, 1895*, p. 191.
103. Cumb. *Weekly Civilian*, Nov. 16 and 23, 1894.
104. N.Y. *Times*, July 6, 1913.
105. Randolph, pp. 525–526.

Chapter 17. The Grasping Corporations (1895–1900)

1. Frank Julian Warne, *Coal-Mine Workers, A Study in Labor Organization*, pp. 213–214; and Chris Evans, *History of UMW of America*, II, 367.
2. Cumb. *Evening Times*, Nov. 12 and 15, 1900.
3. *UMW Journal*, Feb. 26, 1903.

4. From a ballad composed by Robert Simpson, a Frostburg miner, published in Balt. *Sun*, April 12, 1900.

5. *Frost. Mining Journal*, March 28, 1896; *Coal Trade Journal*, April 1. 1896; and Charles E. Beachley, *History of the Consolidation Coal Co.*, p. 30.

6. Beachley, p. 30.

7. *Coal Trade Journal*, March 25 and April 8, 1896.

8. *Ibid.*, July 3, 1893.

9. *Ibid.*, April 8, 1896.

10. Quoted in *Frost. Mining Journal*, March 27, 1897.

11. Cumb. *Daily Times*, July 2 and 3, 1897.

12. W.Va. Commission of Labor, *Report, 1897–1898* (Charleston, 1898), pp. 93–94.

13. Philip Taft, *A.F. of L. in the Time of Gompers* (N.Y., 1957), pp. 137–140; Joseph G. Rayback, *History of American Labor*, pp. 209–210; and Harold U. Faulkner, *Decline of Laissez Faire* (N.Y., 1959), p. 293.

14. Cumb. *Daily Times*, July 9, 1897.

15. *Coal Trade Journal*, July 28, 1897.

16. Cumb. *Daily Times*, Sept. 1, 1897.

17. Md. Bur. Ind. Stat., *Report, 1898*, p. 214.

18. Consolidation Coal Co., *Annual Report of the President and Directors to the Stockholders for the Year Ended Dec. 31, 1898* (Balt., 1899), p. 14.

19. *Coal Trade Journal*, Sept. 14, 1898.

20. *Ibid.*, Feb. 20, 1901.

21. *Ibid.*, Feb. 1, 1899.

22. *Frost. Mining Journal*, May 27, 1899.

23. *Coal Trade Journal*, May 24, 1899.

24. Balt. *Sun*, June 22, 1899.

25. *Frost. Mining Journal*, Aug. 19, 1899.

26. Md. Bur. Ind. Stat., *Ninth Annual Report, 1900* (Balt., 1901), p. 22.

27. *Frost. Mining Journal*, April 1, 1899.

28. Md. Bur. Ind. Stat., *Report, 1900*, p. 22.

29. Cumb. *Evening Times*, June 14, 1900.

30. *Ibid.*, Aug. 31, 1899; and Balt. *Sun*, Aug. 31, 1899.

31. Balt. *Sun*, Sept. 4 and 9, 1899; and Cumb. *Evening Times*, Sept. 9, 1899.

32. *UMW Journal*, Nov. 9, 1899.

33. *Ibid.*, Nov. 9, 16, and 23.

34. *Ibid.*, Dec. 21; Balt. *Sun*, Dec. 12, 13, 14, and 15, 1899; and Cumb. *Evening Times*, Dec. 12, 13, 14, and 15, 1899.

35. *UMW Journal*, Dec. 21, 1899.

36. Md. Bur. Ind. Stat., *Report, 1900*, p. 23.

37. *Ibid.*

38. *Ibid.*, p. 24.

39. Balt. *Sun*, Dec. 30, 1899.

40. *Frost. Mining Journal*, Jan. 6, 1900.

41. *Coal Trade Journal*, Jan. 29 and Feb. 19, 1900.

42. Balt. *Sun*, Feb. 23, 1900.

43. Cumb. *Evening Times*, Feb. 23, 1900; and *Frost. Mining Journal*, Feb. 24, 1900.

44. Cumb. *Evening Times*, March 13, 15, and 16, 1900.

45. *Frost. Mining Journal*, March 17, 1900.

46. Balt. *Sun*, March 23, 1900.

47. *Ibid.*, March 29 and 30.

48. Cumb. *Evening Times*, April 18, 1900.

49. Balt. *Sun*, April 4, 1900.

50. *Ibid.*, April 5, 6, 7, and 9.

51. *Ibid.*, April 6, 7, 9, and 10.

52. *Ibid.*, April 11.

53. Md. Bur. Ind. Stat., *Report, 1900*, p. 30.

54. Balt. *Sun*, April 11, 1900.

55. Md. Bur. Ind. Stat., *Report, 1900*, p. 30.

56. *Ibid.*; and Balt. *Sun*, April 11, 1900.

57. Cumb. *Evening Times*, April 18, 1900.

58. Balt. *Sun*, April 12, 13, 14, 16, 17, 19, 20, 21, and 30, 1900; and Cumb. *Evening Times*, April 26 and May 19, 1900.

59. Md. Bur. Ind. Stat., *Report, 1900*, p. 53.

60. *Ibid.*, p. 54.

61. Consolidation Coal Co., *Annual Report of the President and Directors to the Stockholders of the Consolidation Coal Co. for the Year Ended Dec. 31, 1900* (Balt., 1900), pp. 14–15.

62. Md. Bur. Ind. Stat., *Report, 1900*, p. 35.

63. Balt. *Sun*, April 21, 1900; and Md. Bur. Ind. Stat., *Report, 1900*, p. 32.

64. *UMW Journal*, May 17, 1900.

65. Balt. *Sun*, June 1, 1900; and *UMW Journal*, June 7, 1900.

66. *UMW Journal*, June 21, 1900.

67. Md. Bur. Ind. Stat., *Report, 1900*, p. 36; and Balt. *Sun*, May 19, 21, 22, and 24, 1900.

68. *UMW Journal*, May 31, 1900.

69. Balt. *Sun*, May 25, 1900; and Md. Bur. Ind. Stat., *Report, 1900*, p. 36.

70. David J. Lewis to John Mitchell, March 24, 1911 (Folder 135, Mitchell Papers, Archives of The Catholic University of America, Washington, D.C.).

71. Cumb. *Evening Times*, July 28, 1900; and James W. Thomas and T. J. C. Williams, *History of Allegany County*, I, 358–359.

72. *UMW Journal*, June 28, 1900.

73. Cumb. *Evening Times*, July 5, 1900.

74. Md. Bur. Ind. Stat., *Report, 1900*, p. 42.

75. Balt. *Sun*, July 11 and 12, 1900.

76. *Ibid.*, July 21.

77. Cumb. *Evening Times*, June 16, 1900; and Balt. *Sun*, June 17, 1900.

78. *UMW Journal*, July 5, 1900.

79. Balt. *Sun*, July 17, 1900.

80. Cumb. *Evening Times*, Nov. 15 and 16, 1900.

81. *UMW Journal*, July 19, 1900.

82. Balt. *Sun*, July 19, 1900.

83. *Ibid.*, July 17, 18, and 19.

84. *Ibid.*, July 17.

85. *Ibid.*, July 20.

86. *Ibid.*, July 21, 23, 30, 31, and Aug. 1.

87. *Ibid.*, Aug. 4 and 6; and Md. Bur. Ind. Stat., *Report, 1900*, p. 48.

88. Balt. *Sun*, Aug. 6, 1900; and *UMW Journal*, Aug. 9, 1900.

89. Md. Bur. Ind. Stat., *Report, 1900*, p. 47.

90. Cumb. *Evening Times*, Aug. 10, 1900.

91. *Ibid.*, Oct. 25.

92. *Ibid.*, Oct. 23.

93. *Ibid.*, Dec. 13.

94. *UMW Journal*, Dec. 27, 1900.

95. *Ibid.*, Feb. 21, 1901, quoting Balt. *News;* see also *UMW Journal*, Feb. 14, 1901.

96. Cumb. *Evening Times*, Feb. 7, 1901.

97. Balt. *Sun*, May 8 and 9, 1900.

98. *UMW Journal*, May 24, 1900.

99. *Ibid.*, June 21.

100. Balt. *Sun*, April 19, 20, and 21, 1900; and *UMW Journal*, May 10 and 17, 1900.

101. Lonaconing *Review*, quoted in *UMW Journal*, June 7, 1900.

102. *Ibid.*

103. Md. Bur. Ind. Stat., *Report, 1900*, p. 43; and Balt. *Sun*, July 12, 1900.

104. Balt. *Sun*, May 10, 1900.

105. *Ibid.*, April 11.

106. *Ibid.*, May 11.

107. Md. Bur. Ind. Stat., *Report, 1900*, p. 54.

108. *Ibid.*

109. *Frost. Mining Journal*, March 10, 1900.

110. Cumb. *Evening Times*, April 25, 1900; Balt. *Sun*, April 18, 19, and 20, 1900; and Md. Bur. Ind. Stat., *Report, 1900*, p. 32.

111. Cumb. *Evening Times*, April 17, 1900.

112. Balt. *Sun*, May 28, 1900.

113. Cumb. *Evening Times*, May 7, 1900.

114. *Ibid.*, June 14.

115. Balt. *Sun*, May 18, 1900.

116. *UMW Journal*, Aug. 16, 1900; and Md. Bur. Ind. Stat., *Report, 1900*, p. 48.

117. *Ibid.*

118. Balt. *Sun*, May 9 and 19, 1900.

119. *Ibid.*, May 9, June 11, 13, and 14.

120. *Ibid.*, April 19; and Cumb. *Evening Times*, April 19, 1900.

121. Balt. *Sun*, June 23 and 26, 1900.

122. *Ibid.*, July 20.

123. *Ibid.*, July 20, 21, and 28.

124. *Ibid.*, July 23.

125. Md. Bur. Ind. Stat., *Report, 1900*, table following p. 18.

126. Balt. *Sun*, Aug. 4, 1900.

127. *Ibid.*, Aug. 6.

128. *Ibid.*, Aug. 11.

129. *Ibid.*

130. *Ibid.*, Aug. 15.

131. *Frost. Mining Journal*, Aug. 25, 1900.
132. Balt. *Sun*, Aug. 25, 1900.
133. As reported by *UMW Journal*, Jan. 24, 1901.
134. Md. Bur. Ind. Stat., *Report, 1900*, table following p. 18.
135. *Frost. Mining Journal*, Nov. 23, 1901.
136. Cumb. *Evening Times*, Nov. 13, 1901.

Chapter 18. Reform through Legislation (1888–1910)

1. Md. Bur. Ind. Stat., *Report, 1900*, p. 51.
2. For the history of this bill see *Frost. Mining Journal*, Jan. 30, March 3 and 6, April 10 and 17, May 15, and Sept. 4, 1886; Balt. *Sun*, March 25, April 10, and May 3, 6, and 7, 1886; and Cumb. *Daily Times*, March 3 and April 13, 1886.
3. Md. Bur. Ind. Stat., *Report, 1886–87*, pp. 24–25.
4. Md. Mine Insp., *Report, 1887*, pp. 4–5.
5. *Frost. Mining Journal*, March 17, 1888.
6. *Ibid.*, Jan. 30, 1892.
7. *Ibid.*, Oct. 19 and Nov. 5, 1889.
8. *Ibid.*, Feb. 1 and 8, 1890.
9. See *Frost. Mining Journal*, Feb. 22, 1890; and *Coal Trade Journal*, March 5, 1890.
10. *Frost. Mining Journal*, April 5, 1890.
11. See Balt. *Sun*, March 25, 1886; MGA, House, *Journal of Proceedings, Jan. Session, 1888* (Annapolis, 1888), p. 1154; and MGA, Senate, *Journal of Proceedings, Jan. Session, 1888* (Annapolis, 1888), p. 700.
12. *Frost. Mining Journal*, Feb. 27, 1892.
13. *Ibid.*, Feb. 22, 1896.
14. Cumb. *Evening Times*, March 6, 1896.
15. *Ibid.*
16. *Frost. Mining Journal*, April 11, 1896.
17. Cumb. *Evening Times*, April 9, 1896.
18. *Ibid.*, April 13.
19. Quoted in Cumb. *Evening Times*, Oct. 19, 1901.
20. *Frost. Mining Journal*, Nov. 6, 1897.
21. Cumb. *Evening Times*, Jan. 14, 1898.
22. *Ibid.*, Jan. 19; and *Frost. Mining Journal*, Jan. 22, 1898.

23. Cumb. *Evening Times,* Jan. 14, 1898.

24. *Ibid.,* Feb. 8.

25. *Ibid.,* Feb. 12.

26. *Ibid.,* Feb. 18.

27. *Frost. Mining Journal,* Oct. 22, 1898.

28. Cumb. *Evening Times,* March 2 and 3, 1898.

29. *Ibid.,* March 15.

30. Md. *Laws* 1898, Ch. 34.

31. *Ibid.,* 1878, Ch. 157.

32. *Frost. Mining Journal,* Feb. 21, 1903.

33. *UMW Journal,* Jan. 11, 1900; see also Balt. *Sun,* Dec. 30, 1899.

34. Cumb. *Evening Times,* Feb. 16, 1900.

35. Balt. *Sun,* March 20, 1900.

36. Cumb. *Evening Times,* Feb. 16, 1900.

37. *Ibid.,* March 1.

38. *Ibid.,* March 21.

39. Quoted by Cumb. *Evening Times,* Oct. 19, 1900.

40. *UMW Journal,* Nov. 28, 1901.

41. *Ibid.*

42. *Ibid.*

43. Cumb. *Evening Times,* Nov. 12, 1901; *Frost. Mining Journal,* Nov. 16, 1901; and *UMW Journal,* Nov. 21 and 28, 1901.

44. Cumb. *Evening Times,* Feb. 6 and 19, 1902.

45. *Ibid.,* March 4.

46. Md. *Laws* 1902, Ch. 124.

47. See letter of D. E. Dick, *Frost. Mining Journal,* Feb. 21, 1903; Md. Bur. Ind. Stat., *Report, 1900,* p. 31; and *UMW Journal,* April 18, 1901.

48. Cumb. *Evening Times,* March 28, 1898; and *Frost. Mining Journal,* April 2, 1898.

49. Cumb. *Evening Times,* March 28, 1898.

50. Md. Bur. Ind. Stat., *Report, 1900,* pp. 20–31.

51. *Frost. Mining Journal,* Jan. 9, 1904; *UMW Journal,* May 25, 1905, April 12, 1906, and May 16, 1907.

52. *Frost. Mining Journal,* Jan. 9, 1904.

53. *Ibid.,* Jan. 24, 1903.

54. *Ibid.*

55. See *Frost. Mining Journal,* March 11, 1905; *Coal Trade Journal,* March 1 and Nov. 8, 1905; *UMW Journal,* March 2 and Nov. 9, 1905; and Cumb. *Evening Times,* Oct. 31, 1905.

56. Md. *Laws* 1904, Ch. 243.
57. *Ibid.*, 1910, Ch. 361.
58. Cumb. *Evening Times*, Jan. 29, 1908.
59. Md. *Laws* 1896, Ch. 133.
60. *Ibid.*, 1900, Ch. 453.
61. *Ibid.*, 1898, Ch. 493.
62. *Luman* v. *Hitchens*, 90 Md. 14.
63. Cumb. *Evening Times*, Jan. 27, 1900.
64. Md. *Laws* 1900, Ch. 453.
65. U.S. Dept. of Labor, *Growth of Labor Law in the U.S.* (Washington, 1962), p. 160.
66. Md. *Laws* 1902, Ch. 139.
67. Md. Bur. Ind. Stat., *Eleventh Annual Report*, 1902 (Balt., 1903), pp. 319–327.
68. *Ibid.*, p. 319.
69. *Ibid.*
70. George E. Barnett, "The Md. Workmen's Compensation Act," *Quarterly Journal of Economics*, XVI, 591.
71. "The State Cooperative Accident Insurance Fund of Md.," Bulletin of the [U.S.] Bureau of Labor, No. 57 (Washington, 1905), p. 645.
72. Md. *Laws* 1902, Ch. 412.
73. *UMW Journal*, Nov. 20, 1902.
74. J. G. Schonfarber, Asst. Chief, Md. Bur. Ind. Stat., in *Proceedings of Association of Officials of Bureaus of Labor Statistics*, Twentieth Annual Convention, 1904, pp. 56–60.
75. Barnett, "The Md. Workmen's Compensation Act," p. 594.
76. George E. Barnett, "The End of the Md. Workmen's Compensation Law," *Quarterly Journal of Economics*, XIX, 320–322.
77. Cumb. *Evening Times*, April 3, 1902.
78. *Frost. Mining Journal*, April 16, 1904.
79. MGA, Senate, *Journal of Proceedings, Jan. Session, 1908* (Annapolis, 1908), pp. 323–324.
80. MGA, Senate, *Journal of Proceedings, Jan. Session, 1910* (Annapolis, 1910), pp. 173–174.
81. Cumb. *Evening Times*, Jan. 20, 1910.
82. MGA, Senate, *Journal, 1910*, pp. 619–631; see also Cumb. *Evening Times*, Feb. 9, 1910.
83. Cumb. *Evening Times*, April 8, 1910.
84. *Ibid.*, Jan. 27.

85. Md. *Laws* 1910, Ch. 159.

86. Lindley D. Clark, "Recent Action Relating to Employers' Liability and Workmen's Compensation," Bulletin of the [U.S.] Bureau of Labor, No. 90 (Washington, 1910), p. 678.

87. Malcolm H. Lauchheimer, *Labor Law of Md.* (Balt., 1919), pp. 48–50.

88. Md. Mine Insp., *Report, 1909–10,* p. 61.

89. Md. Bur. Ind. Stat., *Twentieth Annual Report, 1911* (Balt., 1912), p. 158.

Chapter 19. End of an Era (1900–1910)

1. *Coal Trade Journal,* June 20, 1900.

2. Beverley S. Randolph, "History of the Md. Coal Region," p. 527.

3. Cumb. *Evening Times,* Jan. 21, 1902.

4. *UMW Journal,* May 30, 1901.

5. *Ibid.*

6. *Ibid.,* June 6, June 20, and Oct. 24, 1901.

7. *Frost. Mining Journal,* May 18, 1901; see also *Frost. Mining Journal,* June 1, 1901; and *UMW Journal,* June 20, 1901.

8. *UMW Journal,* March 6, 1902.

9. *Frost. Mining Journal,* Sept. 20, 1902; see also Cumb. *Evening Times,* Sept. 17, 1902.

10. *Frost. Mining Journal,* Sept. 27, 1902; see also Cumb. *Evening Times,* Sept. 20, 23, 24, and 27, Oct. 2, Nov. 7, 8, and 10, and Dec. 17, 1902.

11. *Coal Trade Journal,* Jan. 4, 1903; see Charles E. Beachley, *History of the Consolidation Coal Co.,* pp. 35–36; and Consolidation Coal Co., *Annual Report of the President and Directors to the Stockholders of the Consolidation Coal Co. for the Year Ended Dec. 31, 1903* (Balt., 1904), p. 7.

12. Cumb. *Evening Times,* Jan. 7, 1903.

13. *Ibid.,* Feb. 1, 1902, and Jan. 10, 1903; see also J. A. Springer to John Mitchell, Nov. 15, 1901 (Mitchell Papers, Correspondence 1899–1902, Box 1, 1901, Confidential File S-Z).

14. Cumb. *Evening Times,* Jan. 10, 1903.

15. Md. Bur. Ind. Stat., *Eleventh Annual Report, 1902* (Balt., 1903), p. 193.

16. Cumb. *Evening Times,* June 21 and July 7, 1902; and Md. Bur. Ind. Stat., *Report, 1902,* p. 193.

17. Cumb. *Evening Times,* Feb. 11, 1903.

18. *Ibid.,* June 30, 1902.

19. *Coal Trade Journal,* March 18 and 25, 1903; and *Frost. Mining Journal,* March 28, 1903.

20. Cumb. *Evening Times,* March 14, 1903.

21. *Coal Trade Journal,* March 25, 1903.

22. Cumb. *Evening Times,* April 14, 1903.

23. *UMW Journal,* Feb. 26, 1903.

24. As reported by the secretary at the 14th annual convention of the UMW; see *UMW Journal,* Jan. 22, 1903.

25. *UMW Journal,* April 23, 1903.

26. Cumb. *Evening Times,* Feb. 16, 1903; see also Cumb. *Evening Times,* Feb. 21, April 13, May 1, 2, 9, and 20, 1903; *UMW Journal,* May 14, 1903; and *Frost Mining Journal,* Feb. 28, March 14, and April 11, 1903.

27. Cumb. *Evening Times,* May 20, 1903; and *UMW Journal,* May 28, 1903.

28. See Cumb. *Evening Times,* May 20, June 3, 4, 5, 12, and 24, July 18 and 21, Aug. 14 and 15, 1903; *UMW Journal,* June 4 and 25 and Aug. 13, 1903; and *Frost. Mining Journal,* June 13, 1903.

29. *UMW Journal,* Aug. 13, 1903.

30. *Ibid.,* June 4; see also *UMW Journal,* June 25, 1903.

31. Cumb. *Evening Times,* June 4, 1903.

32. *Ibid.,* June 12.

33. *Ibid.,* Aug. 28.

34. *Ibid.,* Sept. 28; see also *Frost. Mining Journal,* Sept. 26, 1903; and *UMW Journal,* Oct. 8, 1903.

35. Cumb. *Evening Times,* Sept. 28, 1903.

36. *Ibid.,* Nov. 30, Dec. 3, 5, 7, 9, and 15; *Coal Trade Journal,* Dec. 2 and 16, 1903; and *Frost. Mining Journal,* Dec. 5 and 12, 1903.

37. *Frost. Mining Journal,* Dec. 5, 1903; and Cumb. *Evening Times,* Dec. 3, 1903.

38. Cumb. *Evening Times,* Dec. 16, 1903.

39. *Ibid.*

40. *Coal Trade Journal,* Dec. 30, 1903, and Feb. 10, 1904.

41. Cumb. *Evening Times,* Dec. 29, 1903.

42. *Coal Trade Journal,* Jan. 20 and July 20, 1904; Cumb. *Evening Times,* April 19, 20, and 21, 1903; and E. Howard Blackburn and

William H. Welfley, *History of Bedford and Somerset Counties, Penn.*, II, 527.

43. Cumb. *Evening Times*, April 19 and 20, 1904.

44. *Coal Trade Journal*, July 20, 1904.

45. Cumb. *Evening Times*, Dec. 7, 1903.

46. *Ibid.*, July 30, 1904.

47. *UMW Journal*, Jan. 21, 1904. Local union officials in reporting to the Md. Bur. Ind. Stat. for the year 1903 claimed nine locals with a membership of 5,000; obviously not all of these were considered "in good standing" by the national organization.

48. *UMW Journal*, Jan. 19, 1905. Only five locals answered the 1904 questionnaire of the Md Bur. Ind. Stat.; they reported a membership of 1,000, again obviously not all "in good standing."

49. *UMW Journal*, Jan. 19, 1905.

50. *Frost. Mining Journal*, April 9, 1904; *Coal Trade Journal*, April 13, 1904; and Cumb. *Evening Times*, April 8, 1904.

51. Cumb. *Evening Times*, April 13, 1904.

52. *Ibid.*, March 11.

53. *Ibid.*, March 16.

54. *Ibid.*, Aug. 5.

55. *Ibid.*, April 15.

56. *Ibid.*, Sept. 6.

57. *Ibid.*, Dec. 1.

58. *Ibid.*, April 8.

59. *Ibid.*, April 21.

60. *Frost. Mining Journal*, April 29, 1905.

61. Letter signed "An Organization Man," Cumb. *Evening Times*, Nov. 6, 1905.

62. *UMW Journal*, March 2, 1905, quoting George's Creek *Press*.

63. Cumb. *Evening Times*, March 1, 1905.

64. *Ibid.*, April 17; see also *Frost. Mining Journal*, April 15, 1905.

65. *Frost. Mining Journal*, Aug. 19, 1905.

66. Cumb. *Evening Times*, May 23, 1905.

67. *UMW Journal*, May 25, 1905.

68. *Ibid.*, Sept. 14.

69. *Frost. Mining Journal*, Aug. 26, 1905; and *UMW Journal*, Sept. 7, 1905.

70. See sworn copies of docket entries furnished to Governor

Edwin Warfield by UMW district president F. J. Drum (Executive Papers, Complaints 1900–1906, MHR).

71. *UMW Journal*, Sept. 14, 1905.

72. Md. *Laws* 1880, Ch. 460.

73. Executive Papers, Special Policemen 1899, 1904–1916, MHR.

74. J. H. Wheelwright to Van Lear Black, Aug. 30, 1905 (Executive Papers, Special Policemen 1899, 1904–1916, MHR).

75. *Ibid.*

76. Executive Papers, General Executive Appointments, 1900–1909, MHR.

77. Warfield to Drum, Oct. 13, 1905 (Executive Papers, Complaints 1900–1906, MHR).

78. Cumb. *Evening Times*, Sept. 18, 1905; see also *Frost. Mining Journal*, Sept. 23, 1905, particularly letter from "A Tax-Payer"; *Coal Trade Journal*, Sept. 27, 1905; and *UMW Journal*, Sept. 28, 1905.

79. Balt. *Sun*, Sept. 12, 1905; see also sworn copies of docket entries furnished the governor (Executive Papers, Complaints 1900–1906, MHR).

80. Finley C. Hendrickson, Cumb. atty., to Governor Warfield, Sept. 13, 1905 (Executive Papers, Complaints 1900–1906, MHR).

81. Allegany Trades Council to Warfield, Sept. 14, 1905 (Executive Papers, Complaints 1900–1906, MHR).

82. Cumb. *Evening Times*, Sept. 18, 1905.

83. Balt. *Sun* and Balt. *American*, Sept. 22, 1905.

84. Rev. A. H. Thompson to Warfield, Sept. 19, 1905 (Executive Papers, Complaints 1900–1906, MHR).

85. Joseph Sprigg to Warfield, Sept. 19, 1905 (Executive Papers, Complaints 1900–1906, MHR).

86. Executive Papers, Complaints 1900–1906, MHR.

87. J. H. Wheelwright to Van Lear Black, Sept. 23, 1905 (Executive Papers, Complaints 1900–1906, MHR).

88. *Frost. Mining Journal*, Oct. 7, 1905.

89. Drum to Warfield, Sept. 20, 1905 (Executive Papers, Complaints 1900–1906, MHR).

90. Governor's secretary to Drum, Sept. 22, 1905 (Executive Papers, Complaints 1900–1906, MHR).

91. William S. Bryan to Warfield, Sept. 26, 1905 (Executive Papers, Complaints 1900–1906, MHR).

92. Drum to Warfield, Oct. 3, 1905 (Executive Papers, Complaints 1900–1906, MHR).

93. Governor's secretary to Bryan, Oct. 4, 1905, and Bryan to Warfield, Oct. 5, 1905 (Executive Papers, Complaints 1900–1906, MHR).

94. Warfield to Drum, Oct. 13, 1905 (Executive Papers, Complaints 1900–1906, MHR).

95. Cumb. *Evening Times,* Oct. 17, 1905.

96. Executive Papers, Complaints 1900–1906, MHR.

97. Cumb. *Evening Times,* Oct. 16, 1905.

98. *Ibid.,* Oct. 2.

99. *Ibid.,* Oct. 18.

100. *Ibid.,* Oct. 31.

101. *UMW Journal,* Oct. 19, 1905.

102. Balt. *Sun,* June 18, 1894.

103. *UMW Journal,* Sept. 28, 1899; and Balt. *Sun,* Dec. 29, 1899.

104. Balt. *Sun,* July 23 and 30 and Aug. 2, 1900.

105. *UMW Journal,* June 28, 1900.

106. Cumb. *Evening Times,* Oct. 7, 1905.

107. *Ibid.,* Oct. 14.

108. *Ibid.*

109. *Frost. Mining Journal,* Nov. 18, 1905.

110. Cumb. *Evening Times,* Nov. 28, 1905.

111. *UMW Journal,* Jan. 18, 1906.

112. *Coal Trade Journal,* Nov. 29, 1905.

113. *Ibid.*

114. Cumb. *Evening Times,* Jan. 17, 1906.

115. *UMW Journal,* July 4, 1907.

116. W. B. Wilson to John Mitchell, Aug. 4, 1909 (Mitchell Papers, File 99).

117. Beachley, p. 70.

118. Cumb. *Evening Times,* May 1, 1906.

119. *Coal Trade Journal,* May 19, 1909.

120. Consolidation Coal Co., *Annual Report of The Consolidation Coal Co. for the Year Ending Dec. 31, 1909* (Balt., 1910), p. 5.

Chapter 20. Sequel

1. *UMW Journal,* Nov. 9 and 16, 1916.

2. *Ibid.,* Nov. 1, 1917; see also UMW, *Proceedings of the Twenty-sixth Consecutive and Third Biennial Convention of the UMWA Held in the City of Indianapolis, Ind., Jan. 15 to 26, 1918, Inclusive* (Indianapolis, 1918), I, 27, 52, 252, 283, 539.

3. *UMW Journal,* Nov. 16, 1916.

4. *Ibid.,* Aug. 23 and Sept. 20, 1917; and William Diamond to H. A. Garfield, Dec. 14, 1917, (NA, Record Group 67, Records of U.S. Fuel Administration [hereinafter referred to as Fuel Admin. Records], Administrative Divison, Bureau of Labor, General Correspondence and Documents on Contracts, Box 1359).

5. William Diamond to Garfield, Dec. 14, 1917 (Fuel Admin. Records, Box 1359).

6. Fuel Admin. Records, Box 1359; a similar resolution from Local 1097 is in this file. See also *UMW Journal,* Jan. 3, 1918; and Cumb. *Evening Times,* Jan. 3, 1918.

7. Unsigned memo to Garfield, dated Jan. 8, 1918 (Fuel Admin. Records, Box 1359).

8. See transcript of convention (Fuel Admin. Records, Box 1359).

9. Unsigned memo dated May 7, 1918 (Fuel Admin. Records, Box 1359).

10. Letters addressed to William Diamond and Francis Drum, representing the miners, and to J. H. Wheelwright and John Brophy, representing Coal Producers of the Md. and Upper Potomac Districts, May 7, 1918 (Fuel Admin. Records, Box 1359).

11. See signed agreement (Fuel Admin. Records, Box 1359); the text of the agreement appears in Md. Mine Insp., *Report, 1917–1918,* pp. 8–12. See also UMW, *Proceedings of the Twenty-seventh Consecutive and Fourth Biennial Convention of the UMWA Held at Cleveland, O., Sept. 9, 1919* (Indianapolis, 1919), I, 36–41, 62; and *UMW Journal,* May 23, 1918.

12. Cumb. *Evening Times,* Oct. 31, Nov. 1, and Nov. 3, 1919; see also Md. Bur. Ind. Stat., *Twenty-eighth Annual Report, 1919* (Balt., n.d.), p. 165.

13. Cumb. *Evening Times,* Dec. 11, 1919.

14. *Ibid.*, April 1, 1922.

15. *Ibid.*, Nov. 20 and 23, 1923; and *UMW Journal*, Dec. 1, 1923.

16. See Cumb. *Evening Times*, June 8, 9, Aug. 9, 22, Sept. 6, 1922; and Aug. 24, 25, 27, Oct. 10, 30, Nov. 20, 1923.

17. *UMW Journal*, April 1 and May 15, 1923.

18. Cumb. *Evening Times*, Aug. 24 and 25, 1923.

19. *Ibid.*, Aug. 24.

20. *Ibid.*, Nov. 23.

21. *Ibid.*, and Md. Bureau of Mines, *First Annual Report of the Md. Bureau of Mines, 1922–23* (Balt., n.d.), p. 9.

22. Cumb. *Evening Times*, March 25, 1922.

23. UMW, *Proceedings of the Twenty-ninth Consecutive and Sixth Biennial Convention of the UMWA Held in the City of Indianapolis, Ind., Jan. 22 to Feb. 2, 1924* (Indianapolis, 1924), I, 193, 195.

24. Cumb. *Evening Times*, June 30, 1930.

25. George H. Love, *An Exciting Century in Coal!* (Princeton, 1955), pp. 13–14.

26. Charles E. Beachley, *History of the Consolidation Coal Co.*, p. 70.

27. W. L. Fairbanks and W. S. Hamill, *Coal-Mining Industry of Md.*, p. 193.

28. Cumb. *Evening Times*, May 20 and June 30, 1930.

29. "Coal: The Industrial Darkness," *Fortune*, March 1947, p. 225.

30. Philip Taft, *Organized Labor in American History* (N.Y., 1964), p. 426.

31. Md. Bureau of Mines, *Eleventh Annual Report of the Md. Bureau of Mines, 1933* (n.p., n.d.), p. 58.

32. Cumb. *Evening Times*, Sept. 29, 1933.

33. *UMW Journal*, April 15, 1934.

34. Cumb. *Evening Times*, editorial, May 4, 1939.

35. *Ibid.*, April 2, May 3, 12, and 14; and Md. Bureau of Mines, *Seventeenth Annual Report of the Md. Bureau of Mines, 1939* (n.p., n.d.), p. 6.

36. Cumb. *Evening Times*, June 1, 7, 21, 22, 24, Oct. 28–31, Nov. 1–9, 1943; see also Md. Bureau of Mines, *Twenty-first Annual Report of the Md. Bureau of Mines, 1943* (n.p., n.d.), p. 36.

37. Cumb. *Evening Times*, April 1, 17, 19, May 10–14, 20–31, and June 3, 1946; see also Md. Bureau of Mines, *Twenty-fourth*

Annual Report of the Md. Bureau of Mines, 1946 (n.p., n.d.), p. 35.

38. Cumb. *Evening Times*, March 15–17, 23, April 3, 5, 7, 13–22, 1948; and Md. Bureau of Mines, *Twenty-sixth Annual Report of the Md. Bureau of Mines, 1948* (n.p., n.d.), p. 46.

39. Cumb. *Evening Times*, July 2, 3, 5, Aug. 17, 18, 24, 31, Sept. 12, 19, Dec. 1, 5, 14, 1949; and Feb. 6, 7, 13, 18, 27, March 4, 6, 1950.

40. Fairbanks and Hamill, p. vii.

41. *Ibid.*

42. Md. Mine Insp., *Report, May 1, 1915, to May 1, 1917* (Baltimore, n.d.), p. 5.

43. Fairbanks and Hamill, p. 68; and Md. University, Bureau of Business and Economic Research, *An Economic Survey of Allegany County, Md.* (College Park, 1947), Table 66, p. 77.

44. "Coal: The Industrial Darkness," p. 121.

45. *Ibid.*, p. 86.

46. "Coal: The 'Pitt Consol' Adventure," *Fortune*, July 1947, p. 104.

47. Fairbanks and Hamill, p. vii.

48. *Ibid.*, pp. 115–116.

49. *Ibid.*, p. 183.

50. *Ibid.*, p. 68; and Md. University, *Economic Survey of Allegany County, Md.*, Table 68, p. 80.

51. Md. Bureau of Mines, *Second Annual Report of the Md. Bureau of Mines, 1924* (Balt., n.d.), pp. 5–6.

52. *Ibid., Third Annual Report of the Md. Bureau of Mines, 1925*, pp. 5–7.

53. *Ibid., Seventh Annual Report of the Md. Bureau of Mines, 1929*, p. 4.

54. Fairbanks and Hamill, p. vii.

55. Md. Bureau of Mines, *Tenth Annual Report of the Md. Bureau of Mines, 1932* (Balt., n.d.), p. 57.

56. *UMW Journal*, Feb. 15, 1933.

57. Md. Bureau of Mines, *Eleventh Annual* Report, p. 57.

58. Love, p. 14.

59. Md. Bureau of Mines, annual reports, 1929–1933.

60. Md. University, *Economic Survey of Allegany County, Md.*, Table 66, p. 77; and Md. Bureau of Mines, annual reports, 1945–1948.

61. Md. Bureau of Mines, annual reports 1942 (p. 30), 1943 (p. 36), and 1945 (p. 27).

62. Md. University, *Economic Survey of Allegany County, Md.*, p. 17.

63. Cumb. *Evening Times,* March 2, 1944; and Moody's Investors Service, *Moody's Manual of Investments*, Part II, *Industrial Securities* (N.Y., 1944), p. 2789.

64. See Md. Bureau of Mines, annual reports 1949 (p. 37), 1950 (p. 24), and 1951 (p. 25).

65. Md. University, Bureau of Business and Economic Research, *Coal in the Md. Economy, 1736–1965* (College Park, 1953), p. 16.

66. Md. Bureau of Mines, annual reports, 1948–1952.

67. Md. *Laws* 1955, Ch. 635.

68. See instances of bond forfeiture, Md. Bureau of Mines reports 1955–1966.

69. Md. *Laws* 1965, Ch. 514.

70. *Ibid.*, 1967, Ch. 144.

71. See Harry M. Caudill, *Night Comes to the Cumberlands* (Boston, 1963).

72. Benjamin Silliman, *Extracts from a Report Made to the Md. Mining Company*, pp. 4–5.

Sources

The manuscript sources, although few in number, have for the most part been rewarding. The Powderly material, for example, provided extensive documentation on the history of the Knights of Labor in this coal region which is nowhere else reported. Perhaps the most interesting information resulted from the rediscovery of the George's Creek Coal and Iron Company Journal (a primary source of the utmost value in the early history of this industry in the United States), which had apparently disappeared after W. M. Richards used it for his thesis in 1950. Locating the document involved a six months' search. To insure the accessibility of this privately owned manuscript for further study, it should be microfilmed before it is "lost" again. The material in the Maryland Hall of Records, on the other hand, was disappointingly scanty, particularly for the period 1879–1899. The various governors took most of their papers with them when they left office. Furthermore, the Executive Papers are not indexed and have been sorted only into broad categories. Any scholar seeking information would do well to search all the material for his time period, no matter how unpromising the label on any particular box or folder.

Newspaper sources must be assessed with an eye to the bias of their editors. Fortunately, the existence of a number of contemporary dailies and weeklies made it possible to confirm questions of fact and to indicate what was presented as opinion. On the one hand the *Coal Trade Journal* and on the other the *United Mine Workers' Journal* provided contrasting accounts of the same events. The Cumberland newspapers for the period after the Civil War were for the most part unexpectedly impartial reporters of labor disputes. It is evident from editorial references in them that several other newspapers were published in this same region during this period—some distinctly labor organs and some violently anti-labor—but regrettably no files of these publications seem to have been preserved.

Labor newspapers of the 1870's and 1880's contain a wealth of Maryland material, but much of this duplicates what is contained in the Cumberland and mining region press. Maryland miners were prolific writers of letters to the editor, but their epistles were almost entirely concerned with immediate strike, wage, or weight situations, or with politics. Unlike the miners of other regions, they did not describe their living conditions.

Concerning the reports of Maryland governmental agencies, two comments should be made. The Bureau of Industrial Statistics was created at the instance of the Knights of Labor, and its early chiefs and assistants had K. of L. affiliations. (The Bureau's statistics in the 1880's bear watching for errors in simple arithmetic.) The reports of the mine inspectors—at least until 1900—were possibly overfavorable to the operators where questions of ventilation were involved.

Regrettably, I was not granted access to the invaluable records of either the Consolidation Coal Company or the United Mine Workers of America. The company, therefore, speaks for itself only through its historian, Beachley, and through available copies of its annual reports to stockholders. For the UMW point of view I have had to rely entirely on the *United Mine Workers' Journal.*

Manuscripts

Allegany County Board of School Commissioners, Cumberland, Maryland
    Journal of Proceedings
The Catholic University of America, Washington, D.C.
    Papers of T. V. Powderly
    Papers of John Mitchell
Maryland Hall of Records, Annapolis, Maryland
    Executive Papers
Maryland Historical Society, Baltimore, Maryland
    J. H. Alexander Papers
    George's Creek Coal and Iron Company Letterbook, 1839–1847

National Archives, Washington, D.C.
Manuscript Population Census, Allegany County, Maryland, 1840–1880
Records of the Chesapeake and Ohio Canal Company
Records of U.S. Fuel Administrator (World War I)
Privately owned
Journal of George's Creek Coal and Iron Company
West Virginia University, Morgantown, West Virginia
Aretus Brooks Fleming Collection

Newspapers

Baltimore
American
Sun
Coal Trade Journal
Cumberland, Md.
Alleganian
Alleganian and Times
Civilian
Civilian and Telegraph
Cumberland Miners' Journal
Daily Times
Democratic Alleganian
Evening Times
Telegraph and Mining Register
Union
Union and Allegany County Gazette
Weekly Civilian
Fincher's Trades Review
Frank Leslie's Illustrated Newspaper
Frostburg [Md.] *Mining Journal*
Journal of the Knights of Labor
Journal of United Labor
Miners' National Record
Missouri Republican
National Labor Tribune
New York
Evening Post
Times
Pottsville, Pa., *Miners' Journal*

United Mine Workers' Journal
Workingman's Advocate

Publications of State and Federal Governments

Maryland Bureau of Industrial Statistics and Information (under its various names). *First Biennial Report, 1884–1885.* Baltimore, 1886.

——. *Second Biennial Report, 1886–87.* Annapolis, 1888.

——. *Third Biennial Report, 1888–89.* Annapolis, 1890.

——. *Fourth Biennial Report, 1890–91.* Annapolis, 1892.

——. *First Annual Report, 1892.* Baltimore, 1893.

——. *Second Annual Report, 1894.* Annapolis, 1894.

——. *Third Annual Report, 1895.* Baltimore, 1895.

——. *Fourth Annual Report, 1896.* Baltimore, 1896.

——. *Fifth Annual Report, 1896.* Baltimore, 1897.

——. *Sixth Annual Report, 1897.* Baltimore, 1898.

——. *Seventh Annual Report, 1898.* Baltimore, 1899.

——. *Ninth Annual Report, 1900.* Baltimore, 1901.

——. *Tenth Annual Report, 1901.* Baltimore, 1902.

——. *Eleventh Annual Report, 1902.* Baltimore, 1903.

——. *Twelfth Annual Report, 1903.* Baltimore, 1904.

——. *Thirteenth Annual Report, 1904.* Baltimore, 1905.

——. *Sixteenth Annual Report, 1907.* Baltimore, 1908.

——. *Seventeenth Annual Report, 1908.* Baltimore, 1908.

——. *Nineteenth Annual Report, 1910.* Baltimore, 1911.

——. *Twentieth Annual Report, 1911.* Baltimore, 1912.

——. *Twenty-eighth Annual Report, 1919.* Baltimore, n.d.

Maryland Bureau of Mines. Annual Reports, 1922–1966.

Maryland Commissioners of Mines. *Report to the Governor of Maryland, Submitted by Wm. H. Smith, Chas. H. Hamill, John Ryan, Commissioners.* Senate Document I. Annapolis, 1880.

Maryland General Assembly, House of Delegates. *Journal of Proceedings, January Session, 1874.* Annapolis, 1874.

——. *Journal of Proceedings, January Session, 1880.* Annapolis, 1880.

——. *Journal of the Proceedings of the House of Delegates of Maryland, January Session, 1888.* Annapolis, 1888.

——. *Message of John Lee Carroll, Governor of Maryland, to the*

General Assembly, at the Regular Session, January 1878. Document A. Annapolis, 1878.

——. *Report of the Select Committee Appointed to Prepare a Statement in Relation to the Resources of Maryland.* Document EE. Annapolis, 1865.

——. *Special Message from His Excellency Wm. T. Hamilton to the General Assembly of Maryland, February 12th, 1880.* Document L. Annapolis, 1880.

Maryland General Assembly, Senate. *Journal of Proceedings, January Session, 1874.* Annapolis, 1874.

——. *Journal of Proceedings, January Session, 1880.* Annapolis, 1880.

——. *Journal of Proceedings, January Session, 1888.* Annapolis, 1888.

——. *Journal of Proceedings, January Session, 1908.* Annapolis, 1908.

——. *Journal of Proceedings, January Session, 1910.* Annapolis, 1910.

Maryland *Geological Survey*, Vol. I. Baltimore, 1897.

——, Vol. V. Baltimore, 1905.

Maryland *Laws* [made and passed at sessions of the General Assembly], 1828–1967.

Maryland Mine Inspector. "First Annual Report of Peter Cain, Mine Inspector for Allegany and Garrett Counties, Maryland, December 30, 1876." Unpublished manuscript, Maryland Historical Society.

——. *Second Annual Report of Peter Cain, Inspector of Mines, to the Governor of Maryland, December 31, 1877.* House of Delegates Document H. Annapolis, 1878.

——. *Report of the Inspector of Mines for Allegany and Garrett Counties, 1878.* N.p., n.d.

——. *Second Annual Report of Owen Riordan, Inspector of Mines, for Allegany and Garrett Counties, for Year Ending December 31st, 1879.* House of Delegates Document J. Annapolis, 1880.

——. *Report of the Inspector of Mines for 1880. Thomas Brown, Inspector of Mines for Allegany and Garrett Counties.* N.p., n.d.

——. *Report of Thomas Brown, Inspector of Mines for Allegany and Garrett Counties, for the Year Ending December 31st,*

1881. House of Delegates Document F. Printed by authority, 1882.

——. *Report of Thomas Brown, Inspector of Mines for Allegany and Garrett Counties, for the Year Ending December 31st, 1883*. Annapolis, 1884.

——. *Report of C. H. Hamill, Inspector of Mines for the Counties of Allegany and Garrett, 1887*. Annapolis, 1888.

——. *Report of R. T. Browning, Inspector of Mines for Allegany and Garrett Counties, 1888*. Frostburg, 1889.

——. *Annual Report of the Inspector of Mines for Allegany and Garrett Counties, for the Years Ending December 31, 1892–93*. Baltimore, 1894.

——. *Annual Report of the Inspector of Mines for Allegany and Garrett Counties, for the Year Ending December 31, 1894*. Baltimore, 1895.

——. *Annual Report for Year Ending Dec. 31, 1898*. N.p., n.d.

——. *Annual Report for Year Ending Dec. 31, 1899*. Cumberland, n.d.

——. *Annual Report of the Mine Inspector for Allegany and Garrett Counties, Maryland, Covering the Period from January 1, 1902, to May 1, 1903*. Midland, 1903.

——. *Annual Report, 1903–04*. Baltimore, 1905.

——. *Annual Report, 1904–05*. Lonaconing, 1905.

——. *Annual Report, 1905–[06]*. Lonaconing, 1907.

——. *Annual Report, 1906–07*. Lonaconing, 1907.

——. *Annual Report, 1908–09*. Cumberland, 1909.

——. *Annual Report, 1909–1910*. Cumberland, 1910.

——. *Annual Report, 1910–1911*. Cumberland, n.d.

——. *Annual Report, 1912–1913*. Cumberland, n.d.

——. *Annual Report, 1913–1914*. Baltimore, n.d.

——. *Annual Report, 1915–1917*. Baltimore, n.d.

Maryland State Board of Education. *Fifteenth Annual Report of the State Board of Education, Shewing the Condition of the Public Schools of Maryland, for the Year Ending September 30, 1881*. House Document I. Annapolis, 1882.

Maryland State Board of Health. *Biennial Report 1888*. Annapolis, 1888.

——. *Twelfth Biennial Report for the Two Years Ending December 31st, 1897*. Baltimore, 1898.

——. *Annual Report, 1904.* Baltimore, 1906.

Pennsylvania Bureau of Statistics of Labor and Agriculture. *First Annual Report, 1872–3.* Harrisburg, 1874.

Pennsylvania Department of Internal Affairs. *Annual Report, 1881–82.* Part III, *Industrial Statistics.* Harrisburg, 1883.

——. *Annual Report, 1884.* Part III, *Industrial Statistics.* Harrisburg, 1885.

——. *Annual Report, 1885.* Part III, *Industrial Statistics.* Harrisburg, 1886.

——. *Annual Report, 1886.* Part III, *Industrial Statistics.* Harrisburg, 1887.

——. *Annual Report, 1888.* Part III, *Industrial Statistics.* Harrisburg, 1889.

——. *Annual Report, 1902.* Part III, *Industrial Statistics.* N.p., 1903.

United States Bureau of the Census (also United States Census Office). *Seventh Census of the United States: 1850. Abstract of the Statistics of Manufactures.* Np., n.d.

——. *Seventh Census of the United States: 1850. History and Statistics of the State of Maryland.* Washington, 1852.

——. *Eighth Census of the United States: 1860.* III, *Manufactures of the United States in 1860.* Washington, 1865.

——. *Eighth Census of the United States: 1860.* I, *Population.* Washington, 1864.

——. *Ninth Census of the United States: 1870.* I, *The Statistics of the Population of the United States.* Washington, 1872.

——. *Ninth Census of the United States: 1870.* III, *The Statistics of the Wealth and Industry of the United States.* Washington, 1872.

——. *Tenth Census of the United States: 1880.* I, *Population of the United States.* Washington, 1883.

——. *Tenth Census of the United States: 1880.* XV, *Report on the Mining Industries of the United States.* Washington, 1886.

——. *Eleventh Census of the United States: 1890.* XV, *Report of the Population of the United States, Part I.* Washington, 1895.

——. *Eleventh Census of the United States: 1890.* XIV, *Mineral Industries in the United States.* Washington, 1892.

——. *Twelfth Census of the United States: 1900.* I, *Population, Part I.* Washington, 1901.

——. *Special Reports, Mines and Quarries: 1902.* Washington, 1905.

——. *Thirteenth Census of the United States: 1910.* II, *Population.* Washington, 1913.

——. *Thirteenth Census of the United States: 1910.* XI, *Mines and Quarries, 1909.* Washington, 1913.

United States Coal Commission. *Report on the Bituminous Mine Workers and Their Homes.* Washington, 1925.

United States Department of Labor. *Growth of Labor Law in the United States.* Washington, 1962.

——. "The State Cooperative Accident Insurance Fund of Maryland," *Bulletin of the Bureau of Labor,* No. 57. Washington, 1905.

United States Industrial Commission. V, *Report on Labor Legislation.* Washington, 1900.

——. XII, *Report on the Relations and Conditions of Capital and Labor Employed in the Mining Industry.* Washington, 1901.

West Virginia Commissioner of Labor. *Report, 1897–1898.* Charleston, 1898.

General Publications

Arnot, Robert P. *A History of the Scottish Miners from the Earliest Times.* London, 1955.

——. *The Miners: A History of the Miners' Federation of Great Britain, 1889–1910.* London, 1949.

Association of Officials of Bureaus of Labor Statistics of America. *Proceedings of Twentieth Annual Convention, 1904.* N.p., n.d.

Barnett, George E. "The Maryland Workmen's Compensation Act," *Quarterly Journal of Economics,* XVI, 591–594.

——. "The End of the Maryland Workmen's Compensation Act," *Quarterly Journal of Economics,* XIX, 320–322.

Beachley, Charles E. *History of the Consolidation Coal Company 1864–1934.* New York, 1934.

Billington, Ray Allen. *Westward Expansion, A History of the American Frontier.* 2d ed. New York, 1960.

Blackburn, E. Howard, and Welfley, William H. *History of Bedford and Somerset Counties, Pennsylvania.* 3 vols. New York, 1906.

Blandi, Joseph G. *Maryland Business Corporations, 1783–1852.* Johns Hopkins University Studies in Historical and Political Science, Series LII, No. 3. Baltimore, 1934.

Bowen, Ele. *Rambles in the Path of the Steam-Horse.* Philadelphia, 1855.

Broehl, Wayne G., Jr. *The Molly Maguires.* Cambridge, 1964.

Brophy, John. *A Miner's Life.* An autobiography edited and supplemented by John O. P. Hall. Madison and Milwaukee, 1964.

Browne, Henry Joseph. *The Catholic Church and the Knights of Labor.* Washington, 1949.

Caudill, Harry M. *Night Comes to the Cumberlands.* Boston, 1963.

Clark, Lindley D. "Recent Action Relating to Employers' Liability and Workmen's Compensation," *Bulletin of the Bureau of Labor,* No. 90. Washington, 1911.

Clark, William Bullock. "Historical Sketch Embracing an Account of the Progress of Investigation Concerning the Physical Features and Natural Resources of Maryland," Maryland *Geological Survey,* Vol. I. Baltimore, 1897.

——. "Origin, Distribution and Uses of Coal," Maryland *Geological Survey,* Vol. V. Baltimore, 1905.

——, and Mathews, Edward B. "Maryland Mineral Industries 1896–1907," Maryland *Geological Survey,* Vol. VIII. Baltimore, 1909.

"Coal: The Industrial Darkness," *Fortune,* March 1947.

"Coal: The 'Pitt Consol' Adventure," *Fortune,* July 1947.

Conant, C. B. "Coal Fever: The Price and Prospects of Anthracite Coal," *Merchants' Magazine and Commercial Review,* May 1865.

Consolidation Coal Company. *Consol News,* 100th Anniversary Edition, Vol. III, No. 1, First Quarter 1964.

——. *Report to the Stockholders, December, 1871.* New York, 1872.

——. *Report to the Stockholders, for Year 1873.* New York, 1874.

——. *Report to the Stockholders for the Year Ending December 31, 1877.* Baltimore, 1878.

——. *Report to the Stockholders for the Year Ending December 31st, 1878.* Baltimore, n.d.

——. *Annual Report of the President and Directors to the Stockholders for the Year Ended December 31, 1898.* Baltimore, 1899.

——. *Annual Report of the President and Directors to the Stockholders of the Consolidation Coal Company for the Year Ended December 31, 1900.* Baltimore, 1900.

——. *Annual Report of the President and Directors to the Stock-*

holders of the Consolidation Coal Company for the Year Ended December 31, 1903. Baltimore, 1904.

——. *Annual Report of the Consolidation Coal Company for the Year Ending December 31, 1909*. Baltimore, 1910.

Cumberland Coal and Iron Company. *Report of the President and Board of Directors of the Cumberland Coal and Iron Company*. New York, 1853.

"Cumberland Coal Region, The," *The Bankers' Magazine and Statistical Register* [New York], IV, 394–399.

Cunz, Dieter. *The Maryland Germans*. Princeton, 1948.

Drum, Francis J. "Labor Relations from the Standpoint of the Union," *Annals American Academy of Political and Social Science*, 1924, pp. 120–124.

Eavenson, Howard N. *The First Century and a Quarter of American Coal Industry*. Pittsburgh, 1942.

Edwards, Richard, ed. *Statistical Gazeteer of the State of Maryland, and the District of Columbia*. Baltimore, 1856.

Erickson, Charlotte. *American Industry and the European Immigrant, 1860–1885*. New York, 1967.

Evans, Chris. *History of United Mine Workers of America from the Year 1860 to 1890*. Vol. I. N.p., n.d.

——. *History of United Mine Workers of America from the Year 1890 to 1900*. Vol. II. N.p., n.d.

Fairbanks, W. L., and Hamill, W. S. *The Coal-Mining Industry of Maryland*. Baltimore, 1932.

Faulkner, Harold U. *The Decline of Laissez Faire*. New York, 1959.

Franklin Coal Company. *First Annual Report of the President and Directors of the Franklin Coal Company of Allegany County*. Baltimore, 1855.

George's Creek Coal and Iron Company. *George's Creek Coal and Iron Company*. N.p., 1836.

——. Untitled report of directors, 1839. In J. H. Alexander Papers, 1839–51, Maryland Historical Society.

——. *George's Creek Coal and Iron Company*. N.p., [1849]. Maryland Historical Society.

——. *Letterbook, 1839–1847*. Maryland Historical Society Acc. No. 58383.

——. "Memoranda and Register, I: Memoranda of Calculations and Determinations for the Construction of the Lonaconing Iron-

Works and Daily Transactions from 21st day of August 1837 to 21st September 1838." Privately owned manuscript, in possession of Mr. Lloyd R. Stallings, Convent, New Jersey.

——. "Memoranda and Register, II: Lonaconing Transactions." Privately owned manuscript, in possession of Mr. Lloyd R. Stallings.

Hamilton, Walton H., and Wright, Helen R. *The Case of Bituminous Coal.* New York, 1925.

Hampshire and Baltimore Coal Company of Virginia. *Report to the Directors.* N.p., 1864.

Hughes, George W. *Extracts from Reports of an Examination of the Coal Measures Belonging to the Maryland Mining Company in Allegany County; and of a Survey for a Railroad from the Mines to the Chesapeake and Ohio Canal, at Cumberland.* Washington, 1837.

Kemmerer, Donald L., and Jones, C. Clyde. *American Economic History*. New York, 1959.

Knights of Labor. *Record of the Proceedings of the Second Regular Session of the General Assembly, Held at St. Louis, Mo., January 14–17, 1879.* N.p., n.d.

——. *Record of the Proceedings of the Third Regular Session of the General Assembly, Held at Chicago, Ill., Sept. 2–6, 1879.* N.p., n.d.

——. *Record of the Proceedings of the Fourth Regular Session of the General Assembly, Held at Pittsburg, Pa., Sept. 7–11, 1880.* N.p., n.d.

——. *Record of the Proceedings of the Fifth Regular Session of the General Assembly, Held at Detroit, Mich., Sept. 6–10, 1881.* N.p., n.d.

——. *Record of the Proceedings of the Sixth Regular Session of the General Assembly, Held at New York City, N.Y., Sept. 5–12, 1882.* N.p., n.d.

——. *Record of the Proceedings of the Seventh Regular Session of the General Assembly, Held at Cincinnati, Ohio, Sept. 4–11, 1883.* N.p., n.d.

——. *Record of the Proceedings of the Eighth Regular Session of the General Assembly, Held at Philadelphia, Pa., Sept. 1–10, 1884.* N.p., n.d.

——. *Record of the Proceedings of the Ninth Regular Session of the General Assembly, Held at Hamilton, Ont., Oct. 5–13, 1885.* N.p., n.d.

——. *Record of the Proceedings of the Special Session of the General Assembly, Held at Cleveland, O., May 25 to June 3, 1886.* N.p., n.d.

——. *Record of Proceedings of the General Assembly of the Knights of Labor of America, Vol. IV. Commencing with Session at Richmond, Virginia, October 4 to 20, 1886.* N.p., 1886.

——. *Proceedings of the General Assembly of the Knights of Labor of America Eleventh Regular Session, Held at Minneapolis, Minnesota, October 4 to 19, 1887.* N.p., 1887.

——. *Proceedings of the General Assembly, Twelfth Regular Session, Held at Indianapolis, Indiana, Nov. 13 to 27, 1888.* Philadelphia, 1889.

Korson, George. *Coal Dust on the Fiddle.* Philadelphia, 1943.

Lauchheimer, Malcolm H. *The Labor Law of Maryland.* Johns Hopkins University Studies in Historical and Political Science, Series XXXVII, No. 2. Baltimore, 1919.

Lloyd, A. L. *Come All Ye Bold Miners.* London, 1952.

Lombardi, John. *Labor's Voice in the Cabinet, A History of the Department of Labor from its Origin to 1921.* Studies in History, Economics and Public Law Edited by the Faculty of Political Science of Columbia University, No. 496. New York, 1942.

Love, George H. *An Exciting Century in Coal!* Princeton, 1955.

Lovejoy, Owen R. "Child Labor in the Soft Coal Mines." In National Child Labor Committee, *Proceedings of the Third Annual Meeting.* New York, 1907.

Lovell, Arthur. *Borden Mining Company, a Brief History.* Frostburg, Md., 1938.

Lowitt, Richard, ed. "Frostburg 1882: German Strikers vs German Strikebreakers." In Society for the History of the Germans in Maryland, *Twenty-eighth Report.* Baltimore, 1953.

Lumsden, James. *American Memoranda: A Short Tour in the Summer of 1843.* Glasgow, 1844.

Macfarlane, James. *The Coal-Regions of America: Their Topography, Geology, and Development.* New York, 1873.

Manakee, Harold R. *Maryland in the Civil War.* Baltimore, 1961.

Maryland Coal Company. *Annual Report of the Board of Directors of the Maryland Coal Company of Allegany County, Maryland, New York, December 13th, 1870.* New York, 1870.

——. *Annual Report of the Maryland Coal Co. of Allegany County, Maryland, for the Year 1873.* New York, 1874.

——. *Annual Report of the Maryland Coal Co. of Allegany County, Maryland, for the Year 1876.* New York, 1877.

Maryland Directory. Baltimore, 1880.

Maryland Directory and State Gazeteer, 1887. Baltimore, 1887.

Maryland Mining Company. *Report to the Stockholders of the Maryland Mining Company, 1847.* New York, n.d.

——. *Report for June, 1848, to the Stockholders of the Maryland Mining Company, Horatio Allen, Pres.* New York, June 26, 1848.

Maryland University, Bureau of Business and Economic Research. *An Economic Survey of Allegany County, Maryland.* Studies in Business and Economics, Vol. I, No. 1. College Park, 1947.

——. *Coal in the Maryland Economy, 1736–1965.* Studies in Business and Economics, Vol. VII, No. 3. College Park, 1953.

Midland Coal and Iron Company. *Report on the Valuable Coal Properties of The Midland Coal & Iron Company and The Manchester Mining and Manufacturing Company, of Alleghany [sic] County, Maryland.* New York, 1856.

Moody's Investors Service. *Moody's Manual of Investments American and Foreign,* Part II, *Industrial Securities.* New York, 1944.

——. *Moody's Industrial Manual, American and Foreign.* New York, 1967.

Munn, Robert F. *The Coal Industry in America. A Bibliography and Guide to Studies.* Morgantown, 1965.

McBride, John. "Coal Miners." In George E. McNeill, ed., *The Labor Movement: The Problem of To-Day.* Boston, 1887.

McDonald, David J. and Lynch, Edward A. *Coal and Unionism, a History of the American Coal Miners' Unions.* Silver Spring, Md., 1939.

Neu, Irene D. *Erastus Corning, Merchant and Financier, 1794–1872.* Ithaca, 1960.

Nevins, Allan. *The War for the Union: The Improvised War, 1861–1862.* New York, 1959.

Nicolls, William J. *Above Ground and Below in the George's Creek Coal Region.* Baltimore, 1898.

Pickell Mining Company. *First Annual Report of the President to the*

Stockholders of the Pickell Mining Company. Baltimore, 1854.

Portrait and Biographical Record of the Sixth Congressional District, Maryland. New York, 1898.

Powderly, T. V. *Thirty Years of Labor, 1859 to 1889*. Columbus, Ohio, 1889.

Randall, J. G. *The Civil War and Reconstruction*. Boston, 1953.

Randolph, Beverley S. "History of the Maryland Coal Region," Maryland *Geological Survey*, Vol. V. Baltimore, 1905.

Rayback, Joseph G. *A History of American Labor*. New York, 1959.

Renwick, James. *Report to George's Creek Coal & Iron Company, 1836*. Np., n.d. Contained in Maryland Historical Society folder labeled "Georges Creek Coal and Iron Company, 1836."

Richards, William M. "An Experiment in Industrial Feudalism at Lonaconing, Maryland, 1837–1860." Master's thesis, University of Maryland, 1950.

Roberts, Peter. *Anthracite Coal Communities*. New York, 1904.

Roy, Andrew. *The Coal Mines*. Cleveland, 1876.

——. *A History of the Coal Miners of the United States from the Development of the Mines to the Close of the Anthracite Strike of 1902*. 3d. ed. Columbus, Ohio, 1907.

——. *Recollections of a Prisoner of War*. Columbus, Ohio, 1909.

Sanderlin, Walter S. *The Great National Project: A History of the Chesapeake and Ohio Canal*. Johns Hopkins University Studies in Historical and Political Science, Series LXIV, No. 1. Baltimore, 1946.

Saward, Frederick E. *Coal Freight Circular*, Vol. I, April 1869–April 1870. N.p., n.d.

——. *Coal Trade Annual*, Vol. V. New York, 1878.

——. *Coal Trade Annual*, Vol. VII. New York, 1880.

Scharf, J. Thomas. *History of Western Maryland*. 2 vols. Philadelphia, 1882.

Schlesinger, Arthur M., Jr. *The Age of Roosevelt*, Vol. I, *The Crisis of the Old Order, 1919–1933*. Boston, 1957.

Shurick, A. T. *The Coal Industry*. Boston, 1924.

Silliman, Benjamin. *Extracts from a Report Made to the Maryland Mining Company*. New York, 1838.

Steiner, Bernard C. *Descriptions of Maryland*. Johns Hopkins University Studies in Historical and Political Science, Series XXII, Nos. 11–12. Baltimore, 1904.

Stockton, N. Allen. "The Coal Mines of Maryland," Maryland *Geological Survey*, Vol. V. Baltimore, 1905.

Suffern, Arthur E. *The Coal Miners' Struggle for Industrial Status.* New York, 1926.

Swank, James M. *History of the Manufacture of Iron in All Ages.* Philadelphia, 1892.

Taft, Philip. *The A. F. of L. in the Time of Gompers.* New York, 1957.

——. *Organized Labor in American History*. New York, 1964.

Thomas, James W., and Williams, T. J. C. *History of Allegany County, Maryland.* 2 vols., 1923.

United Mine Workers of America. *Proceedings of the Twenty-sixth Consecutive and Third Biennial Convention of the United Mine Workers of America Held in the City of Indianapolis, Indiana, January 15 to 26, 1918, Inclusive.* 2 vols. Indianapolis, 1918.

——. *Proceedings of the Twenty-seventh Consecutive and Fourth Biennial Convention of the United Mine Workers of America Held at Cleveland, Ohio, Sept. 9, 1919.* 3 vols. Indianapolis, 1919.

——. *Proceedings of the Twenty-ninth Consecutive and Sixth Biennial Convention of the United Mine Workers of America Held in the City of Indianapolis, Indiana, January 22 to February 2, 1924.* 3 vols. Indianapolis, 1924.

Ware, Norman J. *The Labor Movement in the United States, 1860–1895: A Study in Democracy*. New York, 1929.

Warne, Frank Julian. *The Coal-Mine Workers, a Study in Labor Organization.* New York, 1905.

Wearmouth, Robert F. *Some Working-Class Movements of the Nineteenth Century*. London, 1948.

Webb, Sidney. *The Story of the Durham Miners (1662–1921)*. London, 1921.

Welbourne, E. *The Miners' Unions of Northumberland and Durham.* Cambridge, England, 1923.

Weld, Henry Thomas. *A Report of the Maryland and New York Iron and Coal Company's Lands, in the County of Alleghany [sic] and State of Maryland.* N.p., n.d.

Wieck, Edward A. *The American Miners Association: A Record of the Origin of Coal Miners' Unions in the United States.* New York, 1940.

Index

A page number in italics refers to one of the illustrations, which follow page 34.